Office of Research [illegible] Evaluation

Regression with Graphics

A Second Course in Applied Statistics

Regression with Graphics

A Second Course in Applied Statistics

Lawrence C. Hamilton

University of New Hampshire, Durham

Duxbury Press
An Imprint of Wadsworth Publishing Company
Belmont, California

Duxbury Press
An Imprint of Wadsworth Publishing Company
A division of Wadsworth, Inc.

Printed in the United States of America
10 9 8 7 6

Library of Congress Cataloging-in-Publication Data
Hamilton, Lawrence C.
Regression with graphics : a second course in applied statistics / Lawrence C. Hamilton.
p. cm.
Includes bibliographical references and index.
ISBN 0-534-15900-1
1. Regression analysis. 2. Regression analysis—Graphic methods.
I. Title.
QA278.2.H37 1991 91-25588
519.5′36—dc20 CIP

Sponsoring Editor: *Michael J. Sugarman*
Marketing Representative: *John Moroney*
Editorial Assistants: *Lainie Giuliano and Carol Ann Benedict*
Production Editor: *Penelope Sky*
Manuscript Editor: *Micky Lawler*
Interior and Cover Design: *Katherine Minerva*
Art Coordinator: *Lisa Torri*
Typesetting: *Polyglot Compositors*
Printing and Binding: *R. R. Donnelley & Sons Company*

To my parents

Alicita and Warren Hamilton

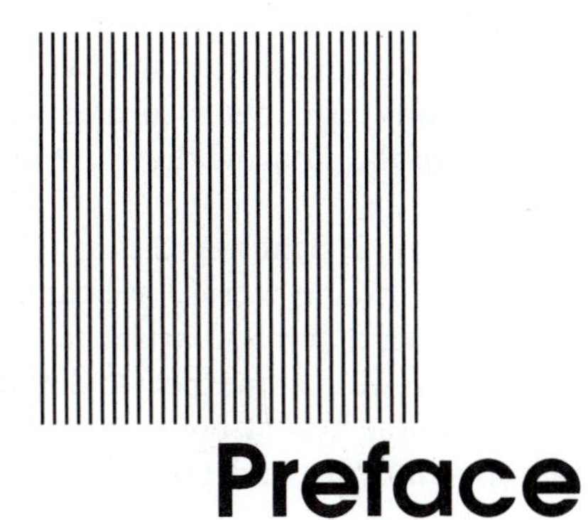

Preface

Regression analysis and graphical displays are basic scientific tools. Regression, the "core technique" of modern statistical analysis, appears regularly in the research journals of many fields. Economists use regression for forecasting; sociologists build regression-based causal models; ecologists employ regression to study the effects of acid rain. Regression encompasses a broad range of methods, from elementary to advanced. Many seemingly unrelated techniques encountered in a first statistics course are actually simple kinds of regression.[1]

Graphing has often been taught as a way to present conclusions, rather than as a tool for analysis and discovery. However, analytical graphics have come into their own in recent decades. During the 1970s, a radical new approach called exploratory

[1] For example, individual means and comparison-of-means procedures (one- and two-sample t-tests, paired-difference tests, analysis of variance), correlation analysis, and trend lines are all special cases of regression.

data analysis (EDA) transformed the field. This transformation accelerated with the computer revolution of the 1980s, which made new kinds of graphs possible and all graphs much easier to draw.

Regression summarizes (or models) complex data in a compact way. Regression models are easy to describe, study, and compare. We can use them to test hypotheses or to make predictions. Such models have many uses but also carry risks: by focusing on the model, we may overlook details of the data and draw oversimplified or false conclusions.

The strengths and weaknesses of graphs complement regression techniques. Graphs are not compact; if they are "worth a thousand words," we cannot easily describe or compare them. Graphs show data rather than summarize them. Modern (EDA-inspired) graphical methods are designed to reveal unexpected details of the data. Using regression and graphs together gives the best of both worlds: compact numerical summaries, checked out and improved by using detailed visual displays of the data.

The graphical and regression methods described in this book form a natural partnership, providing a general, flexible, and quite powerful approach to data analysis. Although they fit well together, these methods have not previously appeared under one cover. Dozens of books describe basic regression, and most general statistics texts contain chapters on this topic. Principal components and factor analysis are also well known. EDA, regression diagnostics, analytical graphics, robust estimation, logit regression, and bootstrapping are newer developments, each with its own specialized literature. In these literatures things are constantly changing, as new ideas or findings supersede the old. General textbooks have not yet assimilated most of the new developments, or the basic philosophical shift they bring.

Computers have changed the underlying strategy of data analysis. When calculation was difficult, analysis was often a one-shot effort: plan what to do, make calculations, and write up the results. Now calculation is easy, so we can afford to view analysis as an interactive process: plan what to do, let the computer calculate, study the results, and think about what to do next. Are our findings stable, when analyzed in several different ways? Are our assumptions plausible, as far as we can tell from the data? Are there unexpected results or complications that we should look into? We may want to try many further analyses before settling down with a "final" conclusion. As analytical strategies have changed, so has the emphasis in learning applied statistics, from a narrow focus on choosing and calculating the "correct" procedure to a broader focus on understanding our data.

Organization and Emphasis

Students enter a second statistics course with varied backgrounds and interests. Some want only practical advice—what method they should use, which program will do it, and how they can interpret the results. Other students need more detailed explanations. I have tried to accommodate student variety by writing a multilevel book:

1. The most technical material in each chapter appears in boxes. This material could be skipped by nonmathematical readers; it is always there to return to later.

2. The boxes contain formal definitions and other equations, using matrix algebra when needed.
3. Least squares regression is introduced in the first four and a half chapters. Some instructors will focus on this material and sample or skip the more advanced topics of Chapters 5–8. In other courses, early chapters can be treated as review, with students proceeding quickly to the later material.
4. Three appendices cover topics that do not fit into the main sequence. Statistical concepts and matrix algebra are briefly reviewed in Appendices 1 and 3; these could be consulted as needed for help in understanding the boxes. Modern computer-intensive methods are illustrated in Appendix 2, which further pursues some issues raised in Chapters 4 and 6.
5. To provide hands-on experience, each chapter ends with a set of exercises that use fresh data.[2]

The overall mathematical level of the book is not high, although inevitably it rises with later chapters. Throughout, I place more emphasis on practical issues and troubleshooting than on statistical theory.

Most of the examples use real data with environmental themes. These data come from diverse disciplines including sociology, economics, geology, medicine, zoology, forestry, and chemistry. I hope these examples will prove interesting and understandable to readers in any field and at the same time will convey the variety of areas in which regression and graphical methods apply.

Acknowledgments

Many people helped in the development of this book. John Moroney provided initial encouragement and John Kimmel got the project under way. Mike Sugarman coordinated reviews and revisions; Penelope Sky supervised final editing and production.

Valuable contributions came from the following reviewers: Patricia M. Buchanan, Penn State University; Daniel C. Coster, Purdue University; Patrick Doreian, University of Pittsburgh; James D. Hamilton, University of Virginia; Jason C. Hsu, Ohio State University; Robert M. Jenkins, Yale University; David Ruppert, Cornell University; and Kent D. Smith, California Polytechnic State University. I also benefited from advice by Kenneth Bollen, Marjorie Flavin, William Gould, Warren Hamilton, Ernst Linder, Adrian Raftery, and John Tukey. The real-data focus of the book would not have been possible without researchers who freely shared or helped locate data, notably Terry G. Bensel, T. Jean Blocker, David K. Cairns, S. Lawrence Dingman, R. W. Furness, Mark Hines, Anthony Iannacchione, W. Berry Lyons, Gary K. Meffe, Michael Rabinowitz, and Carole Seyfrit.

I owe thanks to many students who worked hard as "unclarity detectors" in reading through early drafts. Karen Gartner, Susan Larson, and Les MacLeod completed every exercise except a few (subsequently deleted) that proved to be

[2] Most exercises could be done with any full-featured statistical package. For example, see *Statistics with Stata* (Hamilton, 1990b).

impossible. John Anderson, Sharon Billings, Kimberly Cook, Elizabeth Crepeau, Holley Gimpel, and Beth Jacobsen also made helpful suggestions.

Petr Brym, Joe Danahy, Ralph Draper, Richard England, Marie Gaudard, Betty LeCompagnon, Stuart Palmer, and Deena Peschke facilitated my work at the University of New Hampshire. At home, where most of the writing took place, I counted on the indispensable support of Leslie, Sarah, and David.

Lawrence C. Hamilton

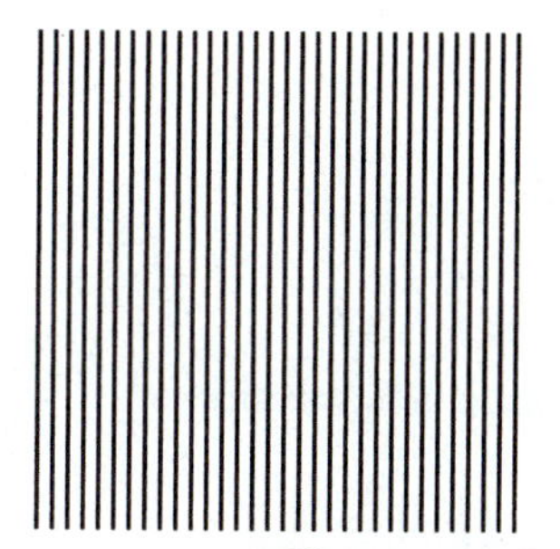

Contents

1 Variable Distributions 1

The Concord Water Study 2
Mean, Variance, and Standard Deviation 2
Normal Distributions 4
Median and Interquartile Range 6
Boxplots 8
Symmetry Plots 10
Quantile Plots 11
Quantile–Quantile Plots 13
Quantile–Normal Plots 15
Power Transformations 17
Selecting an Appropriate Power 19
Conclusion 23

Exercises 23
Notes 26

2 Bivariate Regression Analysis 29

The Basic Linear Model 30
Ordinary Least Squares 32
Scatterplots and Regression 34
Predicted Values and Residuals 37
R^2, Correlation, and Standardized Regression Coefficients 38
Reading Computer Output 41
Hypothesis Tests for Regression Coefficients 42
Confidence Intervals 47
Regression Through the Origin 49
Problems with Regression 51
Residual Analysis 51
Power Transformations in Regression 53
Understanding Curvilinear Regression 57
Conclusion 58
Exercises 59
Notes 63

3 Basics of Multiple Regression 65

Multiple Regression Models 66
A Three-Variable Example 67
Partial Effects 69
Variable Selection 72
A Seven-Variable Example 74
Standardized Regression Coefficients 76
t-Tests and Confidence Intervals for Individual Coefficients 77
F-Tests for Sets of Coefficients 80
Multicollinearity 82
Search Strategies 82
Interaction Effects 84
Intercept Dummy Variables 85
Slope Dummy Variables 88
Oneway Analysis of Variance 92
Twoway Analysis of Variance 95

Conclusion 101
Exercises 101
Notes 105

4 Regression Criticism 109

Assumptions of Ordinary Least Squares 110
Correlation and Scatterplot Matrices 113
Residual Versus Predicted *Y* Plots 116
Autocorrelation 118
Nonnormality 124
Influence Analysis 125
More Case Statistics 130
Symptoms of Multicollinearity 133
Conclusion 136
Exercises 137
Notes 140

5 Fitting Curves 145

Exploratory Band Regression 146
Regression with Transformed Variables 148
Curvilinear Regression Models 148
Choosing Transformations 154
Evaluating Consequences of Transformation 155
Conditional Effect Plots 158
Comparing Effects 161
Nonlinear Models 163
Estimating Nonlinear Models 167
Interpretation 171
Conclusion 173
Exercises 174
Notes 181

6 Robust Regression 183

A Two-Variable Example 184
Goals of Robust Estimation 189
M-Estimation and Iteratively Reweighted Least Squares 190
Calculation by IRLS 195
Standard Errors and Tests for *M*-Estimates 198

Using Robust Estimation 200
A Robust Multiple Regression 203
Bounded-Influence Regression 207
Conclusion 211
Exercises 212
Notes 215

7 Logit Regression 217

Limitations of Linear Regression 218
The Logit Regression Model 220
Estimation 223
Hypothesis Tests and Confidence Intervals 225
Interpretation 229
Statistical Problems 233
Influence Statistics for Logit Regression 235
Diagnostic Graphs 238
Conclusion 242
Exercises 243
Notes 246

8 Principal Components and Factor Analysis 249

Introduction to Components and Factor Analysis 250
A Principal Components Analysis 252
How Many Components? 258
Rotation 259
Factor Scores 263
Graphical Applications: Detecting Outliers and Clusters 267
Principal Factor Analysis 270
An Example of Principal Factor Analysis 273
Maximum-Likelihood Factor Analysis 278
Conclusion 281
Exercises 283
Notes 287

Appendix 1 Population and Sampling Distributions 289

Expected Values 289
Covariance 291
Variance 292

Further Definitions 293
Properties of Sampling Distributions 294
Ordinary Least Squares 296
Some Theoretical Distributions 297
Exercises 300
Notes 301

Appendix 2 Computer-Intensive Methods 303

Monte Carlo Simulation 304
Bootstrap Methods 313
Bootstrap Distributions 314
Residual Versus Data Resampling 318
Bootstrap Confidence Intervals 319
Evaluating Confidence Intervals 323
Computer-Intensive Methods in Research 325
Exercises 326
Notes 329

Appendix 3 Matrix Algebra 333

Basic Ideas 334
Matrix Addition and Multiplication 335
Matrix Inversion 337
Regression in Matrix Form 338
An Example 340
Regression from Correlation Matrices 342
Further Definitions 344
Exercises 345
Notes 346

Appendix 4 Statistical Tables 349

A4.1: Critical Values for Student's *t*-Distribution 350
A4.2: Critical Values for the *F*-Distribution 351
A4.3: Critical Values for the Chi-Square Distribution 354
A4.4: Critical Values for the Durbin–Watson Test for Autocorrelation 355

References 357

Index 361

Regression with Graphics

A Second Course in Applied Statistics

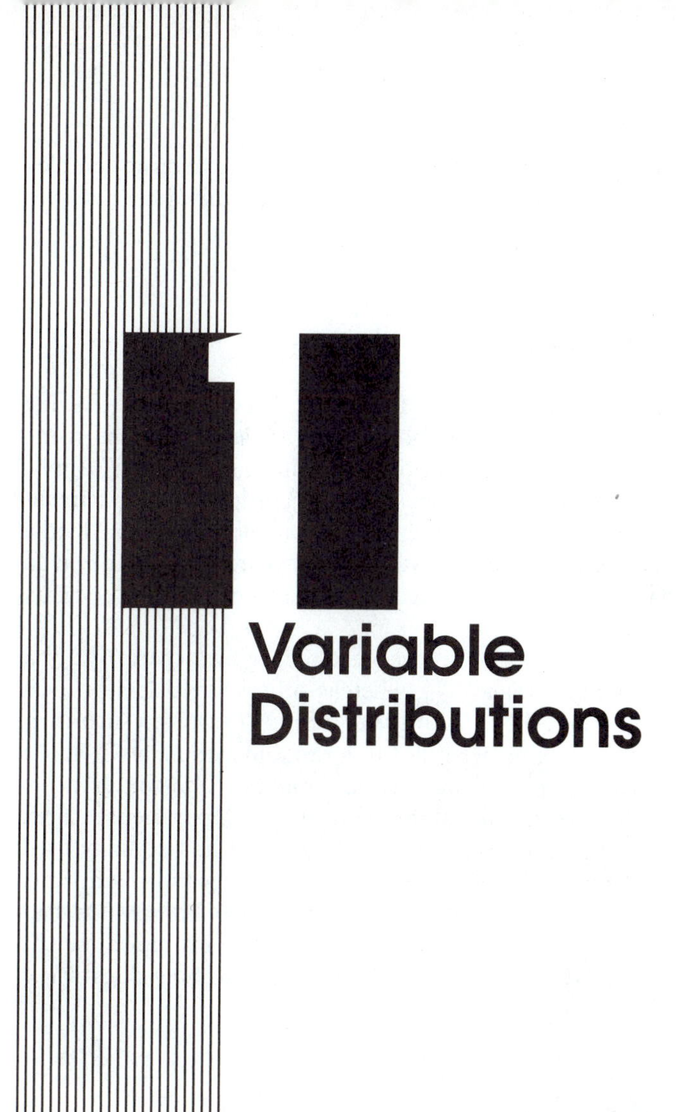

1 Variable Distributions

Not long ago, multiple regression was considered a difficult technique that was seldom practical because it required so much calculation. Computers have changed this picture. Regression is now easy and has become the technique of choice in many fields. The most interesting research questions typically involve relations between variables, and regression provides powerful methods to investigate such relations. For this reason researchers sometimes want to jump straight to regression analysis as soon as they collect their data.

Univariate analysis, or the study of variables one at a time, is usually less interesting to the researcher. Nonetheless, univariate distributions are the foundation upon which *multivariate* (multivariable) analysis rests. Every feature of the univariate distributions will have multivariate implications. Sometimes researchers, encountering unexpected results in their multivariate analysis, must go back to the univariate distributions for an explanation. A better way to proceed is from the ground up: carefully studying each variable first before proceeding to more complex analyses. This chapter illustrates some techniques and issues of univariate analysis.

The Concord Water Study

During the 1970s the city of Concord, New Hampshire, experienced a growing demand for water, despite a roughly stable population. In late 1979 this rising demand, together with unusually dry weather, led to a shortage in water supply. In late summer of 1980, as the shortage worsened, the Concord Water Department and municipal officials began a campaign (described by one observer as a "media blitz") to persuade citizens to use less water. Over the next year, water use declined by about 15%, which was naturally interpreted as evidence of the conservation campaign's success.

An overall 15% decline does not mean that everyone reduced water use by 15%, of course. Some users saved more than 15%, whereas others saved less or even increased their consumption. The 1981 Concord Water Study examined such variations in water savings. Questionnaires went out to a random sample of Concord households, asking about demographic characteristics, opinions, and conservation behavior. These questionnaires were then matched with Water Department records (from meter readings) of the amount of water each household had actually used during the summers of 1980 and 1981, before and after the conservation campaign.

The Concord data, containing information about how 496 households had responded to the scarcity of a basic natural resource, provide examples for the initial chapters of this book. We begin with a simple question: how does water use vary from house to house?

Mean, Variance, and Standard Deviation

During the summer of 1981, at the height of the water shortage, 496 households in the Concord Water Study used a total of 1,140,000 cubic feet of water. Dividing 1,140,000 cubic feet by 496 households, we obtain the *sample mean* ($\bar{Y}$), about 2298 cubic feet of water per household:

$$\begin{aligned}\bar{Y} &= 1{,}140{,}000/496 \\ &= 2298\end{aligned}$$

This calculation illustrates the formal definition of the sample mean as the sum of Y values (ΣY_i) divided by the number of cases in the sample (n):

$$\begin{aligned}\bar{Y} &= (Y_1 + Y_2 + Y_3 + \cdots + Y_n)/n \\ &= \Sigma Y_i/n \qquad [1.1]\end{aligned}$$

The i subscript indexes individual cases: $i = 1, 2, 3, \ldots, n$. Y_1 represents the value of Y for case 1, Y_2 the value for case 2, and so on. In [1.1] and elsewhere in this book, I use the *summation operator* Σ (sigma) with no super- or subscripts to indicate summation over each case in the sample (from $i = 1$ to n). That is, Σ by itself will stand for

$$\sum_{i=1}^{n}$$

Explicit super/subscripts appear only when summation is *not* over each case.

During summer 1980, before the conservation program took effect, the same 496 households used 1,355,072 cubic feet of water, so the mean was 2731. The post-shortage mean is thus $2731 - 2298 = 433$ cubic feet less than the preshortage mean. Means are often used in this manner, to quickly summarize or compare variable distributions.

A mean describes the location of a distribution's center but says nothing about variation around that center. Each case deviates from the mean by a certain amount: $Y_i - \bar{Y}$. Deviations from means possess two important properties:

■ *Zero-sum property*: the sum of deviations from a mean equals zero:

$$\Sigma(Y_i - \bar{Y}) = 0$$

Least squares property: the sum of squared deviations from a mean is lower than the sum of squared deviations from any other value:

$$\Sigma(Y_i - \bar{Y})^2 < \Sigma(Y_i - c)^2 \qquad \text{for any } c \neq \bar{Y}$$ ■

The mean is the simplest member of a family of statistical procedures that share similar properties.

Variation around the mean is commonly measured in squared deviations. The sum of squared deviations is called the *total sum of squares*, or TSS_Y:

$$\text{TSS}_Y = \Sigma(Y_i - \bar{Y})^2 \qquad [1.2]$$

A useful measure of variation based on the total sum of squares is the *sample variance* (s_Y^2):

$$s_Y^2 = \text{TSS}_Y/(n-1)$$
$$= \Sigma(Y_i - \bar{Y})^2/(n-1) \qquad [1.3]$$

Variance is approximately the mean squared deviation. Its units are the variable's natural units squared—cubic feet of water squared, dollars squared, years squared, and so forth.

Taking the square root of the variance returns us to natural units—cubic feet, dollars, years, and so on. The square root of the variance is the *sample standard deviation*:

$$
\begin{aligned}
s_Y &= \sqrt{s_Y^2} \\
&= \sqrt{\Sigma(Y_i - \bar{Y})^2/(n-1)} \qquad [1.4]
\end{aligned}
$$

Like variance, standard deviations measure variation or spread about the mean.

The standard deviation of summer water use in this sample was 1764 cubic feet in 1980 but only 1486 cubic feet in 1981. This tells us that the *sample households varied less* (*were more alike*) in their water use after the conservation program.

Sample mean, variance, and standard deviation can be viewed as *estimators* of corresponding unknown *population parameters*. For example, the sample mean $\bar{Y}$ estimates the population mean μ.

Concept	*Sample statistic*		*Population parameter*	
mean	$\bar{Y}$	[1.1]	$\mu = \mathrm{E}[Y]$	[A1.1],[A1.3]
variance	s_Y^2	[1.3]	$\sigma_Y^2 = \mathrm{Var}[Y]$	[A1.13]
standard deviation	s_Y	[1.4]	$\sigma_Y = \sqrt{\mathrm{Var}[Y]}$	

The Greek letters μ (mu) and σ (sigma) denote the population parameters. E[ ] and Var[ ] are mathematical operations defining these parameters (Appendix 1).

Normal Distributions

The mean describes the center of a distribution, and the standard deviation describes spread around this center. With *normal* or *Gaussian* distributions, mean and standard deviation tell all we need to know.[1]

A *Gaussian* (*normal*) distribution has the probability density function

$$
f(Y) = \frac{1}{\sigma_Y \sqrt{(2\pi)}} e^{-(Y-\mu)^2/(2\sigma_Y^2)} \qquad [1.5]
$$

where μ denotes the population mean and σ_Y the population standard deviation. Since π and e are mathematical constants ($\pi = 3.14159\ldots$, $e = 2.71828\ldots$), only the parameters μ and σ_Y distinguish one normal distribution from another.

Graphing $f(Y)$ against Y produces the *normal curve*, seen in Figure 1.1. This curve is symmetrical around μ, extending from $-\infty$ to $+\infty$. σ_Y is the distance from μ to the points of inflection, where the curve changes from convex to concave. Areas under this curve correspond to normal-distribution probabilities.

Normal distributions simplify the analyst's job in several respects:

1. Comparing normal distributions reduces to comparing only means and standard deviations. If standard deviations are the same, the task becomes even simpler: just compare means. On the other hand, means and standard deviations may be incomplete or misleading as summaries for nonnormal distributions.

2. Some statistical procedures assume that sample data come from a normally distributed population. The Central Limit Theorem suggests that such normality assumptions become less important with larger samples, however.[2]
3. Many other procedures do not require normality but nonetheless work better when applied to normally distributed variables.

To be aware of possible complications, we routinely examine variable distributions at an early stage in any analysis. Could our sample plausibly have come from a normal population distribution?

Unfortunately, normal distributions are the exception rather than the rule with many types of real data. For example, Figure 1.2 shows the sample distribution of household water use as a histogram. Superimposed on this histogram is a normal curve with the same mean (2298) and standard deviation (1486).

The data histogram and theoretical normal curve in Figure 1.2 are notably different:

1. The upper tail of the histogram continues out well past the point where normal-curve probabilities fall to near zero, more than four standard deviations above the mean.
2. The histogram's lower tail is cut off, since no household can use less than zero cubic feet of water. In a normal distribution with $\mu = 2298$ and $\sigma = 1486$, about 6% of the cases would have values below zero, and there is no theoretical lower (or upper) limit.
3. The normal curve is symmetrical and centered around the mean, 2298. In contrast, the histogram is asymmetrical and peaks to the left of this mean.

So nonnormal a distribution, in so large a sample, would be extremely unlikely if the population distribution of water use were normal.[3]

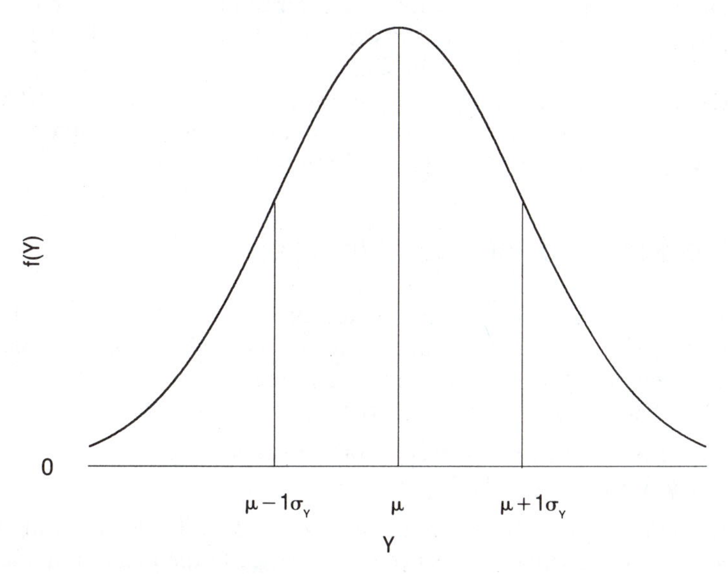

Figure 1.1 Normal distribution with mean μ and standard deviation σ_Y.

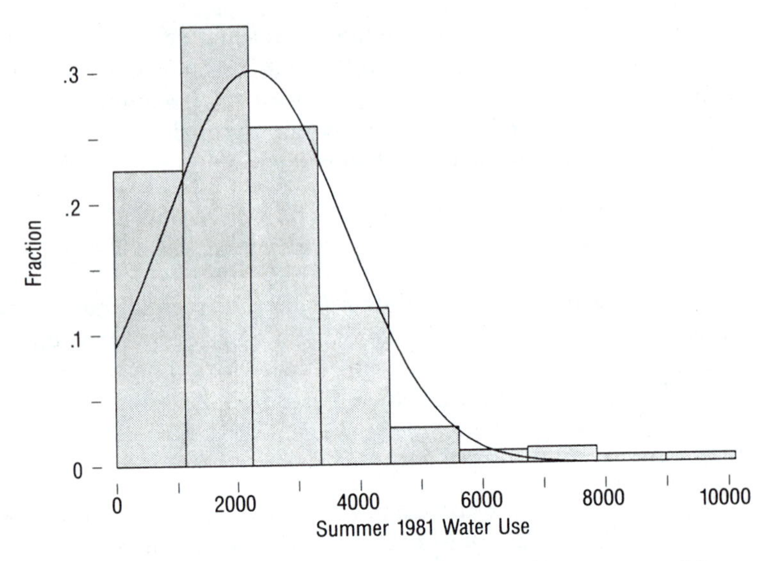

Figure 1.2 Histogram of postshortage household water use, with normal curve.

Distributions with elongated upper tails, like the histogram in Figure 1.2, are called *positively skewed.* (Distributions with elongated lower tails, the reverse of Figure 1.2, are called *negatively skewed.*) Positive skew often occurs with variables that have a lower limit of zero but no definite upper limit.

Visually, skewed sample distributions have one "longer" and one "shorter" tail. More general terms are "heavier" and "lighter" tails. Tail weight reflects not only distance from the center (tail length) but also the frequency of cases at that distance (tail depth, in a histogram). Tail weight corresponds to actual weight if the sample histogram were cut out of wood and balanced like a seesaw on its median (see next section). A positively skewed distribution is heavier to the right of the median; negative skew implies the opposite.[4]

One problem with skewed distributions is that the concept of "center" becomes ambiguous. This makes it trickier to summarize or compare distributions; we cannot rely solely on the mean and standard deviation.

Median and Interquartile Range

Order statistics are based on position in an ordered list of the data. The simplest order statistics are the *low* and *high extremes,* also called the *minimum* and *maximum*: the first and last values in an ordered list.

Another well-known order statistic is the *median* (Md):

■ If n cases are listed in order from lowest to highest, the sample median is the value at position $(n + 1)/2$. ■

The median equals the value of the middle case if n is odd and of the average of the two middle values if n is even. Theoretically the median divides an ordered list in half. Fifty percent of the cases have values below the median, and the other 50%

have values above it. Like the mean, the median is a measure of center, but one with different statistical properties.

Quartiles divide ordered lists into quarters. The *first quartile* is a number (theoretically) greater than the values of 25% of the cases and lower than the values of the remaining 75%. Similarly, the *third quartile* is a number greater than the values of 75% of the cases and lower than the remaining 25%. The *second quartile* equals the median.

Percentiles divide ordered lists into hundredths. One percent of the cases lie below the first percentile, and 99% lie above it. Ten percent lie below the tenth percentile, and so on. Percentiles can define other order statistics:

first quartile (Q_1):	25th percentile
median (Md):	50th percentile
third quartile (Q_3):	75th percentile

Sample order statistics are often just approximations, since the number of cases may not divide evenly into halves, fourths, or hundredths. Several alternative formulas exist that can lead to different results. The following explanation is not the simplest, but it has two advantages: it applies consistently to medians, quartiles, or percentiles; and it adapts to weighted data.

We start with a list of the values of variable Y, in ascending order. Let Y_i represent the ith value in this list and cum(P_i) represent the cumulative percentage of cases up to and including Y_i. *To find the value of the pth sample percentile*:

1. Find the first Y_i such that cum(P_i) $> p$.
2. Then the pth percentile equals:

$$\begin{cases} (Y_{i-1} + Y_i)/2 & \text{if } \mathrm{cum}(P_{i-1}) = p \\ Y_i & \text{otherwise} \end{cases} \qquad [1.6]$$

This formula also finds approximate sample quartiles ($p = 25$ and $p = 75$) and medians ($p = 50$).

With unweighted data, [1.6] obtains the same median as the $(n + 1)/2$ rule given earlier.

Applying [1.6] to the water data, we first sort the 496 sample households in order, from lowest to highest water use. To find the median (50th percentile, or $p = 50$), we locate the first household for which cum(P_i) > 50. This is the 249th household (cum(P_{249}) $= 100 \times (249/496) = 50.2016$), which consumed 2100 cubic feet of water ($Y_{249} = 2100$). The next-lower household, #248, consumed 2000 cubic feet ($Y_{248} = 2000$). Since cum(P_{i-1}) equals 50 (cum(P_{248}) $= 100 \times (248/496) = 50$):

$$\begin{aligned} \mathrm{Md} &= (Y_{i-1} + Y_i)/2 \\ &= (Y_{248} + Y_{249})/2 \\ &= (2000 + 2100)/2 \\ &= 2050 \end{aligned}$$

The median is Md = 2050 cubic feet, whereas we earlier found the mean to be $\bar{Y} = 2298$. Disagreement between Md and $\bar{Y}$ reflects the fact that such skewed distributions (Figure 1.2) do not have a single clear-cut "center." The heavy upper tail, consisting of a handful of high-consumption households, pulls the mean up relative to the median. (Had the lower tail been heavier, it would have pulled the mean down.) We can detect serious skew by comparing mean and median:

■ positive skew (Figure 1.2): $\bar{Y} > \text{Md}$
approximate symmetry: $\bar{Y} \approx \text{Md}$
negative skew: $\bar{Y} < \text{Md}$ ■

The greater the skew, the larger the mean-median difference will be.

The *interquartile range* (IQR) equals the distance between the first and third quartiles:

$$\text{IQR} = Q_3 - Q_1 \qquad [1.7]$$

This distance spans the middle 50% of the data. Following [1.6], the first quartile for the water-use data is $Q_1 = 1200$ and the third quartile is $Q_3 = 2900$. Therefore IQR = 2900 − 1200 = 1700 cubic feet.

Median and IQR measure center and spread, respectively. They are *resistant*, or not easily influenced by a few extreme values. In contrast, mean and standard deviation are *not* resistant; extreme values pull at the mean and inflate the standard deviation.[5] Consequently, median and IQR often provide better summaries when the data are skewed or contain exceptionally high or low values. Like mean and median, standard deviation and IQR provide an informative comparison. In approximately normal distributions, their relationship is

$$s_Y \approx \text{IQR}/1.35$$

We might therefore judge the normality *of symmetrical distributions* by comparing the standard deviation with IQR/1.35:

■ heavier-than-normal tails: $s_Y > \text{IQR}/1.35$
normal tails: $s_Y \approx \text{IQR}/1.35$
lighter-than-normal tails: $s_Y < \text{IQR}/1.35$ ■

Heavier-than-normal tails indicate that more of the distribution's cases are in the tails, far from the mean, than occurs with a normal distribution.

If a distribution is skewed (that is, if it is not true that $\bar{Y} \approx \text{Md}$), a comparison of standard deviation with IQR/1.35 makes less sense, because (1) a skewed distribution cannot be normal, and (2) it will typically have one light and one heavy tail. Thus we should not attempt this comparison with the water-use data. We already know it is skewed, unlike a normal distribution.[6]

Boxplots

Boxplots graphically display the median and interquartile range. Figure 1.3 shows a boxplot of the water-use data graphed earlier as a histogram (Figure 1.2). The cen-

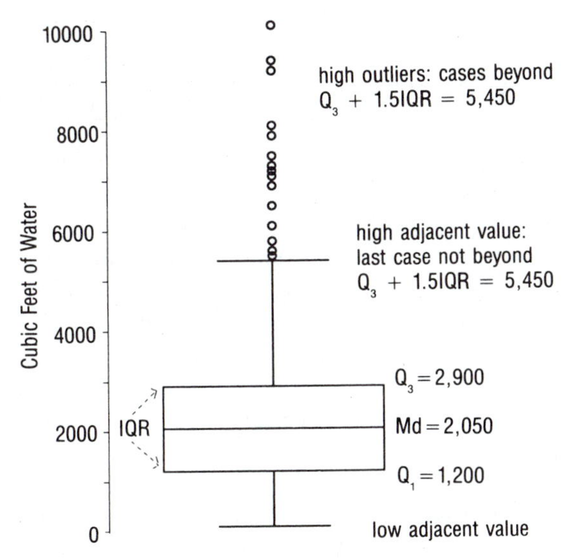

Figure 1.3 Boxplot of household water use.

tral box extends from the first quartile ($Q_1 = 1200$) to the third quartile ($Q_3 = 2900$), so its height equals the interquartile range (IQR = 1700). A horizontal line within the box indicates the median (Md = 2050).

Vertical lines extend from each quartile to *adjacent values,* values of the last cases not more than 1.5IQR beyond the quartiles. That is, adjacent values represent the last cases not less than

$$Q_1 - 1.5\text{IQR} = 1200 - 1.5(1700) = -1350$$

or not more than

$$Q_3 + 1.5\text{IQR} = 2900 + 1.5(1700) = 5450$$

Adjacent values in Figure 1.3 are 100 and 5400.

Farther-out values are called *outliers* and are graphed individually as small circles. *Thus Y_i is an outlier if*

$$Y_i < Q_1 - 1.5\text{IQR}$$

or if

$$Y_i > Q_3 + 1.5\text{IQR}$$

Of course, no households used less than −1350 cubic feet of water, so no low outliers exist. Twenty households used more than 5450 cubic feet, so each of these is a high outlier. The numerous high outliers and lack of low outliers reflect the overall positive skew of this distribution.

Many variants of boxplots exist. Most follow the basic convention of marking the median within a box that spans the interquartile range and indicating tails and outliers beyond the box. Areas of difference include the following:

1. Definitions of the quartiles, or similar statistics called *hinges* or *fourths*. One article noted eight possible definitions.[7]
2. Definitions of the adjacent values and outliers. Some versions use Q $\pm$ IQR instead of Q $\pm$ 1.5IQR as limits for adjacent values, for instance. Others drop the concept entirely, drawing lines out to the extremes of the data rather than stopping at adjacent values. Although simpler, such graphs hide the outliers.
3. A distinction between mild and severe outliers. Typically, *severe outliers* are defined as cases more than 3IQR beyond the first or third quartile. *Mild outliers* are cases more than 1.5IQR, but not more than 3IQR, beyond the quartiles.[8]
4. Embellishment of the box to show confidence intervals ("notched boxplots") or density.[9]
5. Simplification for the sake of clarity when many boxplots are shown together or combined with other graphs. For example, boxplots in the margins of another graph (for example, Figure 1.13) usually omit outliers to avoid confusion.

Boxplots provide information at a glance about center (median), spread (interquartile range), symmetry, and outliers. With practice they are easy to read and are especially useful for quick comparisons of two or more distributions. Sometimes unexpected features such as outliers, skew, or differences in spread are made obvious by boxplots but might otherwise go unnoticed.

Boxplots are a technique for *exploratory data analysis* (EDA), an approach developed by John Tukey (1977). EDA takes an open-minded, exploratory attitude toward the data, employing a modern toolbox of resistant and/or graphic analytical techniques (like boxplots). These EDA techniques require relatively few assumptions but are designed to detect and cope with problematic data. Tukey's approach contrasts with traditional techniques, which often start out with stronger assumptions (for example, Gaussian distributions) and work poorly when these assumptions are false. This book covers both traditional and EDA techniques, but the guiding philosophy leans toward EDA.[10]

Symmetry Plots

Boxplots and other EDA techniques were originally developed for paper-and-pencil work, in the days when computer analysis was difficult and expensive. The computer revolution has since opened up new possibilities, making it simple to construct graphs that would be tedious to do by hand. One example is the *symmetry plot*.

The median divides a distribution in half. If the distribution is exactly symmetrical, then for each value above the median there is another value the same distance below the median. A *symmetry plot* (Figure 1.4) graphs the distance from the median of the *i*th value above the median, against the distance from the median of the *i*th

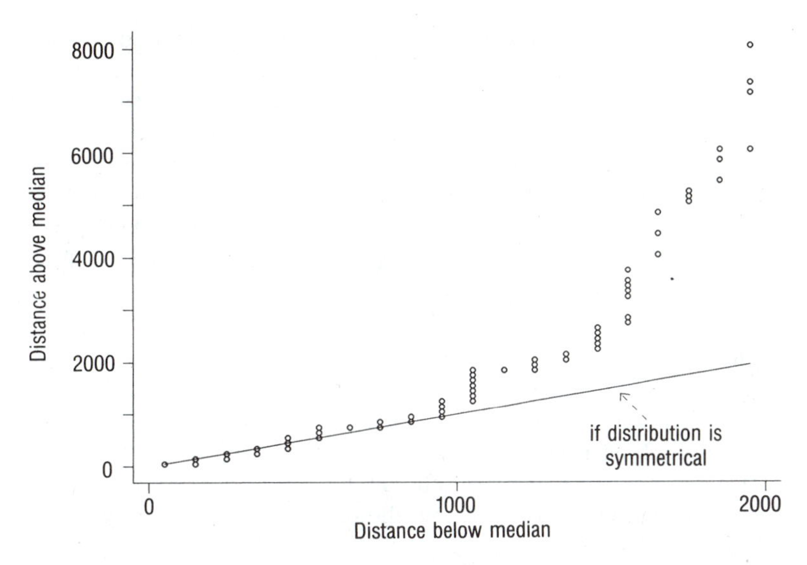

Figure 1.4 Symmetry plot of household water use (positively skewed).

value below the median. Each pair of values defines one point, so a symmetry plot based on n cases contains about $n/2$ points.

If a distribution were perfectly symmetrical, all symmetry-plot points would lie on the diagonal line. Off-line points indicate asymmetry. Points fall above the line when distance above the median is greater than corresponding distance below the median. A consistent run of above-the-line points indicates positive skew; a run of below-the-line points indicates negative skew. Figure 1.4 shows that the distribution of 1981 water use is roughly symmetrical near the median, but positive skew increases as we look farther out and compare the two tails. Figures 1.2 and 1.3 also show this pattern of central symmetry and asymmetrical tails, but Figure 1.4 makes it more obvious.

Quantile Plots

Quantiles are order statistics similar to percentiles but expressed as fractions or proportions instead of percentages. The .25 quantile, for instance, equals the 25th percentile. It is the value that divides the lower .25 of the data from the upper .75. The .63 quantile (or 63rd percentile) divides the lower .63 from the upper .37, and so on.

Sample quantiles are usually the actual data values. When we sort data in ascending order, the ith value, Y_i, becomes the $(i - .5)/n$ quantile. Here $(i - .5)/n$ is the fraction of cases below Y_i, counting half of the ith case itself. For example, when the Concord data are sorted in ascending order of water use, the 400th household used 3200 cubic feet of water; 3200 is therefore the $(400 - .5)/496 = .81$ quantile. That is, about 81% of the households used less than 3200 cubic feet.

Quantile plots graph quantiles (data values) against corresponding fractions, as illustrated in Figure 1.5. The 400th household plots at coordinates (.81, 3200). Each of the other 495 households likewise appears as a (fraction, quantile) point in this graph.

In a *uniform* or *rectangular* distribution, all possible values of the variable are equally common. A histogram shows a flat, rectangular block. A quantile plot of a uniform distribution would place all points along a straight diagonal line, like the line drawn in Figure 1.5. The actual data points do not follow the line, because this water-use distribution is not uniform.

Sets of cases with identical values produce flat areas in a quantile plot. Where successive values do not differ by much, the points climb slowly. Where larger differences or gaps separate successive values, the points climb more steeply. Thus the relatively flat or slow-climbing areas occur where data are most dense, and the steep-climbing areas show where data are sparse. In Figure 1.5 we see a high density of cases (that is, many households) between about 1000 and 3000 cubic feet and much lower density (fewer households) at high values, especially above 4000 cubic feet.

We can estimate the value of any order statistic from a quantile plot, as illustrated in Figure 1.6. Look along the horizontal axis for the appropriate fraction, and then read the height of the line of plotted points above. For example, the .05 quantile (5th percentile) equals the height of the line above the .05 fraction (500 cubic feet); the .25 quantile (first quartile) equals its height above the .25 fraction (1200); the .5 quantile or median is the height above .5 (2050), and so forth. The interquartile range equals the rise in the line of points between .25 and .75 fractions.

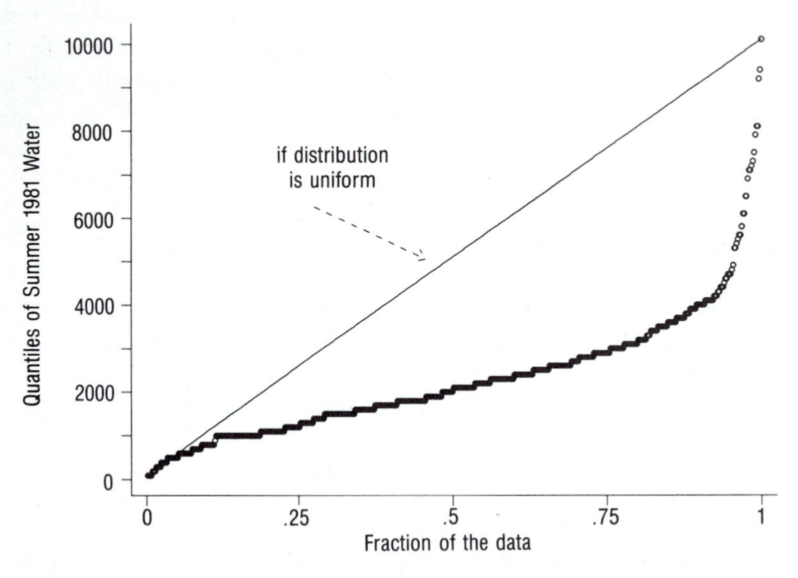

Figure 1.5 Quantile plot of household water use.

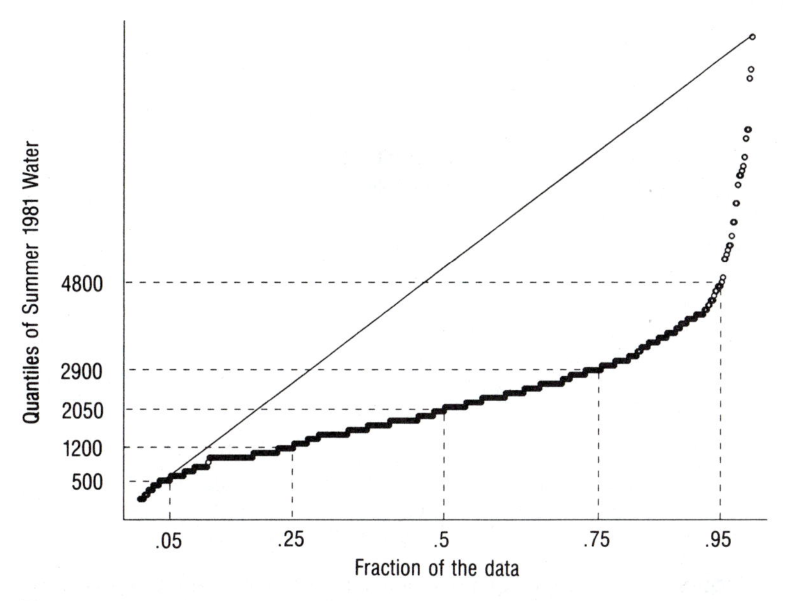

Figure 1.6 Order statistics as read from a quantile plot.

Alternatively, we can read quantile plots to find what fraction of the cases lie below any given value—for instance, what fraction used less than 3000 cubic feet of water (about .80 or 80%, judging from Figure 1.5). Unlike histograms and boxplots, which simplify the data and hence lose information, quantile plots display every single case. Consequently, quantile plots contain more information than histograms or boxplots, although they require more thinking to read.

Quantile-Quantile Plots

Quantile-quantile plots graph quantiles of one variable against quantiles of a second variable. Since sample quantiles are the data values in ascending order, this amounts to graphing the sorted values of Y against the sorted values of X. That is, we graph a set of points with coordinates (X_i, Y_i), where X_i is the ith-from-lowest value of X and Y_i is the ith-from-lowest value of Y. Two uses for quantile-quantile plots are:

1. comparing two empirical distributions;
2. comparing an empirical distribution with a theoretical distribution (Gaussian, for example).

Figure 1.7 shows an empirical quantile-quantile plot: quantiles (sorted values) of summer 1981 water use versus quantiles of summer 1980 water use. If the two distributions were identical, all points would lie on the diagonal line. At low levels of water use, below about 1000 cubic feet, the postshortage (1981) distribution is similar but slightly below the preshortage (1980) distribution. At higher levels of

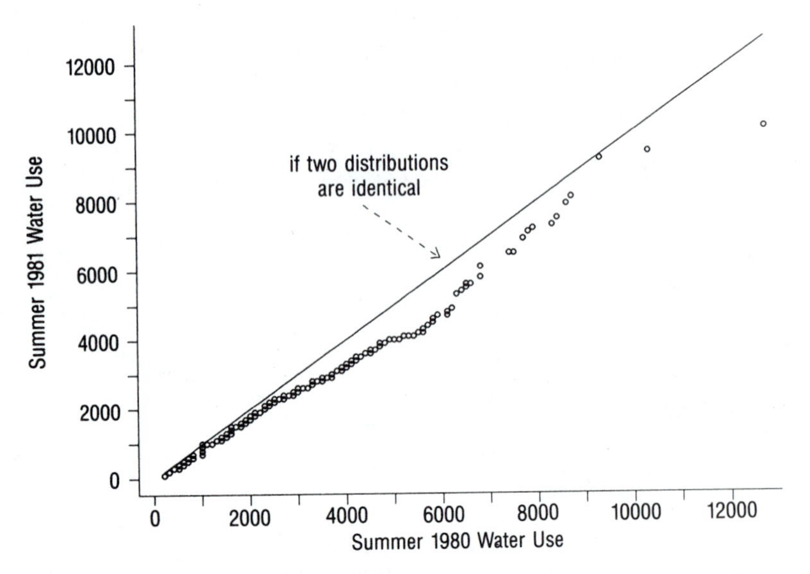

Figure 1.7 Quantile-quantile plot showing postshortage versus preshortage household water use.

water use, the points fall farther below the diagonal line. This indicates that quantiles of 1981 water use are systematically lower than corresponding quantiles of 1980 water use; less water was used after the shortage. The two distributions differ most near the middle of this plot, where points (not counting the two highest quantiles) fall farthest below the line.

Figure 1.8 contains two additional quantile-quantile plots. At left is a plot of summer 1980 use against summer 1979 use. Apart from the four highest quantiles, the distributions of water use in these two preshortage years are nearly identical—

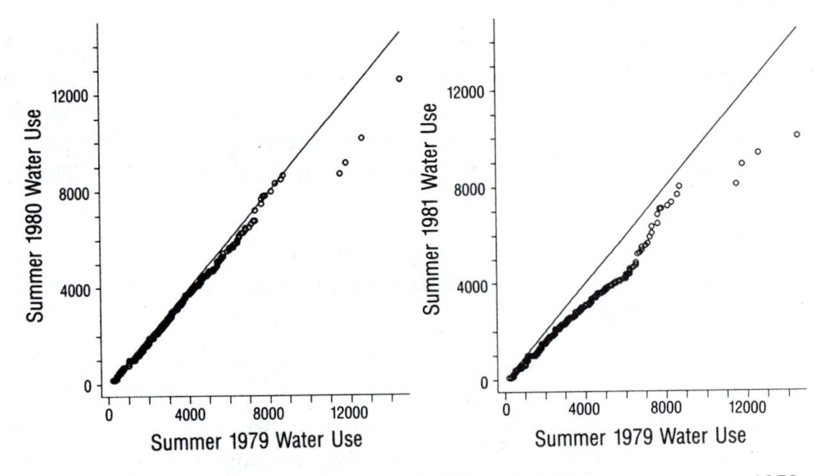

Figure 1.8 Quantile-quantile plots of 1980 and 1981 water use versus 1979 water use.

we see nothing like the systematic reduction that followed the water shortage (Figure 1.7). Reductions in the highest quantiles may reflect the *cost* of high water consumption, which rose due to a rate increase in fall of 1979. The same might apply to the highest quantiles in Figure 1.7 and at right in Figure 1.8: for high-use households, water bills alone provide an incentive for conservation, with or without a shortage. The right-hand plot in Figure 1.8 graphs quantiles of 1981 water use against quantiles of 1979 water use. The pattern is similar to that seen in Figure 1.7: postshortage water use is systematically lower.

Quantile-quantile plots convey details about how two distributions differ:

1. If quantile points lie along the diagonal line, the two distributions are *similar in center, spread, and shape.*
2. If the points follow a straight line that is parallel to the diagonal, but above or below it, the two distributions are *similar in spread and shape but have different centers.*
3. If the points follow a straight line not parallel to the diagonal, the distributions are *similar in shape but have different spreads.* They may also have different centers.
4. If the points do not follow a straight line, the distributions have *different shapes.* They may also differ in center and spread.

Distributions often differ in more than one way. For example, Figures 1.7 and 1.8 reveal that the distribution of postshortage water use (1981) differs from that of preshortage water use (1980 or 1979) in three respects:

1. Postshortage use has a lower center (points are systematically below the diagonal).
2. Postshortage use has less spread or variation (points get farther below the line as water use increases).
3. The two distributions have slightly different shapes (points do not follow a straight line).

Quantile-Normal Plots

Figures 1.7 and 1.8 show *empirical* quantile-quantile plots, which make detailed comparisons between two empirical distributions. *Theoretical* quantile-quantile plots, in contrast, allow comparisons between an empirical distribution and a theoretical distribution. Figure 1.9 shows a *quantile-normal plot,* sometimes called a *normal probability plot,* which graphs quantiles of 1981 water use (vertical axis) against corresponding quantiles of a theoretical Gaussian (normal) distribution with the same mean and standard deviation.

If water use followed a normal distribution, points would lie on the diagonal line in Figure 1.9. In the central region they approximately do, but the tails exhibit marked curvature, or nonnormality. At upper right the points rise steeply: the upper tail is heavier than normal. There are more high-water-use households than we would expect if water use were normally distributed. At lower left the points

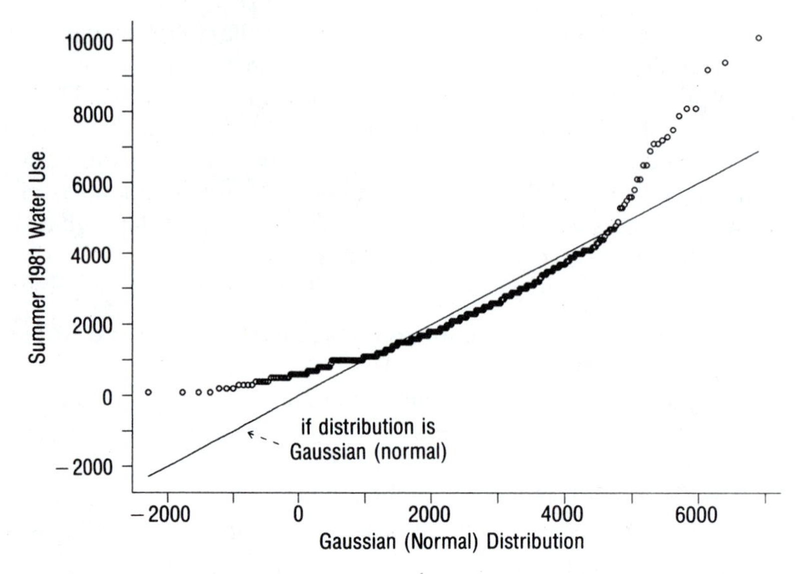

Figure 1.9 Quantile-normal plot of household water use (positively skewed).

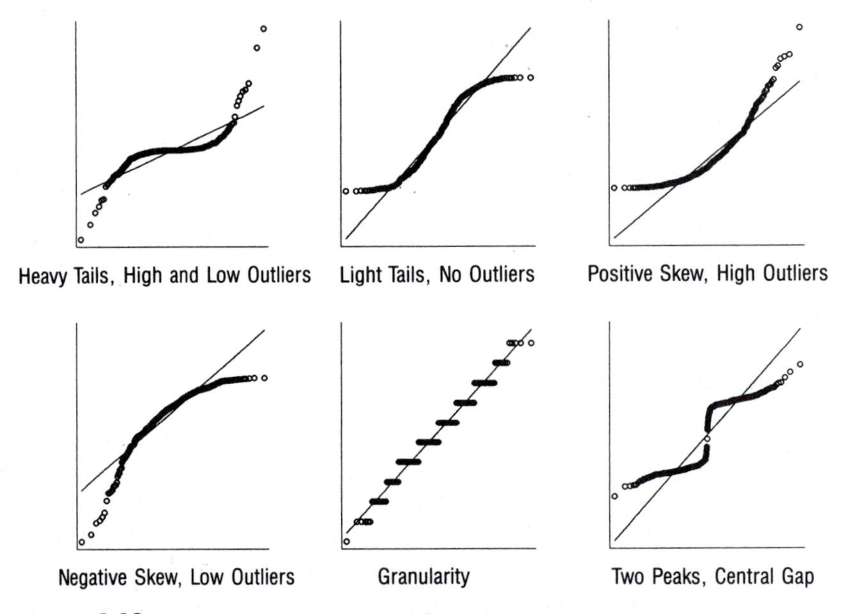

Figure 1.10 Quantile-normal plots reflect distribution shape.

rise slowly, indicating that the lower tail is lighter than nornal. There are fewer households with low water use (and none with negative water use) than would be expected if the distribution were normal.

These differences between the empirical water-use distribution and a theoretical normal distribution also appear in the histogram and normal curve of Figure 1.2. Quantile-normal plots allow more detailed analysis, however. They show how two distributions compare to every quantile, or every sample value.

Figure 1.10 illustrates how quantile-normal plots reflect distribution shape:

1. Heavy-tailed distributions are steepest at the top and bottom.
2. Light-tailed distributions are least steep at the top and bottom.
3. Skewed distributions have one heavy and one light tail. This gives plots of positively skewed distributions a downward-bowed appearance, like that of water use in Figure 1.9. Negatively skewed distributions bow upward.
4. Outliers appear as points toward the upper right or lower left, *vertically* separated from the rest of the distribution.
5. *Granularity* means that certain discrete values occur repeatedly in the data. These appear as plateaus separated by gaps.
6. Since less-steep areas indicate higher-than-normal data density, two less-steep areas separated by a gap or steep climb (lower-than-normal density) indicate that the distribution has two peaks (is *bimodal*).

Power Transformations

Skew and outliers create problems even for simple statistics like the mean. Their potential for mischief grows as we move to more complicated analysis. Fortunately skew can often be reduced, and outliers pulled in, by *power transformation*, which refers to a family of simple transformations:

Y^q	$q > 0$
$\log Y$	$q = 0$
$-(Y^q)$	$q < 0$

We seek to reduce statistical problems by choosing an appropriate value for q.[11]

Log Y denotes the logarithm of Y, which takes the place of zero among power transformations.[12] Two kinds of logarithms appear in this book:

■ *base 10 logarithms*, written $\log_{10} Y$. The base 10 logarithm of Y ($Y > 0$) is the power to which 10 must be raised to yield Y.

natural logarithms, written $\log_e Y$. The natural logarithm of Y ($Y > 0$) is the power to which e ($e = 2.71828\ldots$) must be raised to yield Y. ■

Base 10 and base e logarithms have identical effects on distributional shape. For general statements that apply to either kind, I write simply log Y.

Since power transformations are undefined for some values, we may need to add a positive constant before transformation. For example, log Y is undefined for

$Y \leq 0$. If the lowest data value of Y is 0, we modify the $q = 0$ transformation to $\log(Y + 1)$.

The *ladder of powers* (Tukey, 1977) is a graduated series of power transformations:

Y^3	$q = 3$
Y^2	$q = 2$
Y^1	$q = 1$
$Y^{.5}$	$q = .5$
$\log Y$	$q = 0$
$-(Y^{-.5})$	$q = -.5$
$-(Y^{-1})$	$q = -1$

Higher or lower values of q could be added to this series of round-number powers.

Power transformations systematically change distributional shape:

$q > 1$: Powers greater than 1 shift weight to the upper tail of the distribution and thereby *reduce negative skew*. The higher the power $(2, 3, \ldots)$, the stronger this effect.

$q = 1$: the raw data.

$q < 1$: powers less than 1 pull in the upper tail and may therefore *reduce positive skew*. The lower the power $(.5, 0, -.5, \ldots)$, the stronger this effect. To preserve order, add minus signs after raising to powers less than zero.

By selecting an appropriate power transformation, we may be able to pull in outliers and make a skewed distribution more symmetrical. (These transformations

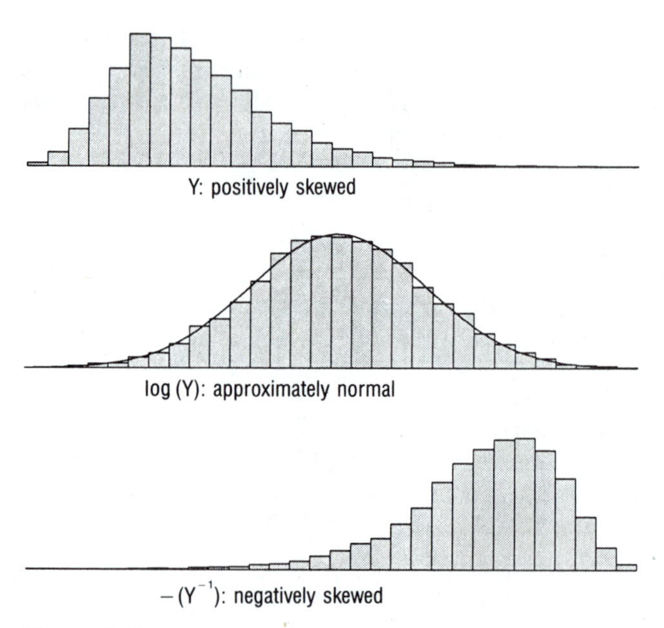

Figure 1.11 How power transformations affect a positively skewed variable Y.

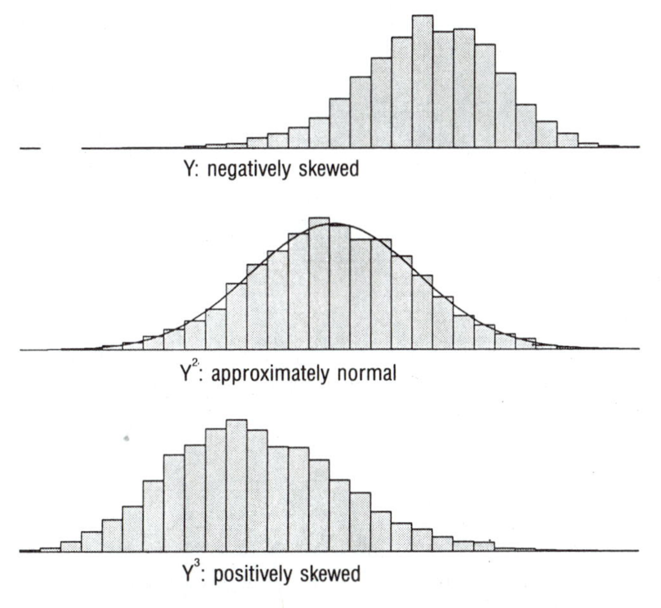

Figure 1.12 How power transformations affect a negatively skewed variable Y.

also have other beneficial effects, which will be seen in later chapters.) Analysis then proceeds using the transformed data.

Powers less than 1, and particularly logarithms ($q = 0$), are used widely because positive skew is so common. In some fields researchers take logarithms routinely, as the first analytical step.

Figure 1.11 illustrates how transformations affect positive skew. Since the raw data (top) are positively skewed, we should choose $q < 1$. For this example, logarithms ($q = 0$) produce an approximately normal distribution (middle). Transformations between $q = 1$ and $q = 0$ would not have been strong enough; some positive skew would remain. Still stronger transformations like negative reciprocals ($q = -1$, bottom) go too far and create *negative* skew.

Figure 1.12 shows an opposite progression. The raw data (top) are negatively skewed, so we choose $q > 1$. Squaring ($q = 2$) works well, but cubing ($q = 3$) goes too far and creates *positive* skew. Inappropriate transformations, such as squaring a distribution that is already symmetrical or positively skewed, usually worsen statistical problems.

Selecting an Appropriate Power

Since the distribution of household water use (Figures 1.2–1.9) is positively skewed, it might be made more symmetrical by a transformation with $q < 1$. Logarithms ($q = 0$) are a good first choice. Figure 1.13 is a graph of $\log_e Y$ against Y, showing

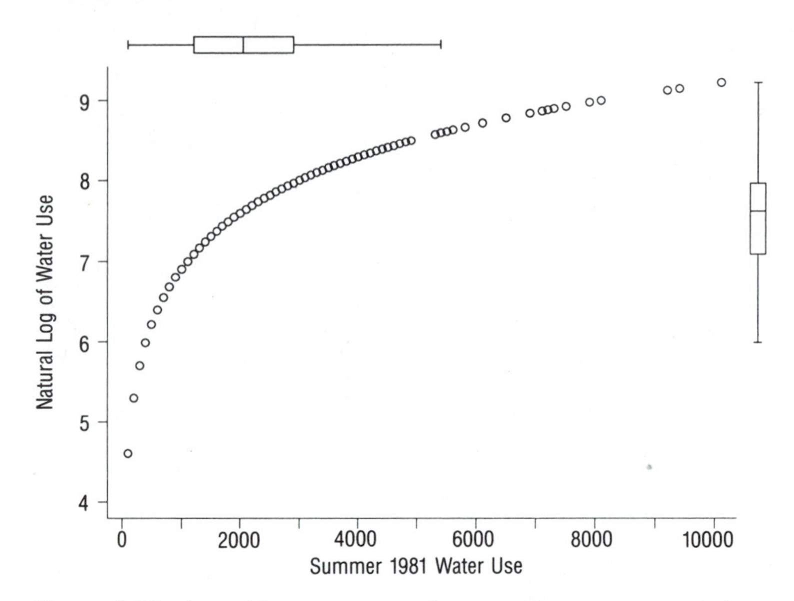

Figure 1.13 Logarithms versus raw data, postshortage water use (scatterplot with marginal boxplots).

the effects of taking natural logs of water use. High values are compressed, pulling in the upper tail of this distribution. For example, in logarithms the difference between 1000 cubic feet ($\log_e(1000) \approx 6.9$) and 3000 cubic feet ($\log_e(3000) \approx 8.0$) is much greater than the difference between 7000 cubic feet ($\log_e(7000) \approx 8.9$) and 9000 cubic feet ($\log_e(9000) \approx 9.1$).

The horizontal boxplot at top in Figure 1.13 shows the distribution of water use, and the vertical boxplot at right shows log water use. As we already know, the water-use distribution is positively skewed. It appears, however, that log water use is *negatively* skewed. The log transformation is *too powerful*; we should try other transformations between $q = 1$ (raw data, no change) and $q = 0$ (logs, too much change).

Compare Figures 1.14 and 1.15. Figure 1.14 recapitulates the graphs of raw data on water use, seen earlier; all four graphs reveal positive skew and high outliers. Figure 1.15 shows a similar set of graphs for the logarithm of water use. These four graphs, in contrast, reveal negative skew and low outliers.

Since the raw data ($q = 1$) are positively skewed (Figure 1.14) and logarithms ($q = 0$) are negatively skewed (Figure 1.15), the ladder of powers suggests that we should next try $q = .5$, or square roots. Figure 1.16 shows a symmetry plot of $Y^{.5}$, the square root of water use. If we had achieved symmetry, all points would lie on the diagonal line, which they do not. The tails remain positively skewed, although less so than with the raw data (lower left in Figure 1.14).

We now know that the .5 power is not strong enough, because $Y^{.5}$ is still positively skewed. Although the ladder of powers directs us to first try round-number powers, we evidently need something between $q = .5$ and $q = 0$. A value that works well is $q = .3$. Four graphs of the distribution of the .3 power of water use, $Y^{.3}$,

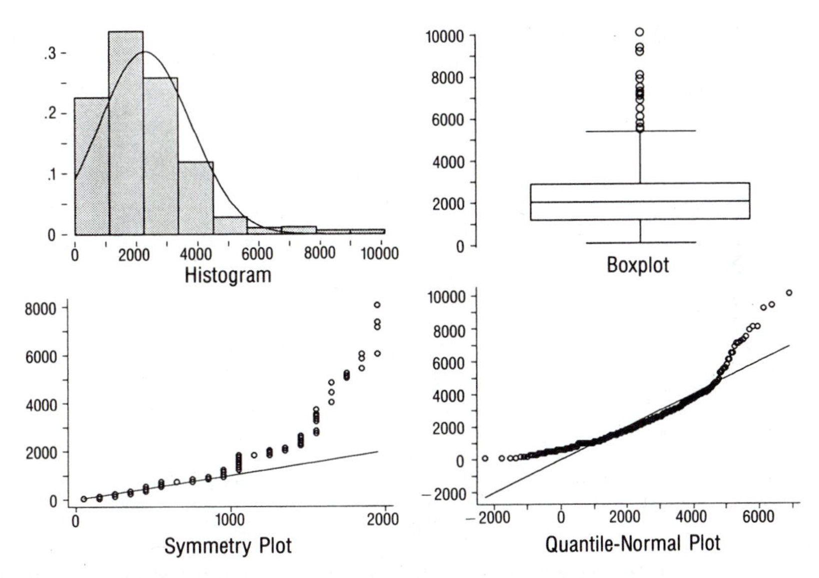

Figure 1.14 Four views of the distribution of household water use (positively skewed).

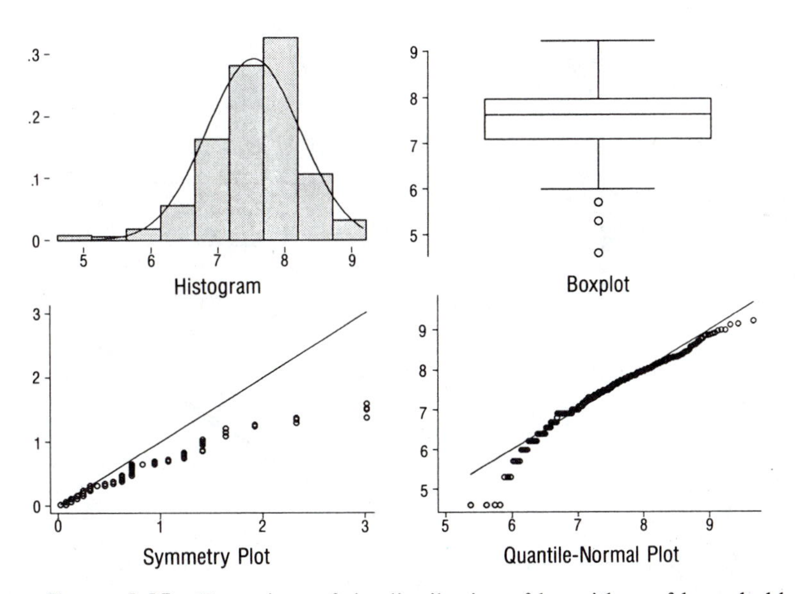

Figure 1.15 Four views of the distribution of logarithms of household water use (negatively skewed).

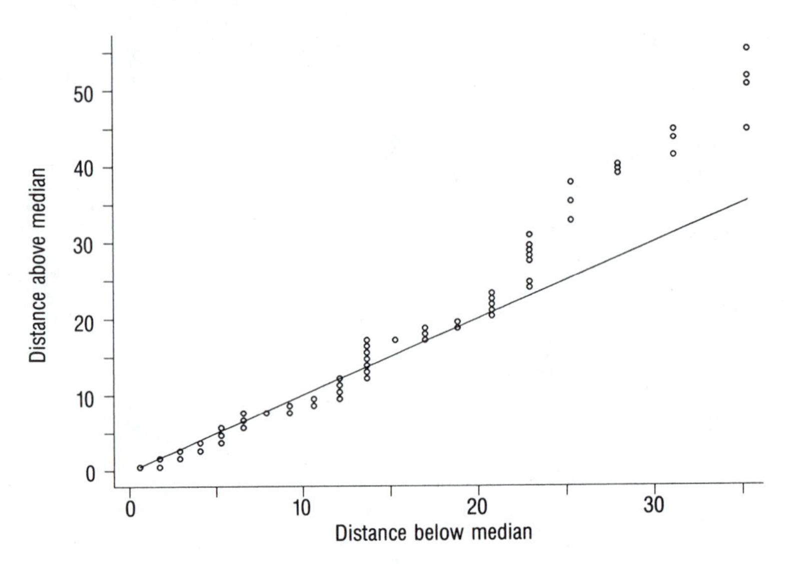

Figure 1.16 Symmetry plot of distribution of *square roots* of household water use (mild positive skew).

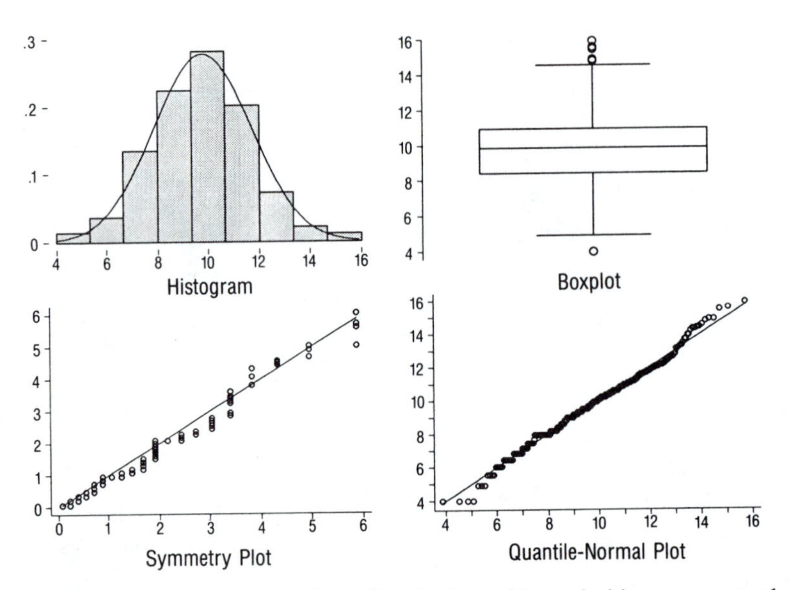

Figure 1.17 Four views of the distribution of household water use to the 0.3 power (approximately normal).

appear in Figure 1.17. We see that:

1. This distribution is approximately symmetrical, because the symmetry-plot points lie close to the diagonal line. Also, the histogram and boxplot look symmetrical.
2. Furthermore, the distribution is approximately normal, because the quantile-normal-plot points lie close to the diagonal line. Also, the histogram's outline resembles that of the normal curve.

Remember that normality and symmetry are not the same thing. All normal distributions are symmetrical, but not all symmetrical distributions are normal. With water use we were able to transform the distribution to be approximately symmetrical *and* normal, but often symmetry is the most we can hope for. For practical purposes, symmetry (with no severe outliers) may be sufficient. Transformations are not a magic wand, however. Many distributions cannot even be made symmetrical.

Figures 1.14, 1.15, and 1.17 might be used for reference as examples of how these four types of graphs appear with positively skewed (Figure 1.14), negatively skewed (Figure 1.15), and approximately normal (Figure 1.17) distributions.[13]

Conclusion

In reviewing the graphing and transformation techniques presented in this chapter, some readers may suspect that we are trying to swat flies with a hammer. The tools seem heavier than the job requires. If all we needed to do was summarize a single variable, this would probably be true. We could present the median, the interquartile range, and the histogram of water use and be done with it. But what if we want to examine how postshortage water use is related to education, controlling for income, number of residents in a household, and preshortage water use? A median or histogram is no longer enough. As research questions grow more complex, we must turn to more powerful analytical methods, which unfortunately can be very sensitive to details of distributional shape. Consequently, techniques for inspecting and transforming distributions become increasingly important.

Exercises

In 1989, as Cold War tensions relaxed, Soviet scientists for the first time published data about nuclear weapons tests conducted during 1961–1972 at sites in Soviet Central Asia. Western scientists had previously estimated the size of these explosions from their seismic magnitudes. Publication of the Soviet data provided an opportunity to check the accuracy of Western magnitude estimates. The following table gives magnitude estimates and reported yields for 19 of these tests (data from Vergino, 1989).

Magnitude estimates and yields of 19 Soviet weapons tests

Test Date	Western Seismologists' Magnitude Estimate	Soviet Reported Yield in Kilotons
November 21, 1965	5.6	29
February 13, 1966	6.1	125
March 20, 1966	6.0	100
May 7, 1966	4.8	4
September 22, 1967	5.2	10
September 29, 1968	5.8	60
July 23, 1969	5.4	10
November 30, 1969	6.0	125
December 28, 1969	5.7	40
April 25, 1971	5.9	90
June 6, 1971	5.5	16
October 9, 1971	5.3	12
October 21, 1971	5.5	23
February 10, 1972	5.4	16
March 28, 1972	5.1	6
August 16, 1972	5.0	8
September 2, 1972	4.9	2
November 2, 1972	6.1	165
December 10, 1972	6.0	140

Later we will examine the relationship between estimated magnitude and yield of these Soviet tests (Chapter 2, Exercises 11–13). As a preliminary step, Exercises 1–6 look at the variables' univariate distributions.

1. Find the mean, standard deviation, median, and IQR of the Western magnitude estimates. Compare mean with median, and compare standard deviation with IQR/1.35. What do these comparisons suggest about the shape of this distribution?
2. Sort the magnitude estimates from low to high. For each data value, calculate and list the corresponding sample quantile, or $(i - .5)/n$. Indicate the approximate locations of median, quartiles, and IQR with respect to this list.
3. Find the mean and median of reported test yields. What do these indicate about the distribution? Why should we not bother to compare standard deviation with IQR/1.35?
4. Examine these graphs of the yield distribution, and describe what they tell us about these 19 Soviet weapons tests:
   a. Histogram
   b. Boxplot
   c. Symmetry plot
   d. Quantile-normal plot (compare with Figure 1.10)
5. Positively skewed distributions may be made more symmetrical by power transformations with $q < 1$. Calculate three new variables by applying these transformations (from the ladder of powers) to the yield data:
   a. $X^{.5}$
   b. $\log_{10} X$
   c. $-(X^{-.5})$

6. For each of the transformed variables from Exercise 5, construct a symmetry plot. Describe what you see. Which transformation works best? (Especially in small samples like this one, we should not expect to see perfect symmetry or "normality.")

The next table presents data on 14 samples of indoor air from a house near a metal-pipe foundry in Phillipsburg, New Jersey (adapted from Lioy, Waldman, Greenberg, Harkov, and Pietarninen, 1988). Investigators suspected that Benzo(a)pyrene, or BaP, from the pipe foundry might be contaminating household air. The measures are concentrations of BaP-containing particles no larger than 10 micrograms (μg). Exercises 7–8 refer to these data.

Airborne Benzo(a)pyrene particles

Sample	BaP in indoor air ($\mu g/m^3$)	BaP in outdoor air ($\mu g/m^3$)
1	10	24
2	10	35
3	25	41
4	40	65
5	40	27
6	45	56
7	45	67
8	55	50
9	55	25
10	70	20
11	75	78
12	90	25
13	220	38
14	285	40

7. Construct a symmetry plot and boxplot of indoor air BaP. Describe what you see.
8. Try to symmetrize the indoor-air BaP distribution, using several appropriate transformations from the ladder of powers. Construct symmetry and boxplots to evaluate these efforts. Which transformation works best?
9. Use a computer program's pseudorandom number function to generate three sets of artificial data: samples of $n = 8$, $n = 25$, and $n = 100$ cases from a Gaussian population with $\mu = 1.41$ and $\sigma = .58$.
   a. Compare your sample means with μ and your sample standard deviations with σ.
   b. For each sample, construct a histogram, boxplot, symmetry plot, and quantile-normal plot. Describe what you see.
   c. As a class project, compare your graphical results with others'. Do you see any systematic differences in apparent "normality" among $n = 8$, $n = 25$, and $n = 100$ samples from normal populations?
10. As a further class project, extending Exercise 9, compile a *distribution* of sample means from all the $n = 8$ samples. What is the mean and standard deviation

of this distribution? Compare with the distributions of sample means from all the $n = 25$ and all the $n = 100$ samples. Are your findings consistent with the Central Limit Theorem (Note 2)? Discuss what that theorem means.

11. Following are estimates of populations of 27 Scottish seabird (cormorant) colonies, from Furness and Birkhead (1984). Use histograms, boxplots, symmetry plots, and quantile-normal plots, together with numerical summaries (mean-median comparisons, skewness statistics), to describe the shape of this distribution and to investigate which ladder-of-powers transformation most improves symmetry or normality.

Colony		Number of breeding pairs
1	Hascosay	36
2	Whalsay	56
3	Muckle Green	65
4	St. Ninians	95
5	S. W. Unst	136
6	Noss	141
7	Bay of Bursay	146
8	Rousay	150
9	S. E. Yell	154
10	Wats Ness	156
11	Burra	191
12	Muckle Roe	232
13	Noup	246
14	P. Westray	260
15	Uyea	275
16	Stronsay	285
17	Hoy	310
18	Papa Stour	348
19	Eday & Calf	354
20	Sumburgh	371
21	Deerness	436
22	Rapness	468
23	Fetlar	500
24	S. Ronaldsay	521
25	Fair Isle	1530
26	N. W. Unst	1696
27	Foula	2000

Notes

1. The word *normal* carries a misleading implication that such distributions are "usual" or "ordinary." To avoid this implication, some statisticians prefer to call such distributions *Gaussian*, after the German scientist Carl Friedrich Gauss, 1777–1855.

2. The Central Limit Theorem states that, as sample size (n) becomes large:
   a. the sampling distribution of the mean becomes approximately normal, regardless of the shape of the variable's frequency distribution.
   b. the sampling distribution will be centered around the variable's population mean μ.
   c. the standard deviation of this sampling distribution, called its *standard error*, approaches $\sigma/\sqrt{n}$, the variable's population standard deviation divided by the square root of the sample size.
3. Large samples like the Concord Water Study ($n = 496$) usually provide a good picture of the general shape of the population distribution. Small-sample distributions, on the other hand, may not closely resemble their population source; they contain less information for checking distributional assumptions.
4. Weight, rather than length alone, determines how the tails of a skewed distribution affect the mean. Furthermore, tail "length" is not useful in describing many theoretical distributions, which have infinitely long tails.
5. Standard deviation and variance, based on *squared* deviations from the mean, have even less resistance to extreme values than the mean does. Many other statistics, also based on squared deviations, share this weakness.
6. Many computer programs calculate statistics called *skewness* and *kurtosis*. The skewness statistic is based on the third power of deviations around the mean. Positive values indicate positive skew, and so on. Because it cubes deviations from the mean, this statistic can be greatly affected by a single far-from-the-mean value.

   The kurtosis statistic is based on the fourth power of deviations around the mean, so it is even less resistant. High kurtosis values indicate heavier-than-normal tails, and low kurtosis indicates lighter-than-normal tails.

   Comparisons of mean with median, and (if symmetrical) of standard deviation with IQR/1.35, provide more resistant ways to assess skew and kurtosis. You should consult graphs when interpreting any shape statistic.
7. See Frigge, Hoaglin, and Iglewicz (1989). Quartiles derived from [1.6] correspond to their definition #5, which is the default in the SPSS, Statgraphics, and Stata statistical programs (and an option in SAS).
8. Hoaglin, Iglewicz, and Tukey (1986) used Monte Carlo simulation to study the occurrence of outliers in random samples from a Gaussian population. They found that *mild outliers*,

$$Q_1 - 3\text{IQR} \le Y_i < Q_1 - 1.5\text{IQR} \qquad \text{or} \qquad Q_3 + 1.5\text{IQR} < Y_i \le Q_3 + 3\text{IQR}$$

   which make up about 0.7% of a Gaussian population, occur often in random samples of any size. (About 20% of their $n = 10$ samples contained mild outliers, as did 52.9% of the $n = 100$ and 85.2% of the $n = 300$ samples.) On the other hand *severe outliers*,

$$Y_i < Q_1 - 3\text{IQR} \qquad \text{or} \qquad Y_i > Q_3 + 3\text{IQR}$$

which make up about .0002% (two per million) of a Gaussian population, rarely occur in small to moderate-sized Gaussian samples. (Only .362% of Hoaglin et al.'s $n = 10$ samples, and .001% of their $n = 300$ samples, contained severe outliers.) Judging from these simulation results we might take the presence of *any severe outliers*, in samples of $n = 10$ to at least 300, as sufficient evidence for rejecting a normality hypothesis at the 5% (or even 1%) significance level.

9. Velleman and Hoaglin (1981) suggest notched boxplots with intervals around the medians:

$$\text{Md} \pm 1.58(\text{IQR})/\sqrt{n}$$

Medians of two distributions whose intervals do not overlap may be said to be significantly different at approximately the .05 level (see pp. 73–75 and 79–81). This is a graphical aid rather than a formal test and does not take into account the number of comparisons made.

For more ideas about embellishing the box of a boxplot, see Benjamini (1988).

10. A widely cited proposal on improving mathematics education (Moore, 1990) argues for EDA in teaching high school students to think about numerical data. Creative researchers have extended EDA by developing more resistant statistics (Chapter 6) and new computer-graphics analytical methods like point cloud spinning (examples in Cleveland and McGill, 1988). Such developments mark the growing influence of Tukey's EDA philosophy on mainstream statistical thought. See Hoaglin, Mosteller, and Tukey (1983, 1985) for a readable, fairly complete introduction to EDA.

11. For some purposes a more general formulation, called the *Box-Cox* transformation, is preferred:

$$\frac{Y^q - 1}{q} \qquad q \neq 0$$

$$\log_e Y \qquad q = 0$$

Box-Cox transformations have the same effect on distributional shape as simpler power transformations but fit with more formal schemes for selecting the appropriate power q.

12. For any nonzero Y, $Y^0 = 1$, so the true zero power is useless as a transformation. Logs "take the place" of the zero power because, as q approaches 0, the distributional effects of Y^q (or, if $q < 0$, of $-[Y^q]$) approach those of the $\log Y$ transformation.

13. For more details about symmetry plots, quantile plots, and other modern graphical methods, see Chambers, Cleveland, Kleiner, and Tukey (1983). An excellent book about the theory and practice of graphing in general is Tufte (1983).

2 Bivariate Regression Analysis

When evaluating evidence for a theoretical claim that X causes Y, scientists consider three points:

1. *Time ordering*: the value of X at one specific time affects later values of Y.
2. *Covariation*: X and Y vary together in a systematic (nonchance) way.
3. *Nonspuriousness*: the covariation between X and Y does not result entirely from their relations with other variable(s).

Bivariate analysis addresses mainly the second point: whether X and Y vary together. But, if they do covary, this does not prove that X causes Y. Perhaps Y causes X, not vice versa (point 1). Or perhaps a third variable causes both X and Y and hence makes them appear related (point 3).

Bivariate analysis alone thus provides insufficient evidence for causality. Multivariate analysis (Chapters 3–8) may provide better, though still incomplete, evidence for causality because it allows checking for possible spuriousness (point 3).

Even when research interest focuses on complicated multivariate hypotheses, however, basic bivariate work and univariate work remain crucial. They lay a solid foundation for multivariate analysis, in which troubleshooting becomes harder and uncorrected problems more damaging.

This chapter begins with the linear model that underlies two-variable regression analysis. A widely used method called *ordinary least squares* (OLS) fits linear models to data. OLS encompasses a system of techniques for describing sample data and extending conclusions to a larger population. Certain statistical problems can undermine the validity of OLS results, however. Experienced analysts learn to watch out for and remedy such problems wherever possible.

The Basic Linear Model

To summarize the relationship between two measurement variables X and Y, we might use a *model* or mathematical expression. One simple possibility is a *linear model*:

$$Y_i = \beta_0 + \beta_1 X_i \qquad [2.1]$$

In this expression β_0 and β_1 are constants; β_0 is the *Y-intercept*, and β_1 is the *slope*. Equation [2.1] shows X as a *predictor* of Y. In causal terms, [2.1] depicts X as the cause and Y as its effect. As previously, the i subscripts index individual cases or observations. Graphically, [2.1] defines a straight line:

■ β_0 (*Y-intercept*) is the height at which the line crosses the vertical Y-axis.

β_1 (*slope*) is the steepness of the line, expressed as rise/run. The line rises β_1 units with each 1-unit increase in X. (If $\beta_1 < 0$, the line actually falls as X increases; if $\beta_1 = 0$, the line is horizontal.) ■

Figure 2.1 shows two examples. At left is a line with Y-intercept $\beta_0 = 3$ and slope $\beta_1 = 1.5$:

$$Y = 3 + 1.5X$$

This line crosses the Y-axis at $(0, 3)$ and rises 1.5 units vertically (Y) with each 1-unit horizontal (X) increase. At right is a line with Y-intercept $\beta_0 = 10$ and slope $\beta_1 = -3$:

$$Y = 10 - 3X$$

This line crosses the Y-axis at $(0, 10)$ and falls 3 units vertically with each 1-unit increase in X.

If we know β_0 and β_1, [2.1] predicts what Y value corresponds to any given value of X. Furthermore, any two data points are sufficient to determine the parameters β_0 and β_1, just as two points (and a ruler) suffice to draw a straight line.

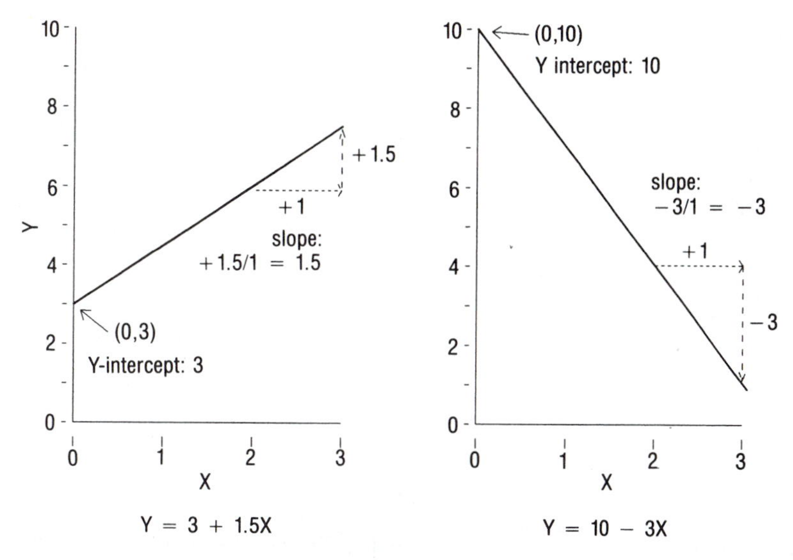

Figure 2.1 Two lines.

Relations in real data are rarely this simple, though. More realistically, we might claim only that the *expected value* (population mean) of Y, rather than exact individual values, changes linearly with X:

$$E[Y_i] = \beta_0 + \beta_1 X_i \tag{2.2}$$

β_0 equals the population mean of Y when $X = 0$. With each 1-unit increase in X, the population mean of Y changes by β_1 units.[1] Such linear models have many practical applications.

Other things besides X cause individual Y_i to vary around $E[Y_i]$. We represent these "other things" with an *error term*, ε_i (epsilon):

$$\begin{aligned} \varepsilon_i &= Y_i - (\beta_0 + \beta_1 X_i) \\ &= Y_i - E[Y_i] \end{aligned} \tag{2.3}$$

Actual Y_i equals expectation plus error:

$$\begin{aligned} Y_i &= E[Y_i] + \varepsilon_i \\ &= \beta_0 + \beta_1 X_i + \varepsilon_i \end{aligned} \tag{2.4}$$

Analysis cannot proceed far without some assumptions about ε. Common assumptions (discussed more fully in Chapter 4) are the following:

1. Errors have *identical distributions*, with zero mean and the same variance, for every value of X.
2. Errors are *independent*: unrelated to X variables or to the errors of other cases.
3. Errors are *normally distributed.*

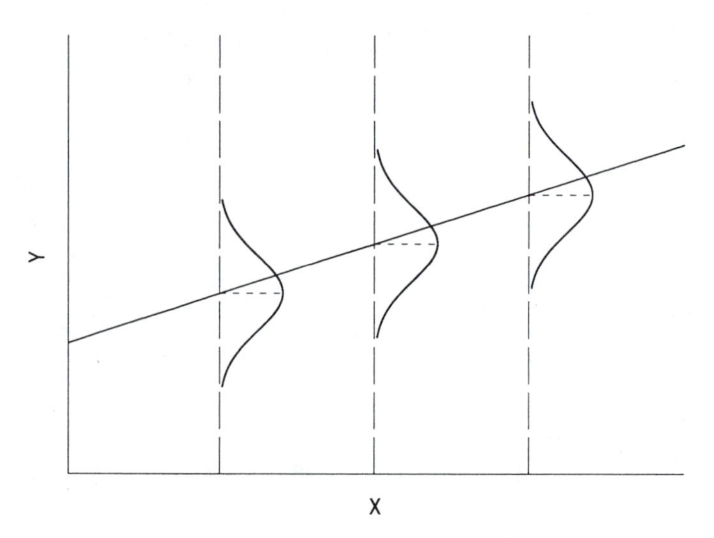

Figure 2.2 Errors normally, independently, and identically distributed at every level of X.

Figure 2.2 illustrates the idea of normal, independent, and identical error distributions (*normal i.i.d.*, for short). Errors ought to be normal i.i.d. if they represent the sum of many small random influences.

We generally do not know if these assumptions are true, nor do we know the true values of the population parameters β_0 and β_1. The remainder of this chapter discusses how we might estimate them using sample data. The process of estimating parameters for linear models like [2.2] is called *regression analysis.*

Ordinary Least Squares

We begin with a linear model ([2.2]) because substantive theory suggests it or, more often, because theory is vague but we wish to start simply. Either way, we want the best possible estimates of the parameters β_0 and β_1. Calling our estimates b_0 and b_1, the sample counterpart of [2.2] is

$$\hat{Y}_i = b_0 + b_1 X_i \qquad [2.5]$$

where $\hat{Y}_i$ ("Y-hat") denotes the *predicted value of* Y for case i.

Residuals (e_i), or sample counterparts of error terms (ε_i), equal differences between actual and predicted Y:

$$\begin{aligned} e_i &= Y_i - \hat{Y}_i \\ &= Y_i - b_0 - b_1 X_i \end{aligned} \qquad [2.6]$$

(As in Chapter 1, Greek letters like β and ε stand for population parameters, whereas lowercase Roman letters like b and e denote sample statistics.) Residuals measure prediction error. If actual Y is higher than predicted ($Y_i > \hat{Y}_i$), the residual is positive; if actual Y is lower ($Y_i < \hat{Y}_i$), the residual is negative. A perfect prediction ($Y_i = \hat{Y}_i$) results in a zero residual.

The *sum of squared residuals*, RSS, reflects the overall accuracy of our predictions.[2]

$$\begin{aligned} \text{RSS} &= \Sigma e_i^2 \\ &= \Sigma(Y_i - \hat{Y}_i)^2 \end{aligned} \tag{2.7}$$

The closer the fit between predictions and data, the lower the RSS. This suggests a criterion for the "best" values of b_0 and b_1:

■ Select b_0 and b_1 such that the sum of square residuals will be as low as possible. ■

The least-squares criterion leads to a technique called *ordinary least squares* (OLS).

Calculus obtains values of b_0 and b_1 that minimize the residual sum of squares:

$$\begin{aligned} \text{RSS} &= \Sigma e^2 = \Sigma(Y - \hat{Y})^2 = \Sigma(Y - b_0 - b_1X)^2 \\ &= \Sigma Y^2 - 2b_0\Sigma Y - 2b_1\Sigma XY + nb_0^2 + 2b_0b_1\Sigma X + b_1^2\Sigma X^2 \end{aligned}$$

(case subscripts dropped for simplicity). Taking the first partial derivative of RSS with respect to b_0, and setting this equal to zero, solve for b_0:

$$\frac{\partial \text{RSS}}{\partial b_0} = -2\Sigma Y + 2nb_0 + 2b_1\Sigma X = 0$$

$$nb_0 = \Sigma Y - b_1\Sigma X$$

$$b_0 = \bar{Y} - b_1\bar{X} \tag{2.8}$$

Similarly, take the derivative of RSS with respect to b_1, set it equal to zero, and solve for b_1:

$$\frac{\partial \text{RSS}}{\partial b_1} = -2\Sigma XY + 2b_0\Sigma X + 2b_1\Sigma X^2 = 0$$

Substituting $\Sigma Y/n - b_1\Sigma X/n$ (from [2.8]) for b_0, we get

$$-2\Sigma XY + \frac{2\Sigma X\Sigma Y}{n} - \frac{2b_1(\Sigma X)^2}{n} + 2b_1\Sigma X^2 = 0$$

$$b_1 n\Sigma X^2 - b_1(\Sigma X)^2 = n\Sigma XY - \Sigma X\Sigma Y$$

$$b_1 = \frac{n\Sigma XY - \Sigma X\Sigma Y}{n\Sigma X^2 - (\Sigma X)^2} \tag{2.9a}$$

Three alternative arrangements are equivalent to [2.9a]:

$$b_1 = \frac{\Sigma(X - \bar{X})(Y - \bar{Y})}{\Sigma(X - \bar{X})^2} \quad [2.9b]$$

$$= \frac{s_{XY}}{s_X^2} \quad [2.9c]$$

$$= r\frac{s_Y}{s_X} \quad [2.9d]$$

where s_{XY} is the covariance ([2.15]) and r the correlation between X and Y ([2.16]).

No other line, applied to the same data, will have an RSS as low as the line obtained by ordinary least squares ([2.8]–[2.9]). In this respect, OLS guarantees a solution that "best fits" the sample data.

OLS has several further attractions. It readily extends to models with two or more X variables, as we will see in Chapter 3. At a simpler level, the arithmetic mean is essentially OLS with *no* X variables:

$$\hat{Y}_i = b_0$$

for which the least-squares solution is $b_0 = \Sigma Y/n = \bar{Y}$. Thus OLS unifies our approach to simple and complex problems. It also has important theoretical advantages (discussed in Chapter 4).

OLS is just one of many techniques for regression analysis, although it is by far the most often used. Its theoretical advantages depend on conditions rarely found in practice. The farther we depart from these conditions, the less we can trust OLS. Because of its simplicity, generality, broad usefulness, and ideal-data properties, however, OLS has become the core technique of modern statistical research.

Scatterplots and Regression

In Chapter 1 we examined the sample distribution of household water use. Figure 2.3 shows the joint (bivariate) distribution of water use and income among the same 496 households. Each household appears as a point. Such *scatterplots* are basic tools for understanding relationships between two measurement variables.

Boxplots for income and water use appear in the margins. The bivariate scatter reflects the shape of these univariate distributions. Both income and water use have positively skewed distributions with high outliers. Consequently the points scatter most densely toward the lower left (low income, low water use) and thin out to a few outliers toward the upper (high water use) and right-hand (high income) parts of the graph.

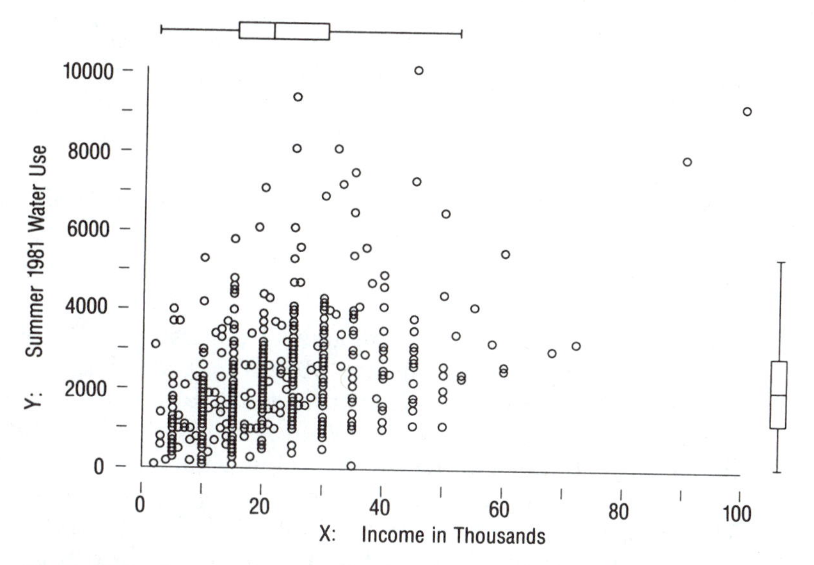

Figure 2.3 Scatterplot of water use versus income, with marginal boxplots.

Scatterplots follow the graphing conventions described earlier with Figure 2.1. The vertical axis is a scale for the Y variable; the horizontal axis is a scale for X. Water use, which might be caused or predicted by household income, is the Y variable in Figure 2.3.

The scatter of points in a two-variable plot may be summarized by a line, as shown in Figure 2.4. Such lines show how Y tends to change with X. The up-to-right pattern indicates that average water use increases with income—wealthier

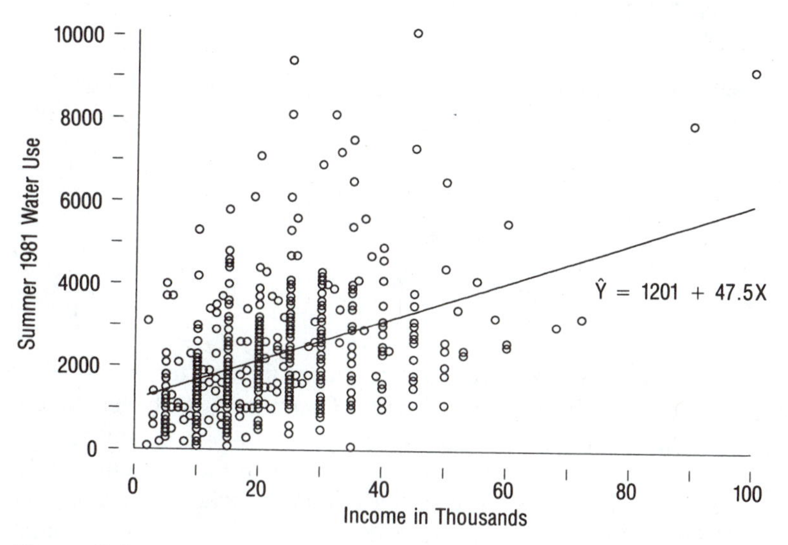

Figure 2.4 Scatterplot with regression line.

households tend to consume more water. Of course, we see exceptions—households with high income and low use (lower right) or with low income and high use (upper left). But the line summarizes the overall pattern of this two-dimensional distribution, much as a mean (a single point) summarizes the one-dimensional distribution of one variable.

Regression lines have the form $\hat{Y}_i = b_0 + b_1 X_i$. Figure 2.4 shows

$$\hat{Y}_i = 1201 + 47.5X_i \qquad [2.10]$$

where X stands for income (thousands of dollars) and $\hat{Y}$ is predicted water use (cubic feet). The height of this line when $X = 0$, at the left side of the graph, is $b_0 = 1201$. Thus [2.10] implies that households with zero income consumed, on average, 1201 cubic feet of water. When X values of zero lie outside the range of the data (as they do here), the Y-intercept may not make substantive sense.[3]

More often than Y-intercepts, slopes can be given a substantive interpretation. The slope, $b_1 = 47.5$, indicates that, with each $1000 increase in income, mean water use goes up by 47.5 cubic feet. We could also say that mean water use rises 475 cubic feet with each $10,000 increase in income, or 4750 cubic feet with a $100,000 increase.

Y-intercept $b_0 = 1201$ and slope $b_1 = 47.5$ were obtained by applying equations [2.8]–[2.9] to these data. Consequently the line $\hat{Y}_i = 1201 + 47.5X_i$ has a lower sum of squared residuals than any other line that might be drawn onto the scatterplot.

Equation [2.10] may fit best, but *how well* does it fit? One indication is the *residual standard deviation*, s_e:

$$s_e = \sqrt{\frac{\text{RSS}}{n - K}} \qquad [2.11]$$

where n is sample size and K the number of estimated parameters. We estimated two parameters, β_0 and β_1. Since RSS = 902,418,143, the residual standard deviation is

$$s_e = \sqrt{\frac{902{,}418{,}143}{496 - 2}}$$
$$= 1351.6$$

The residual standard deviation measures scatter or spread around a regression line—hence the badness of its fit. Compare s_e with the standard deviation of Y, s_Y, which measures spread around the mean of Y (s_e and s_Y have the same units—for example, cubic feet of water). Since here the residual standard deviation ($s_e = 1351.6$) is almost as large as the standard deviation of Y ($s_Y = 1486.1$), the line in Figure 2.3 does not fit the scatter very closely. That is, income does not strongly predict household water use. The next section looks more closely at prediction, residuals, and overall measures of fit.

Predicted Values and Residuals

The 88th household in our sample reports an income of X_{88} = \$30,000. Equation [2.10] predicts that this middle-income household used a middling amount of water, 2626 cubic feet:

$$\begin{aligned}\hat{Y}_{88} &= 1201 + 47.5X_{88} \\ &= 1201 + 47.5(30) \\ &= 2626\end{aligned}$$

The prediction is wrong. Household #88 actually used 6900 cubic feet of water:

$$Y_{88} = 6900$$

The prediction error or residual for household #88 is the difference between its actual and predicted water use:

$$\begin{aligned}e_{88} &= Y_{88} - \hat{Y}_{88} \\ &= 6900 - 2626 \\ &= 4274\end{aligned}$$

Household #88 used 4274 cubic feet more water than we would predict on the basis of income.

In contrast, household #254 earned above-average income, \$50,000, so we predict above-average water use:

$$\begin{aligned}\hat{Y}_{254} &= 1201 + 47.5(50) \\ &= 3576\end{aligned}$$

Their actual use was low, $Y_{254} = 1100$, so the residual is negative:

$$e_{254} = 1100 - 3576 = -2476$$

Predicted values and residuals are defined in this fashion for each case in the data. Squaring each residual, and then adding them all up, obtains the RSS.

Graphically, the predicted value for case i is the height of the regression line directly above X_i. The residual is the *vertical* distance between predicted $(X_i, \hat{Y}_i)$

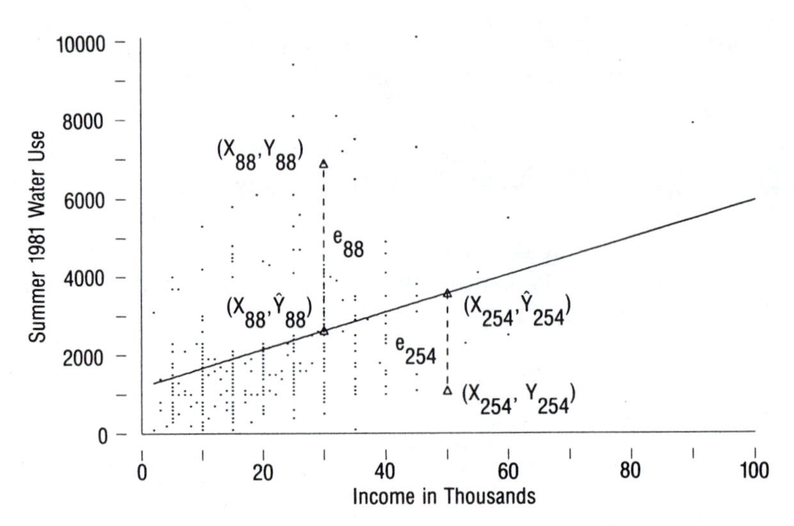

Figure 2.5 Graphical meaning of predicted values ($\hat{Y}$) and residuals (e) for cases 88 and 254.

and actual (X_i, Y_i) data points (Figure 2.5). The residual standard deviation measures vertical variation around the regression line, whereas the standard deviation of Y measures vertical variation around a horizontal line at $\bar{Y}$.

R^2, Correlation, and Standardized Regression Coefficients

Slopes, intercepts, and residual standard deviations refer to variables' natural units—dollars or cubic feet, for instance. For some purposes *standardized* statistics, which do not depend on the variables' units, are more convenient. One such statistic, the *coefficient of determination*, or R^2, gives the proportion of the variance of Y explained by X:

$$R^2 = \text{explained variance}/\text{total variance}$$

$$= s_{\hat{Y}}^2/s_Y^2 \qquad [2.12a]$$

The variance of $\hat{Y}$ ($s_{\hat{Y}}^2$) is termed *explained variance.* R^2 ranges from zero (if $s_{\hat{Y}}^2 = 0$, no variance is explained) to one (if $s_{\hat{Y}}^2 = s_Y^2$, all of Y's variance is explained), providing a simple summary of fit.

For the water-use regression (Figure 2.4), $s_Y^2 = 2{,}208{,}563$ and $s_{\hat{Y}}^2 = 385{,}496$, so

$$R^2 = 385{,}496/2{,}208{,}563$$

$$= .1745$$

A linear relationship with income explains about 17.45% of the sample variance of household water use. If income explains 17.45%, the remaining 82.55% of the variance is unexplained or residual. The total variance of Y equals the sum of explained and residual variances.[4]

The *explained sum of squares* (ESS) measures variation of predicted Y around $\bar{Y}$:

$$\text{ESS} = \Sigma(\hat{Y}_i - \bar{Y})^2 \qquad [2.13]$$

The coefficient of determination equals the ratio of explained to total sums of squares:

$$R^2 = \frac{\text{ESS}}{\text{TSS}_Y} = \frac{\Sigma(\hat{Y}_i - \bar{Y})^2}{\Sigma(Y_i - \bar{Y})^2} \qquad [2.12b]$$

$1 - R^2$ equals the ratio of residual (RSS) to total sums of squares:

$$1 - R^2 = \frac{\text{RSS}}{\text{TSS}_Y} = \frac{\Sigma e_i^2}{\Sigma(Y_i - \bar{Y})^2} \qquad [2.14]$$

Sample covariance measures how much two variables vary together. We define Y's variance (s_Y^2) as

$$s_Y^2 = \frac{\Sigma(Y_i - \bar{Y})(Y_i - \bar{Y})}{n - 1}$$

Similarly, the covariance of X and Y (s_{XY}) is

$$s_{XY} = \frac{\Sigma(X_i - \bar{X})(Y_i - \bar{Y})}{n - 1} \qquad [2.15]$$

Variance equals the covariance of a variable with itself ($s_Y^2 = s_{YY}$).

To standardize covariance, divide by the product of the two variables' standard deviations. This yields the *correlation* (r), a standardized measure of the strength of a bivariate linear relationship:[5]

$$r = \frac{s_{XY}}{s_X s_Y} \qquad [2.16]$$

Correlations theoretically range from -1 (perfect negative relationship) to $+1$ (perfect positive relationship). The sign of a correlation is the same as that of the corresponding regression slope and has the same meaning:

■ negative: *high values of* Y tend to occur with *low values of* X, and low Y with high X;
positive: *high values of* Y tend to occur with *high values of* X, and low Y with low X. ■

("High" and "low" mean above and below average.) A zero correlation, like a zero slope, indicates no linear relationship. The closer r is to ± 1, the more tightly data points cluster around the regression line. At $r = \pm 1$, all points lie on the line.[6]

The correlation between water use and income is abour $r = .42$ (more precisely, $r = .417787$). We could describe this as a "moderate positive relationship." Correlations provide simple summaries of bivariate relationship. Because they are standardized, we need not think about variables' units and can compare relationships involving different variables. *For bivariate regression only*, the square of the correlation equals the coefficient of determination:

$$r^2 = R^2 \qquad [2.17]$$

Standardized regression coefficients (b^*), also called *beta weights*, are defined as follows:

$$b_1^* = b_1 \frac{s_X}{s_Y} \qquad [2.18]$$

where b_1 is the unstandardized regression coefficient, s_X is the standard deviation of X, and s_Y is the standard deviation of Y. Like correlations, standardized regression coefficients theoretically range from -1 to $+1$ (but see Note 6, which applies to either). Interpret b_1^* as the number of standard deviations $\hat{Y}$ changes with each 1-standard-deviation increase in X.

For the water-use regression, $b_1 = 47.5$, $s_X = 13.06$, and $s_Y = 1486$, so the standardized regression coefficient is

$$b_1^* = 47.5(13.06/1486)$$
$$= .42$$

Predicted household water use increases by about .42 standard deviations (about $.42 \times 1486 = 624$ cubic feet) with each 1-standard-deviation (13.06-thousand-dollar) increase in income. This parallels the unstandardized regression equation [2.10]. *For bivariate regression only*, the standardized regression coefficient equals the correlation:

$$b_1^* = r \qquad [2.19]$$

Standardized statistics like R^2, r, and b_1^* are easily read, which makes them popular and often overemphasized by novice analysts. Sometimes these statistics get confused with substantive research goals, so models are evaluated solely by R^2, or relationship strength judged entirely by b_1^*. Natural-units statistics like s_e and b_1 provide a better foundation for substantive understanding.

Reading Computer Output

Every statistical program presents regression results differently, but most provide the same basic information. Table 2.1 contains an example of *Stata* output, showing the regression of 1981 water use (Y) on income (X):

$$\hat{Y}_i = 1201.124 + 47.54869X_i \qquad [2.20]$$

Values for the intercept ($b_0 = 1201.124$) and slope ($b_1 = 47.54869$) for [2.20] appear in the "Coefficient" column of Table 2.1. Equation [2.10] and Figure 2.4 gave rounded-off versions of Equation [2.20].

Other statistics in Table 2.1 include:

R-square, coefficient of determination: $R^2 = .1745$
Model SS, explained sum of squares: ESS = 190,820,566
Residual SS, residual sum of squares: RSS = 902,418,143
Total SS, total sum of squares: $\text{TSS}_Y = 1.0932e + 09$, scientific notation for 1.0932×10^9 or 1,093,200,000.

Note that, within round-off limits, $\text{TSS}_Y = \text{ESS} + \text{RSS}$ and $R^2 = \text{ESS}/\text{TSS}_Y$.

Below R-square in Table 2.1 is the *adjusted* R^2 (R_a^2):

$$R_a^2 = R^2 - \frac{K-1}{n-K}(1 - R^2) \qquad [2.21]$$

TABLE 2.1 Regression of 1981 water use on household income

Source	SS	df	MS	
				Number of obs = 496
				$F(1, 494)$ = 104.46
Model	190820566	1	190820566	Prob > F = 0.0000
Residual	902418143	494	1826757.38	R-square = 0.1745
Total	1.0932e + 09	495	2208563.05	Adj R-square = 0.1729
				Root MSE = 1351.6

Variable	Coefficient	Std. Error	t	Prob > |t|	Mean
water81					2298.387
income	47.54869	4.652286	10.221	0.000	23.07661
_cons	1201.124	123.3245	9.740	0.000	1

where n is sample size and K is the number of parameters in the model. Since Table 2.1 shows a regression with 496 cases and two estimated parameters ($b_0 =$ 1201.124 and $b_1 = 47.54869$),

$$R_a^2 = .1745 - \frac{2-1}{496-2}(1 - .1745)$$
$$= .1729$$

The adjustment to R^2 takes into account the complexity of the regression model relative to the complexity of the data.[7] R_a^2 is often preferred to R^2 when describing more complex models (for reasons explained in Chapter 3).

Hypothesis Tests for Regression Coefficients

We began with a linear model:

$$E[Y_i] = \beta_0 + \beta_1 X_i$$

Ordinary least squares, applied to sample data on water use and income (Table 2.1), yields estimates $b_0 = 1201.124$ and $b_1 = 47.54869$. Thus the relationship among the 496 households of our sample is

$$\hat{Y}_i = 1201.124 + 47.54869X_i$$

On the basis of these results, what should we conclude about the relationship in the larger population of 10,000 or so Concord households? That is, what inferences can we make about the true values of population parameters β_0 and β_1?

If we drew another sample of 496 households, we would almost certainly obtain different estimates of β_0 and β_1. If we drew many samples, we would obtain a distribution of estimates. The distribution of estimates from all possible size-n random samples is called a *sampling distribution.*

Assuming independent and identically distributed (i.i.d.) errors, the sampling distribution of b_1 should be approximately normal with mean β_1 and standard deviation σ_{b_1}. Similarly, the sampling distribution of b_0 should be approximately normal with mean β_0 and standard deviation σ_{b_0}. Standard deviations of sampling distributions, like σ_{b_1} and σ_{b_0}, are called *standard errors.* A small standard error indicates little sample-to-sample variation, so most sample b's are close to β. A larger standard error indicates the opposite: the b's vary more widely, and many sample estimates are far from β. Figure 2.6 illustrates this idea.

The true standard errors, like the true values of β_1 and β_0, are usually unknown and must be estimated from the sample. Such estimates appear in regression output. For the water-use regression, the "Std. Error" column in Table 2.1 gives the

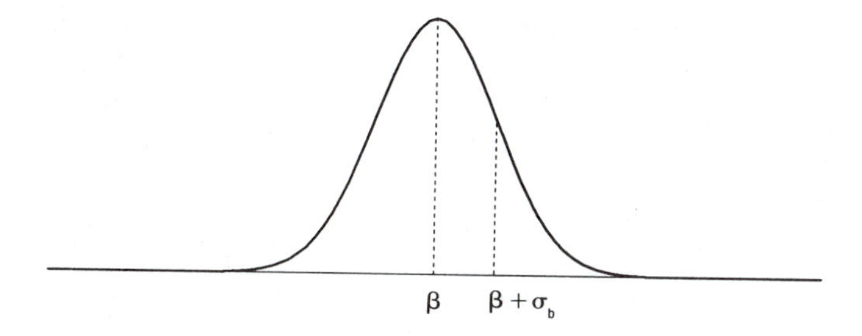

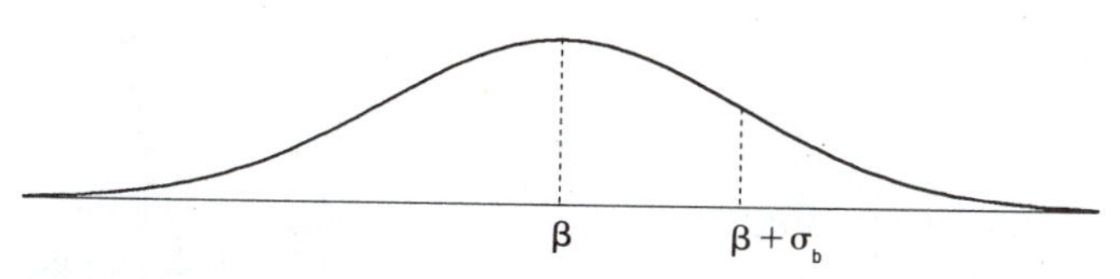

Figure 2.6 Sampling distribution of b_1, with larger and smaller (top) standard errors σ_{b_1}.

estimated standard error of the slope, b_1, as $SE_{b_1} = 4.652286$. The estimated standard error of the intercept, b_0, is $SE_{b_0} = 123.3245$.

We estimate the standard error of a bivariate regression slope as follows:

$$SE_{b_1} = \frac{s_e}{\sqrt{TSS_X}} \qquad [2.22]$$

where s_e is the residual standard deviation (Equation [2.11]) and TSS_X is the total sum of squares for X: $TSS_X = \Sigma(X_i - \bar{X})^2$.

We estimate the standard error of a bivariate regression intercept as follows:

$$SE_{b_0} = s_e\sqrt{\frac{1}{n} + \frac{\bar{X}^2}{TSS_X}} \qquad [2.23]$$

Assuming normal i.i.d. errors, the *t-statistic*,

$$t = \frac{b - \beta}{SE_b} \qquad [2.24]$$

follows a theoretical *t-distribution* with $n - K$ degrees of freedom. We test hypotheses by substituting the hypothesized value of β into [2.24]. The usual *null hypothesis* asserts that there exists no linear relationship between $E[Y_i]$ and X in the population—in other words, that $\beta_1 = 0$. Stated formally:

$$H_0: \beta_1 = 0$$

The alternative to this null hypothesis is that the population slope is not zero:

$$H_1: \beta_1 \neq 0$$

If $\beta_1 = 0$, as claimed by the null hypothesis, [2.24] simplifies to

$$t = \frac{b}{\text{SE}_b} \qquad [2.25]$$

The "t" column in Table 2.1 contains such t-statistics: coefficients divided by their estimated standard errors. For example, $b_1 = 47.54869$ and $\text{SE}_{b_1} = 4.652286$, so

$$t = 47.54869/4.652286$$
$$= 10.221$$

The t-statistics found by [2.24] or [2.25] are compared with a theoretical t-distribution (Table A4.1) with $n - K$ degrees of freedom. From the t-distribution we obtain a probability, called the *P-value*:

■ The *P-value* equals the estimated probability of obtaining these sample results, or results more favorable to H_1, if the sample were drawn randomly from a population where H_0 is true. ■

A low P-value means it is unlikely that such a sample would come from a population where H_0 is true.

In Table 2.1 we have $t = 10.221$, with $n - K = 496 - 2 = 494$ degrees of freedom. Table A4.1 shows the corresponding probability (two-sided test) to be $P < .001$.[8] In fact, it is much less than .001, and the computer printout in Table 2.1 gives this P-value in the "Prob > $|t|$" column as "0.000." (The column heading means "the probability of a greater absolute value of t.") The probability is not actually zero, but it is quite low—low enough to round off to zero to three decimal places. It is very unlikely that we would see such a sample if there were no relationship between income and water use in the population.

Researchers often follow a decision rule such as

"reject H_0 (and believe H_1 instead) if $P < .05$."

Any coefficient for which the obtained P-value is less than .05 is then said to be *statistically significant*. The cutoff point, denoted α (alpha), is arbitrary; there is rarely much substantive difference between findings in which $P = .049$ ("statistically significant at $\alpha = .05$") and findings in which $P = .051$ ("not significant at $\alpha = .05$"). Although the $\alpha = .05$ cutoff is most popular, other choices are equally arbitrary.[9] Reporting actual P-values bypasses this arbitrariness and conveys more information than simply saying which coefficients are significant.

Usually, t-statistic P-values reported with regression output refer to *two-sided tests*, in which the alternative hypothesis does not specify direction:

$$H_0: \beta = 0$$

$$H_1: \beta \neq 0$$

Sample b values either above or below zero would favor H_1 over H_0.

One-sided tests are needed when H_1 does specify direction. Theory or substantive knowledge may strongly suggest the sign of a relationship *before* we look at the data. If we expect a positive relationship, we should test:

$$H_0: \beta \leq 0$$

$$H_1: \beta > 0$$

Only positive b values could support this directional H_1. Alternatively, we might expect negative values:

$$H_0: \beta \geq 0$$

$$H_1: \beta < 0$$

One-sided t-test probabilities equal half the corresponding two-sided probabilities. *If* b lies in the direction specified by H_1, find one-sided P-values by dividing the usual two-sided P-values by 2. If b lies in the direction specified by H_0, do not bother with the test; H_0 cannot be rejected.

Regression output often includes another test, based on an *F-statistic*:

$$F = \frac{\text{ESS}/(K - 1)}{\text{RSS}/(n - K)} \qquad [2.26]$$

where ESS is the explained sum of squares, RSS is the residual sum of squares, n is sample size, and K is the number of parameters estimated ($K = 2$ for bivariate regression). Assuming normal i.i.d. errors, [2.26] follows an *F-distribution* (Table A4.2) with $\text{df}_1 = K - 1$ and $\text{df}_2 = n - K$ degrees of freedom.

From Table 2.1:

$$F = \frac{190{,}820{,}566/(2 - 1)}{902{,}418{,}143/(496 - 2)}$$

$$= 104.46$$

Table 2.2 Regression of 1981 water use on household income, with annotations

Source	SS	df	MS	
Model	190820566[1]	1[4]	190820566[7]	Number of obs = 496[10]
Residual	902418143[2]	494[5]	1826757.38[8]	$F(1, 494)$ = 104.46[11]
				Prob > F = 0.0000[12]
				R-square = 0.1745[13]
Total	1.0932e + 09[3]	495[6]	2208563.05[9]	Adj R-square = 0.1729[14]
				Root MSE = 1351.6[15]

Variable	Coefficient	Std. Error	t	Prob > $\|t\|$	Mean
water81					2298.387[24]
income	47.54869[16]	4.652286[18]	10.221[20]	0.000[22]	23.07661[25]
_cons	1201.124[17]	123.3245[19]	9.740[21]	0.000[23]	1

[1] explained sum of squares, ESS (equation [2.13])
[2] residual sum of squares, RSS (equation [2.7])
[3] total sum of squares, TSS_Y (equation [1.2])
[4] $K - 1$
[5] $n - K$
[6] $n - 1$
[7] $ESS/(K - 1)$
[8] $RSS/(n - K)$
[9] $TSS_Y/(n - 1)$
[10] sample size, n
[11] F statistic for null hypothesis $\beta_1 = 0$ (equation [2.26])
[12] P-value for F statistic
[13] coefficient of determination, R^2 (equation [2.12])
[14] adjusted R^2, R_a^2 (equation [2.21])
[15] standard deviation of residuals (equation [2.11])
[16] regression coefficient or slope, b_1 (equation [2.9])
[17] regression constant or Y-intercept, b_0 (equation [2.8])
[18] estimated standard error of b_1, SE_{b_1} (equation [2.22])
[19] estimated standard error of b_0, SE_{b_0} (equation [2.23])
[20] t statistic for null hypothesis $\beta_1 = 0$ (equation [2.25])
[21] t statistic for null hypothesis $\beta_0 = 0$ (equation [2.25])
[22] P-value for two-sided t test of H_0: $\beta_1 = 0$
[23] P-value for two-sided t test of H_0: $\beta_0 = 0$
[24] mean of Y, $\bar{Y}$
[25] mean of X, $\bar{X}$

With $df_1 = 1$ and $df_2 = 494$ degrees of freedom, this F-statistic leads to a P-value well below .001. Again the P-value ("Prob > F," read as "probability of a greater F") appears rounded off as "0.0000," although it is not actually zero.[10]

In bivariate regression the F- and two-sided t-tests are redundant. Their relationship is

$$F = t^2$$

The obtained P-value from the F-test equals that for a two-sided t-test of

$$H_0: \beta_1 = 0$$

$$H_1: \beta_1 \neq 0$$

In multiple regression, however, F-statistics can test more complex hypotheses regarding *sets* of coefficients.

Table 2.2 annotates the computer output of Table 2.1.

Confidence Intervals

To construct a confidence interval for a regression parameter, find

$$b \pm t(\mathrm{SE}_b) \qquad [2.27]$$

with t chosen from the theoretical t-distribution (Table A4.1) with $n - K$ degrees of freedom.

For a 95% confidence interval, with df $= n - K = 496 - 2 = 494$ (approximately "infinite" degrees of freedom), the appropriate t-value is $t = 1.96$. A 95% confidence interval for the coefficient relating water use to income (Table 2.2) is:

$$47.54869 \pm 1.96(4.652286)$$

$$47.54869 \pm 9.118481$$

$$38.430209 \leq \beta_1 \leq 56.667171$$

Assuming normal i.i.d. errors, over many random samples 95% of the intervals constructed in this manner should contain β_1. For 99% confidence we need a wider interval ($t = 2.576$):

$$47.54869 \pm 2.576(4.652286)$$

$$35.564401 \leq \beta_1 \leq 59.532979$$

We may wish to construct confidence intervals around predicted values of Y. Use the estimated standard error of $\hat{Y}_i$, $\mathrm{SE}_{\hat{Y}_i}$, and find

$$\hat{Y}_i \pm t(\mathrm{SE}_{\hat{Y}_i}) \qquad [2.28]$$

choosing t for the desired degree of confidence, with df $= n - K$. There are two applications for [2.28]:

1. Confidence intervals for the *mean (expected) value of* Y, *when* $X = X_i$; estimate $\mathrm{SE}_{\hat{Y}_i}$ by [2.29].
2. *Prediction intervals*, or confidence intervals for *an individual case's* Y *value, when* $X = X_i$; estimate $\mathrm{SE}_{\hat{Y}_i}$ by [2.30].

We can predict population means with more precision than we can predict individual cases, so [2.29] yields a smaller standard error and narrower interval than [2.30].

Estimated standard error of the *mean value of* Y, when $X = X_i$:

$$SE_{\hat{Y}_i} = s_e \sqrt{\frac{1}{n} + \frac{(X_i - \bar{X})^2}{TSS_X}} \quad [2.29]$$

Substituting $X_i = 0$ into [2.29] obtains [2.23].

Estimated standard error of *predicted Y for an individual case*, when $X = X_i$:

$$SE_{\hat{Y}_i} = s_e \sqrt{\frac{1}{n} + \frac{(X_i - \bar{X})^2}{TSS_X} + 1} \quad [2.30]$$

Confidence or prediction intervals around $\hat{Y}_i$ are narrowest at $X_i = \bar{X}$, forming bands with an hourglass shape when graphed. Figure 2.7 illustrates 99% confidence and prediction bands for water use given income (applying [2.28]–[2.30]).

Confidence intervals and hypothesis tests are related ideas. The popularity of hypothesis tests reflects their simplicity—especially their support for yes/no decisions on significance. Confidence intervals, which focus more directly on issues of substantive importance (relationship strength, prediction), can also test hypotheses; t or F testing H_0: $\beta_1 = 0$ versus H_1: $\beta_1 \neq 0$, at significance level $\alpha = .05$, is equivalent to:

Reject H_0 if the null-hypothesis value of β_1 lies outside of a 95% confidence interval around b_1.

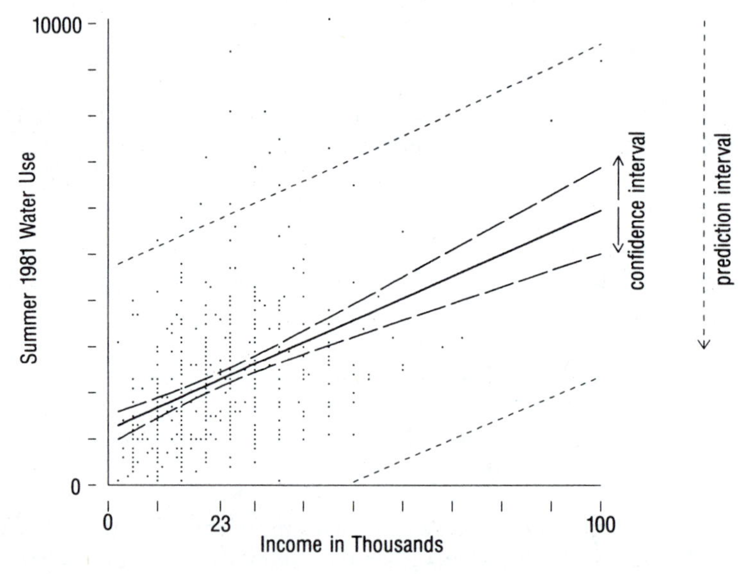

Figure 2.7 Confidence and prediction intervals around regression line.

Intervals and t-tests both rely on the same standard errors, sampling distributions, and statistical theories.

Regression Through the Origin

Sometimes we know beforehand that a Y-intercept must be zero. For example, consider the following data on the number of accidental oil spills at sea and the amount of oil lost in those spills (millions of metric tons) for the years 1973–1985:[11]

Year	Number of Spills	Amount of Oil Lost
1973	36	84.5
1974	48	67.1
1975	45	188.0
1976	29	204.2
1977	49	213.1
1978	35	260.5
1979	65	723.5
1980	32	135.6
1981	33	45.3
1982	9	1.7
1983	17	387.8
1984	15	24.2
1985	8	15.0

Regressing oil loss (Y) on number of spills (X) yields the following prediction equation:

$$\hat{Y}_i = -44 + 7X_i$$

But $b_0 = -44$ implies that, when there are no oil spills, -44 million gallons of oil will be lost. Common sense dictates that, if there are no oil spills, zero gallons of oil will be lost in spills. Therefore the true Y-intercept must be $\beta_0 = 0$, and we should not allow nonzero estimates. We want an equation with the form

$$\hat{Y}_i = b_1 X_i \qquad [2.31]$$

Regression with $b_0 = 0$, called *regression through the origin*, specifies a line through the point (0, 0). The least-squares solution is

$$b_1 = \frac{\Sigma X_i Y_i}{\Sigma X_i^2} \qquad [2.32]$$

Equation [2.32] does not adjust for means, as ordinary regression does (compare with [2.9b]).

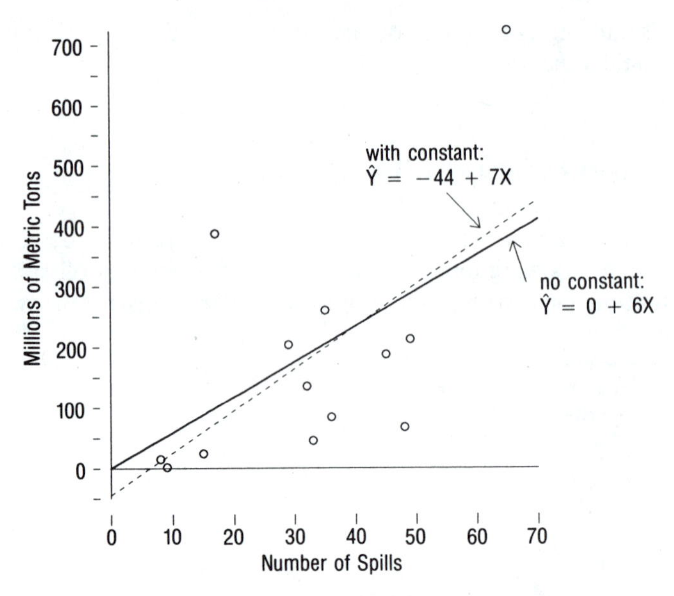

Figure 2.8 Regression with and without a constant.

Computer programs for regression permit the user to specify $b_0 = 0$ ("no constant") when needed. From the oil-spill data we then obtain

$$\hat{Y}_i = 6X_i$$

Figure 2.8 shows both lines.

To estimate a Y-intercept, computer programs implicitly regress Y on one or more X variables *and* one constant, usually 1. In the bivariate case:

$$\begin{aligned}\hat{Y}_i &= b_0 1 + b_1 X_i \\ &= b_0 + b_1 X_i\end{aligned}$$

(See Appendix 3 for the extension to more than one X variable.) Regression through the origin involves the regression of Y on X variables and *no* constant, so we obtain only slope coefficients.

Use caution when reading computer output from a no-constant regression. The R^2 and F-statistics do not have their usual meanings in this context. Ordinarily (with a constant), R^2 and overall F-tests reflect improvement over predicting that *all* Y_i *equal* $\bar{Y}$. With no constant, R^2 and F are commonly redefined to reflect improvement over predicting that *all* Y_i *equal 0*. $Y_i = 0$ is a more naive prediction than $Y_i = \bar{Y}$, so we should not be impressed if the no-constant regression yields a much higher R^2 or F-statistic than regression with a constant. It can do so even though the no-constant fit is objectively worse.[12]

Problems with Regression

Some common statistical problems detract from a regression's validity. These include:

Omitted variables. If other variables affect both X and Y, our sample regression coefficient b_1 may substantially overstate or understate the true relation between X and Y. Omitted variables particularly damage causal interpretations.

Nonlinear relationships. OLS finds the best-fitting straight line, but this is misleading if $E[Y_i]$ is actually a nonlinear function of X.

Nonconstant error variance. If the variance of errors changes with the level of X (a condition called *heteroscedasticity*), then the usual standard errors, tests, and confidence intervals are untrustworthy.

Correlation among errors. The usual standard errors, tests, and confidence intervals assume no correlation among errors (no *autocorrelation*). Error correlations often do occur (for example, when cases are adjacent in time or space).

Nonnormal errors. The usual t and F procedures assume normally distributed errors. Nonnormal errors may invalidate these procedures (especially with small samples) and increase sample-to-sample variation of estimates.

Influential cases. OLS regression is *not resistant*: a single outlier can pull the line up or down and substantially influence all results. (Chapter 1 noted the nonresistance of mean, variance, and standard deviation. OLS belongs to the same mathematical family.)

The following sections illustrate ways to diagnose and deal with some of these problems. Later chapters return to them in more detail.

Residual Analysis

Scatterplots of Y versus X reveal obvious problems and should be consulted at an early stage. Residual graphs also provide useful diagnostic information. Regression assumptions focus on errors (we often assume that ε_i are normal i.i.d.), and sample residuals help assess the plausibility of these assumptions.

Figure 2.9 graphs residuals versus predicted values, from the regression of water use on income. Such e-versus-$\hat{Y}$ plots are a general-purpose diagnostic tool. OLS residuals always have a mean of zero, indicated by the horizontal line. Households that used more water than predicted have positive residuals and are graphed above the $e = 0$ line. Households that used less water than predicted ($e < 0$) appear below the line. Marginal boxplots show the univariate distributions of e (right) and $\hat{Y}$ (top).

As with other analytical graphs, reading e-versus-$\hat{Y}$ plots becomes easier with

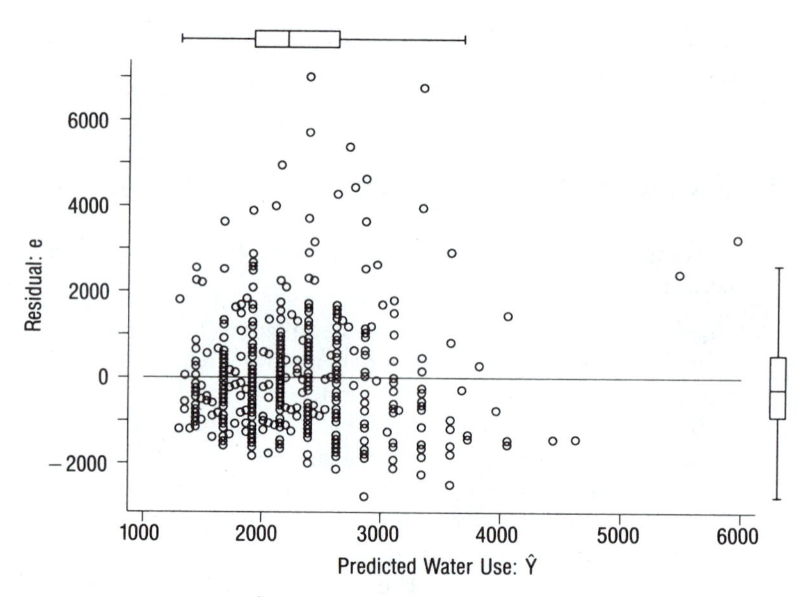

Figure 2.9 e-versus-$\hat{Y}$ plot for water-use regression.

practice. The best situation is an "all clear" plot like that seen in Figure 2.10 (artificial data). The distribution of e appears approximately normal and equally spread out at all levels of X—consistent with assumptions that population errors are normal and identically distributed. Figure 2.10 contains no evidence of influential cases or nonlinearity. Figure 2.11, in contrast, illustrates how four kinds of trouble might appear in e-versus-$\hat{Y}$ plots.

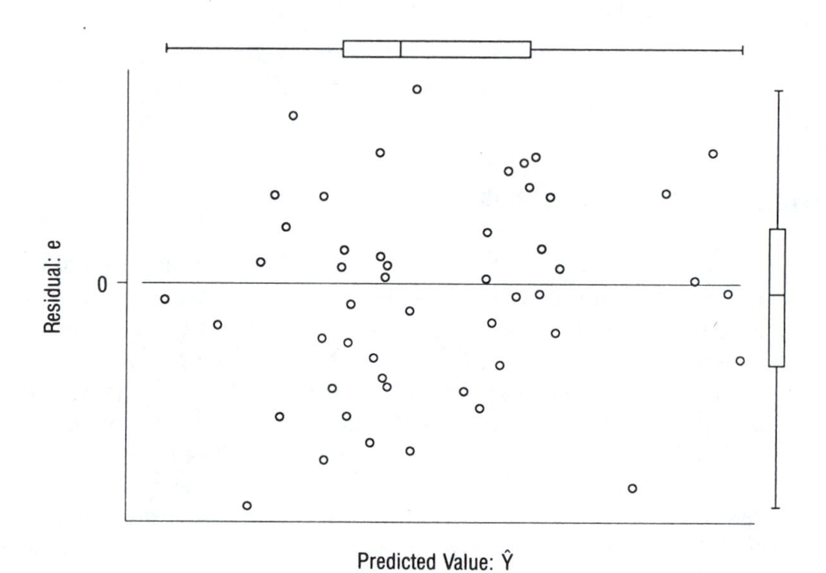

Figure 2.10 "All clear" e-versus-$\hat{Y}$ plot (artificial data).

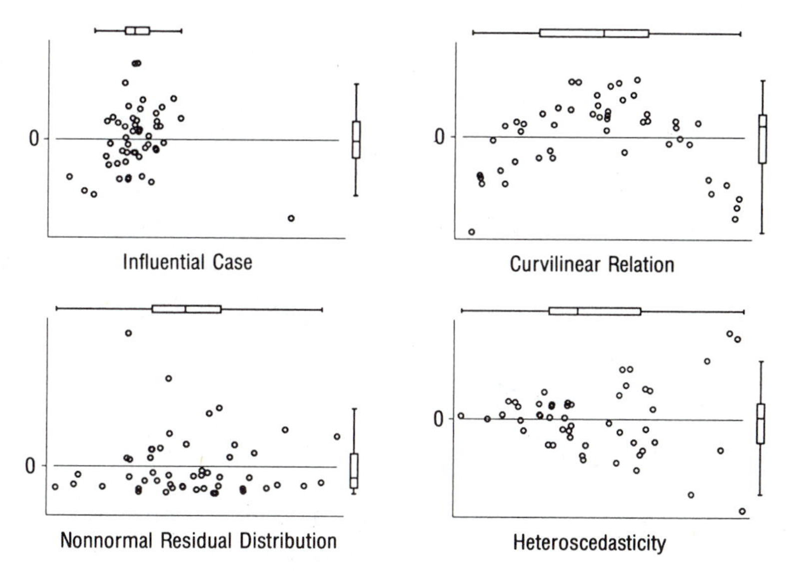

Figure 2.11 Examples of trouble seen in e-versus-$\hat{Y}$ plots (artificial data).

Comparing the e-versus-$\hat{Y}$ plot for water use and income (Figure 2.9) with the artificial-data "all clear" (Figure 2.10) and "trouble" (Figure 2.11) plots leads to the following observations:

1. Two households with high predicted water use (far right in Figure 2.9) might be influential.
2. The scatter in Figure 2.9 fans out left to right, indicating that water use is less predictable at higher incomes (heteroscedasticity).
3. The distribution of residuals in Figure 2.9 is positively skewed, as shown by the right-hand boxplot and the concentration of cases with small negative residuals (below the $e = 0$ line).

Skewed residuals call the normal-errors assumption into question. A quantile-normal plot of the water-use residuals (Figure 2.12) confirms their skewness and nonnormality (compare with the third plot in Figure 1.10).

The residual graphs in Figures 2.9 and 2.12 give us several reasons to distrust the original analysis. The next section considers how to improve it.

Power Transformations in Regression

Power transformations often correct the kinds of problems seen in Figures 2.9 and 2.12. We earlier determined that the distribution of water use is skewed, but the .3 power of water use has a symmetrical, approximately normal distribution. A similar analysis of household income begins in Figure 2.13. Income, too, is positively

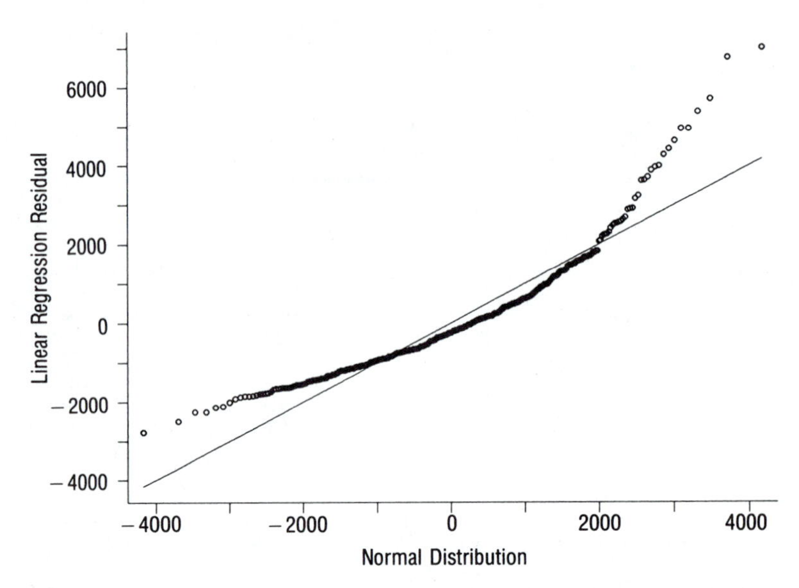

Figure 2.12 Quantile-normal plot of residuals from water-use regression (positively skewed).

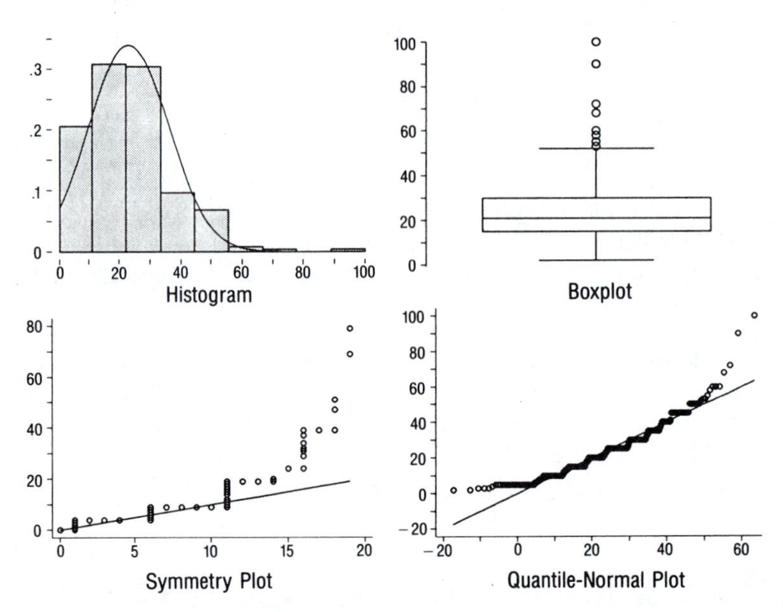

Figure 2.13 Four views of the distribution of household income (positively skewed).

skewed.[13] Regression requires no assumptions about the distribution of X variables, but in practice skewed X distributions are often associated with statistical problems such as influence and heteroscedasticity.

Transformations with $q < 1$ tend to reduce positive skew. It happens that $q = .3$ works as well for income as it does for water use. Figure 2.14 shows that income to the .3 power has a symmetrical, roughly normal distribution.

The univariate distributions of water use and income are improved by these transformations. Table 2.3 shows a regression using the transformed variables

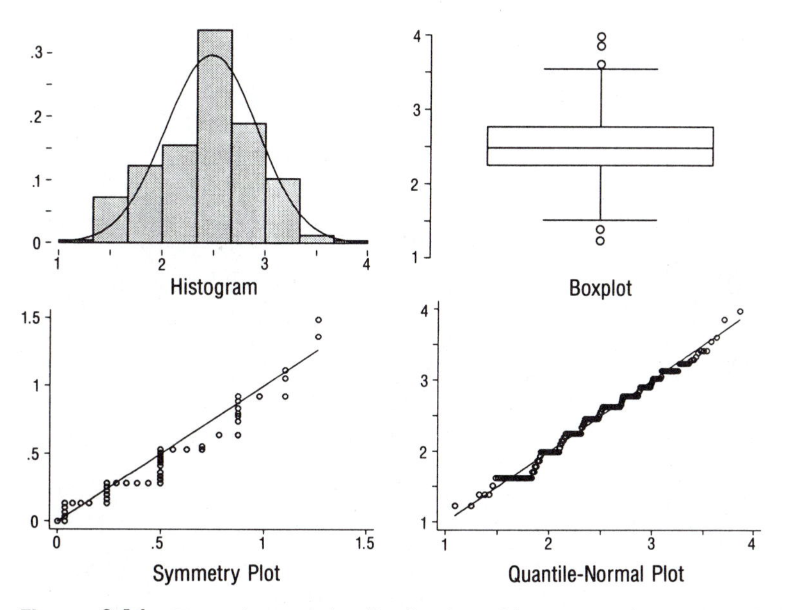

Figure 2.14 Four views of the distribution of income to the .3 power (symmetrical).

Table 2.3 Regression of .3 power of 1981 water use (wtr81_3) on .3 power of household income (inc_3)

Source	SS	df	MS	
Model	370.337058	1	370.337058	Number of obs = 496 $F(1,494)$ = 126.22 Prob > F = 0.0000
Residual	1449.41668	494	2.93404187	R-square = 0.2035
Total	1819.75374	495	3.67627019	Adj R-square = 0.2019 Root MSE = 1.7129

Variable	Coefficient	Std. Error	t	Prob > |t|	Mean
wtr81_3					9.776982
inc_3	1.934535	.1721913	11.235	0.000	2.474998
_cons	4.989011	.4330577	11.520	0.000	1

(named "wtr81_3" and "inc_3"). The resulting equation is

$$\hat{Y}_i^* = 4.989011 + 1.934535X_i^* \quad [2.33a]$$

where $Y^* = Y^{.3}$ and $X^* = X^{.3}$. Expressing [2.33a] in terms of the original variables:

$$\hat{Y}_i^{.3} = 4.989011 + 1.934535X_i^{.3} \quad [2.33b]$$

The relationship between transformed water use ($Y^{.3}$) and transformed income ($X^{.3}$) is linear, as seen in Figure 2.15.

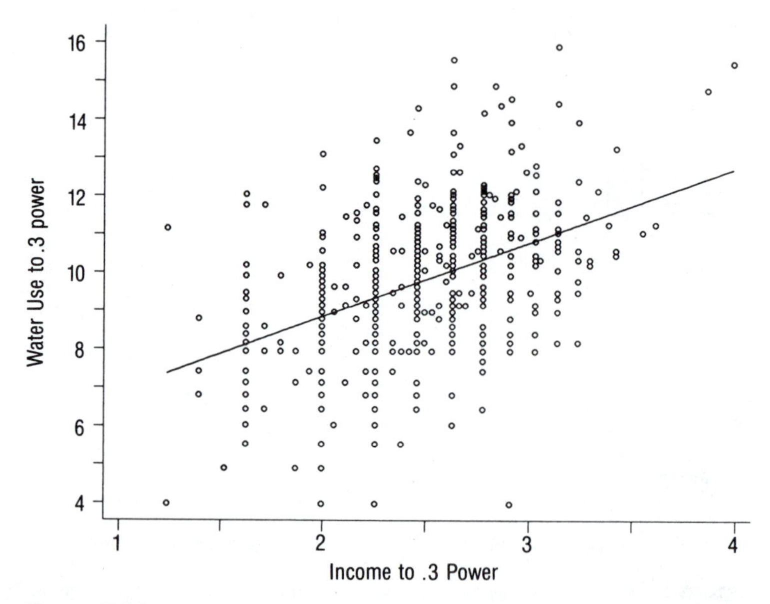

Figure 2.15 Transformed water use versus transformed household income.

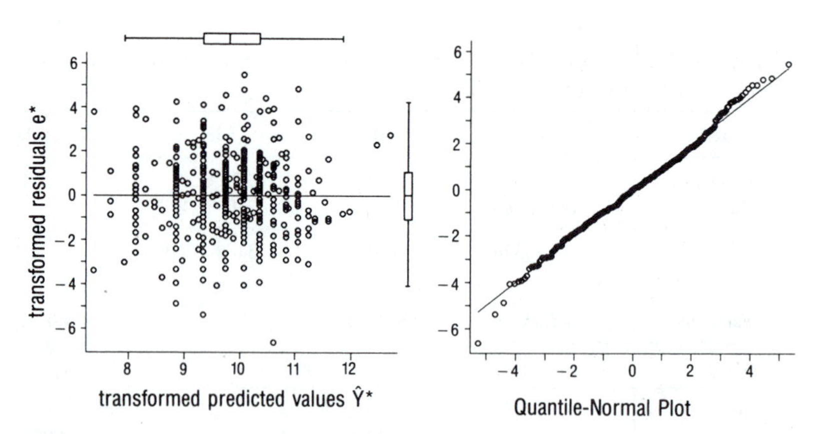

Figure 2.16 e-versus-$\hat{Y}$ plot (left) and quantile-normal plot (right) showing residuals from transformed-variables regression.

At left in Figure 2.16 is an e-versus-$\hat{Y}$ plot for the transformed-variables regression. Unlike the raw-data version (Figure 2.9), this plot shows no evidence of outliers or heteroscedasticity. At right in Figure 2.16, a quantile-normal plot shows a close-to-normal residuals distribution. Basic regression assumptions appear more plausible regarding the transformed variables than they did with the raw data. In this example the same transformations ($q = .3$) that normalize the univariate distributions also improve bivariate analysis.

Both X and Y were transformed in the same way here, but this need not always be the case. Any mixture of transformations (or no transformation) can be tried, depending on the specific curves or distributional problems encountered.

Understanding Curvilinear Regression

The transformed-variables regression equation,

$$\hat{Y}_i^* = 4.989011 + 1.934535X_i^*$$

or

$$\hat{Y}_i^{.3} = 4.989011 + 1.934535X_i^{.3}$$

has statistical advantages but is harder to understand. It asserts that the predicted .3 power of water use increases by 1.93 with every one-unit increase in the .3 power of income. But what does this mean in real-world terms? Graphs help to visualize the implications of transformed-variables regression. First we obtain predicted values ($\hat{Y}^*$) from the transformed-variables equation. Then we do the following:

1. Convert transformed predicted values, $\hat{Y}^*$, back into the natural units of Y (obtaining $\hat{Y}$).
2. Graph $\hat{Y}$ against X.

We perform step 1 only if Y was transformed. *If Y was not transformed, we simply graph $\hat{Y}$ against X.*

Step 1, returning to natural units, requires *inverse transformation.* With power transformations this usually involves raising transformed predicted values to the inverse of the original power:

Transformation	Inverse Transformation	
$Y^* = Y^q$	$Y = Y^{*1/q}$	$q > 0$
$Y^* = \log_e Y$ or $Y^* = \log_{10} Y$	$Y = e^{Y^*}$ or $Y = 10^{Y^*}$	$q = 0$
$Y^* = -(Y^q)$	$Y = (-Y^*)^{1/q}$	$q < 0$

For example, since we transformed water use (Y) by raising to the .3 power—$Y^* = Y^{.3}$—the appropriate inverse transformation, applied to the predicted .3 power of water use ($\hat{Y}^*$), is

$$\hat{Y} = (\hat{Y}^*)^{1/.3}$$

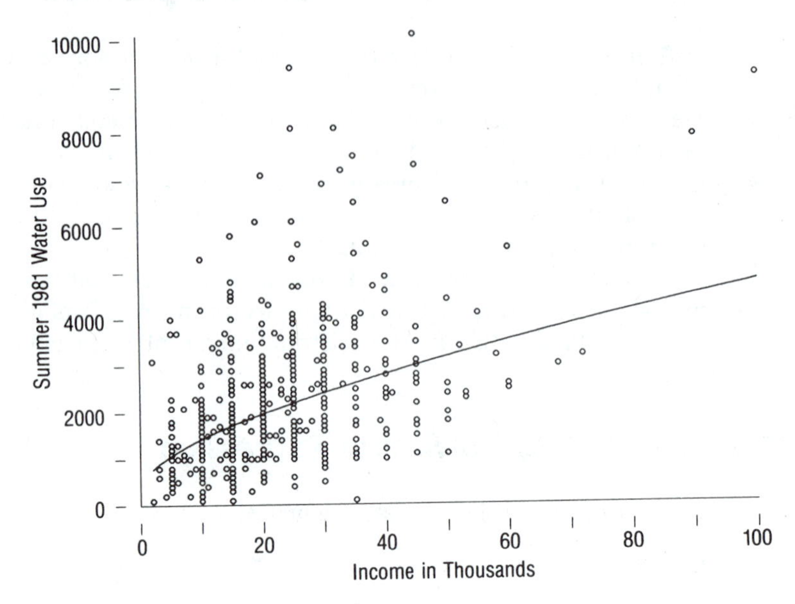

Figure 2.17 Curvilinear relation of water use to income.

Unlike $\hat{Y}^*$, $\hat{Y}$ is measured in the natural units of Y—cubic feet of water.

When $\hat{Y}^*$ values from Equations [2.33] are returned, by inverse transformation, to the natural units of Y and graphed against income, they trace out a gentle curve (Figure 2.17). Regression equations like [2.33], which are linear in terms of power-transformed variables, are curvilinear in terms of the original variables.

The curvilinear regression (Table 2.3, Figure 2.17) differs from the linear regression (Table 2.1, Figure 2.4) in several respects:

1. Nonnormality and heteroscedasticity undermine hypothesis tests and standard errors in Table 2.1.
2. Two high-water-use/high-income households steepen the line in Figure 2.4. With a few other high-use households, they pull the line up so that it lies above a majority of the cases. Curvilinear regression (Figure 2.17) better fits most of the data, concentrated at lower left.
3. The curve in Figure 2.17 implies that mean water use rises more steeply from low to middle income than it does from middle to high income. The straight line in Figure 2.3 implies that mean water use rises by a constant amount per dollar of income, whether we look at rich households or at poor ones.

Although differences between Figures 2.4 and 2.17 are not dramatic, transformations made statistical and substantive improvements.

Conclusion

Bivariate regression analysis seeks the line or curve that best fits a two-dimensional scatter of data. The most widely used regression technique, ordinary least squares

(OLS), defines "best fitting" as that line with the lowest sum of squared residuals. We view the sample intercept and slope as estimates of corresponding population parameters. OLS provides optimal estimates of these parameters *if* certain assumptions are true.

Scatterplots and residual analysis help in diagnosing potential problems. Power transformations can alleviate several kinds of problems, notably curvilinearity, heteroscedasticity, nonnormality, and influence. Models that originally appear inappropriate may become more reasonable after transformations. Transformed-variables regression is implicitly curvilinear, with implications that are most easily understood from graphs.

Exercises

Burning fuels produce air pollutants that return to earth as acid rain, affecting wildlife in lakes and streams. Declines in aquatic life have occurred downwind of industrial regions in parts of Europe, the Rocky Mountains, and the northeastern United States.

The following table lists data from 15 tributaries of Millers River in north-central Massachusetts. For each stream we know the average pH and the number of fish species observed during the summer of 1983.[14] pH measures acidity: the more acidic the stream, the lower its pH. Pure water has a pH of 7.0; vinegar is about 3.0.

Acidity and number of fish species in 15 streams

Stream	Average Summer pH	Number of Fish Species
Moss	6.3	6
Orcutt	6.3	9
Ellinwood	6.3	6
Jacks	6.2	3
Riceville	6.2	5
Lyons	6.1	3
Osgood	5.8	5
Whetstone	5.7	4
Upper Keyup	5.7	1
West	5.7	7
Boyce	5.6	4
Mormon Hollow	5.5	4
Lawrence	5.4	5
Wilder	4.7	0
Templeton	4.5	0

Exercises 1–6 explore the relationship between acidity and fish species.

1. Regress fish species (Y) on average pH (X). Write out the regression equation, and interpret the intercept, slope, and R^2.

2. Is the slope in Exercise 1 statistically significant at $\alpha = .05$, using a two-sided test? What is the P-value for a one-sided test? Which test seems more appropriate, and why?
3. For the fish/acidity regression of Exercise 1:
   a. Construct a 95% confidence interval for the slope. How does this confirm your two-sided test of Exercise 2?
   b. Construct a 90% confidence interval for the mean number of fish species in streams with pH = 6.0.
   c. Construct a 90% prediction interval for the number of fish species in a stream with pH = 6.0.
4. List all X, Y, $\hat{Y}$, and e values. Explain the calculation and interpretation of $\hat{Y}$ and e for the two streams with the largest absolute residuals.
5. Compare the residual mean with median, and residual standard deviation with IQR/1.35, to describe this distribution. Is it approximately symmetrical? Is it light or heavy tailed?
6. Construct a scatterplot, and draw in the regression line. (All predicted values lie on the regression line, so we reveal the line just by graphing $\hat{Y}$ against X. To include the data points, graph both Y and $\hat{Y}$ against X.) Which streams are the two unusual data points at lower left? Repeat the regression *without* these two cases, and add a second line to your scatterplot. Discuss how the two cases affect your results. Do you think we should keep them in or leave them out in drawing our final conclusions about fish and pH?

Is there any relationship between sales of leaded gasoline and lead burden in the bodies of newborn infants? The following table lists monthly gasoline lead sales in Massachusetts and mean lead concentrations in umbilical-cord blood of babies born at a major Boston hospital over 14 months in 1980–1981 (Rabinowitz, Needleman, Burley, Finch, and Rees, 1984). Exercise 7 (and further exercises in Chapters 4 and 8) examines these data.

Month	Year	Statewide Lead Sold in Gasoline (Metric Tons)	Mean Lead in Umbilical-Cord Blood (μg/dl)
3	1980	141	6.4
4	1980	166	6.1
5	1980	161	5.7
6	1980	170	6.9
7	1980	148	7.0
8	1980	136	7.2
9	1980	169	6.6
10	1980	109	5.7
11	1980	117	5.7
12	1980	87	5.3
1	1981	105	4.9
2	1981	73	5.4
3	1981	82	4.5
4	1981	75	6.0

7. Regress lead in umbilical-cord blood (Y) on gasoline lead sales (X). Write out the regression equation, and interpret:
   a. slope and intercept.
   b. R^2.
   c. t-test regarding regression slope.

Building upon seabird-colony population studies by R. W. Furness and T. R. Birkhead (1984), David K. Cairns (1989) analyzed the relation between population and foraging area. The following table presents their data for 22 black-legged kittiwake (a northern gull) colonies of Scotland's Shetland and Orkney Islands. Exercises 8–10 use these data.

Kittiwake Colonies of Shetland and Orkney Islands

	Colony	Foraging Area (km²)	Population: # of Breeding Pairs
1	W. Unst	208	311
2	Hermaness	1,570	3,872
3	N. E. Unst	1,588	495
4	W. Yell	126	134
5	Buravoe	353	485
6	Fetlar	931	372
7	Out Skerries	1,616	284
8	Noss	1,317	10,767
9	Moussa	614	1,975
10	Dalsetter	60	970
11	Sumburgh	1,273	3,243
12	Fitful Head	596	500
13	St. Ninians	106	250
14	S. Havra	242	925
15	Reawick	111	970
16	Vaila	302	278
17	Papa Stour	809	1,036
18	Foula	2,927	5,570
19	Eshaness	1,069	2,430
20	Uyea	898	731
21	Gruney	565	1,364
22	Fair Isle	3,957	17,000

8. Regress population on foraging area, and graph the resulting line. Discuss your findings.
9. Construct an e-versus-$\hat{Y}$ plot, and describe what it shows.
10. Is there a significant relationship between the *natural logarithms* of seabird-colony population and foraging area? Transform both variables, and carry out the regression. Construct a second e-versus-$\hat{Y}$ plot, and compare it with Exercise 9.
11. Return to the nuclear-test data at the end of Chapter 1 (p. 24). Construct two scatterplots:
    a. estimated magnitude (Y) versus yield (X).
    b. estimated magnitude versus $\log_{10} X$.

You will see that magnitude is linearly related to log yield (as by definition it should be).[15] This implies a curvilinear relation between magnitude and yield.

12. Regress nuclear-test magnitude on $\log_{10} X$. By doing so, you fit a curve to the scatterplot in Exercise 11a and a line to the scatterplot in 11b. To see the resulting curve and line, calculate predicted values and then plot:
    a. both Y and $\hat{Y}$ (on one graph) versus X.
    b. both Y and $\hat{Y}$ (on one graph) versus $\log_{10}(X)$.
13. The relationship between magnitude estimates and log yield (Exercise 12) is strong but not perfect. Research continues on possible explanations for prediction errors—different rock types, locations, hole depths of the nuclear explosions, and so on. Construct an e-versus-$\hat{Y}$ plot. Do you see any systematic patterns?

Rathbun (1988) reported on experiments in surveying manatees along Florida's Crystal and Indian Rivers by observation from airplanes and helicopters. He initially suspected that helicopters, due to their slower speed, would provide higher and more accurate counts. Helicopter counts were intended to provide a "truth" count for judging the airplane counts. Data failed to support this expectation, however, partly because helicopters frighten manatees. Rathbun concluded that "there is no significant advantage" in using helicopters, which are much more expensive. Exercises 14–15 refer to these data.

	Manatee Count	
Day	**From Airplane**	**From Helicopter**
1	24	30
2	31	30
3	32	33
4	39	38
5	47	58
6	47	58
7	35	48
8	76	75
9	95	85
10	85	55

14. Graph airplane count (Y) versus helicopter count (X), and draw in the regression line. Interpret intercept, slope, and R^2. Carry out two t-tests regarding the slope β_1: the usual test of H_0: $\beta_1 = 0$ and also a test of H_0: $\beta_1 = 1$. Which test seems more relevant to this study?
15. If the helicopter count really were accurate, and airplane observers counted no imaginary manatees (although they might miss some real ones), the relation between these two counts should be a regression through the origin. Conduct a no-constant regression of airplane count on helicopter count, and graph the result. Is the slope in this graph significantly different from 1?

Notes

1. This notation assumes *fixed X*: we could draw repeated samples with different Y_i but the same set of X_i values, like repeating an experiment. If X values are not fixed, but can vary from sample to sample (*random X*), Equation [2.3] becomes

   $$E[Y_i | X_i] = \beta_0 + \beta_1 X_i$$

   where $E[Y_i | X_i]$ is the *conditional expectation* of Y_i given X_i.

   The fixed-X assumption simplifies presentation. Chapter 4 explores the implications of this and other "usual" regression assumptions.
2. For any straight line through the point $(\bar{X}, \bar{Y})$, the residuals add up to zero. Consequently we cannot measure accuracy by the sum of residuals. Squaring eliminates negative values, so RSS ≥ 0. This reasoning parallels the use of squared deviations in variance and standard deviation to work around the zero-sum property of the mean (Chapter 1).
3. Nonsensical Y-intercepts illustrate the more general hazards of prediction beyond the range of observed X values. Some X values are impossible and hence lead to meaningless predictions. Furthermore, the $Y-X$ relationship might change at very high or very low X values—become nonlinear, for instance. A linear relation over the observed X range provides no assurance against such possibilities.
4. By definition, $Y = \hat{Y} + e$. The variance of Y is therefore

   $$\text{Var}[Y] = \text{Var}[\hat{Y}] + \text{Var}[e] + 2\text{Cov}[\hat{Y}, e]$$

   (see [A1.17]). *OLS residuals are always, by construction, uncorrelated with predicted values*:

   $$\text{Cov}[\hat{Y}, e] = 0$$

   Consequently:

   $$\text{Var}[Y] = \text{Var}[\hat{Y}] + \text{Var}[e]$$

5. Equation [2.16] is sometimes called a *Pearson correlation* or *product moment correlation.*
6. A perfect positive relationship, or $r = 1$, requires that the univariate X and Y distributions have exactly the same shape. A perfect negative relationship, $r = -1$, requires mirror-image shapes. Consequently, unless both variables have symmetrical and identically shaped distributions, correlations cannot actually range from -1 to $+1$.
7. If you try these calculations yourself, you will obtain $R_a^2 = .1728$ instead of .1729. The computer's calculation is based on a more precise value of R^2, to 16 decimal places rather than the four places shown in this example. Such discrepancies often arise when comparing high-precision machine calculations with calculations rounded off at intermediate stages.

8. To obtain approximate P-values for t-statistics using Table A4.1, read down the left-hand column to locate the row with appropriate degrees of freedom (df). In this example, df $= n - K = 496 - 2 = 494 \approx \infty$. Next read across that row to see where your sample t-statistic falls; here our sample statistic is $t = 10.221$, which falls off the right-hand side (because it is larger than 3.291) of the table. Since our t-statistic is larger than 3.291, its P-value must be lower than the P-value of 3.291, which for a two-sided test is .001 (given at top of table). We conclude that $P < .001$. If, for example, our sample t-statistic had been $t = 2$, we would instead have obtained $.02 < P < .05$. Study the graphs at the top of Table A4.1 if this reasoning seems unclear.
9. Theoretically, if our decision rule is to reject H_0 when $P < .05$, we have a 5% chance of being wrong. That is, about 5% of the time this rule will lead us to reject H_0 when in fact it is true. A lower cutoff point (such as $\alpha = .01$ or .001) results in a lower probability of this type of error (called *Type I error*). But, other things being equal, a lower α also *raises* the likelihood of the opposite error: failing to reject H_0 when in fact it is false (*Type II error*). The .05 significance level is popular (but not inevitable) as a compromise between these two risks.
10. To find approximate P-values for F-statistics using Table A4.2, read down the left-hand column to locate the denominator degrees of freedom (df_2). In this example, $\text{df}_2 = 494 \approx \infty$. Next read across the top row of the table to locate the numerator degrees of freedom (df_1). Here we want $\text{df}_1 = 1$, so we focus on the bottom-left cell ($\text{df}_1 = 1$, $\text{df}_2 = \infty$) of Table A4.2. Obtain a P-value by fitting the sample F-statistic among the F-values in that cell. Since our sample F-statistic, $F = 104.46$, exceeds the largest F-value in that cell (10.83), we obtain $P < .001$. If, for example, our sample F-statistic had been $F = 4$, we would instead have obtained $.01 < P < .05$.
11. Data from *Statistical Abstract of the United States*, 1987.
12. Potential confusion arises when the analyst supplies a constant as one of the X variables and asks for no-constant estimation. Although such regressions really do include a constant, computer programs may not automatically recognize this fact and so may print misleadingly high R^2 and F values (Uyar and Erdem, 1990).
13. The *granularity* of the income measure is apparent in the symmetry and quantile-normal plots of Figure 2.13. Many of the recorded incomes are multiples of $10,000. Such granularity does not in itself present serious statistical problems, but it indicates the presence of "rounding-off" measurement error.
14. Data from D. Halliwell, reprinted by the Committee on Monitoring and Assessment of Trends in Acid Deposition (1986).
15. Magnitude increases linearly with $\log_{10}$ of the amplitude, or height, of earthquake waves passing through the earth and recorded by seismographs at various distances. (A one-unit increase in magnitude corresponds to a tenfold increase in wave amplitude.) Amplitude, in turn, increases with the square root of the energy of the earthquake or nuclear blast. We therefore expect a linear relationship between magnitude (corrected for the attenuation of energy passing through rock) and the logarithm of energy or blast yield.

3 Basics of Multiple Regression

Multiple regression refers to regression with two or more X variables. For the most part this is a straightforward extension of bivariate regression. The same computer programs do either, producing similar-looking output that can be understood in much the same way. The potential problems, diagnostic methods, and possible solutions discussed in Chapter 2 apply to multiple regression as well. Of course, multiple regression is more complex, because it involves more variables and encounters some new statistical problems.

This chapter introduces basic ideas of multiple regression but leaves diagnostics, problem solving, and other complications until later. We begin with a general statement of the multiple regression model.

Multiple Regression Models

Bivariate regression models view expected Y as a linear function of X:

$$E[Y_i] = \beta_0 + \beta_1 X_i \qquad [3.1]$$

Expected (population mean) Y equals β_0 when $X = 0$ and changes by β_1 with each 1-unit increase in X. Actual Y equals expected Y plus a random error:

$$Y_i = E[Y_i] + \varepsilon_i$$

Multiple regression models can include more X variables. For example, with two X variables, X_1 and X_2:

$$E[Y_i] = \beta_0 + \beta_1 X_{i1} + \beta_2 X_{i2} \qquad [3.2]$$

(X_{i1} denotes "the ith value of variable X_1.") Expected Y equals β_0 when $X_1 = X_2 = 0$. The *coefficient on* X_1, β_1, reflects change in mean Y per 1-unit increase in X_1, *if* X_2 *stays the same*. Similarly the *coefficient on* X_2, β_2, reflects change in mean Y per 1-unit increase in X_2, if X_1 stays the same.

With $K - 1$ X variables, the general model becomes

$$E[Y_i] = \beta_0 + \beta_1 X_{i1} + \beta_2 X_{i2} + \beta_3 X_{i3} + \cdots + \beta_{K-1} X_{i,K-1} \qquad [3.3]$$

Here K stands for the *number of parameters* (β's) in the model, usually one more than the number of X variables. β_0 equals mean Y when $X_1 = X_2 = X_3 = \cdots = X_{K-1} = 0$. β_k, the coefficient on X_k, tells us the change in mean Y for each 1-unit increase in X_k, if the other X variables do not change. Under [3.2] or [3.3], as under simpler models, actual Y equals expected Y plus error:

$$Y_i = E[Y_i] + \varepsilon_i \qquad [3.4]$$

Matrix algebra permits a compact statement of the general regression model for any number of X variables:

$$\mathbf{Y} = \mathbf{X}\boldsymbol{\beta} + \boldsymbol{\varepsilon} \qquad [3.5]$$

$\mathbf{Y}$ is column of $\mathbf{Y}$ values, $\mathbf{X}$ a matrix of X values (including an initial column of 1's), $\boldsymbol{\beta}$ a column of coefficients, and $\boldsymbol{\varepsilon}$ a column of errors. See Appendix 3 for a review of matrix algebra.

Ordinary least squares applies equally well to bivariate and multiple regression. We find coefficient estimates (b's) for a sample version of [3.3]:

$$\hat{Y}_i = b_0 + b_1X_{i1} + b_2X_{i2} + b_3X_{i3} + \cdots + b_{K-1}X_{i,K-1} \qquad [3.6]$$

such that the sum of squared residuals,

$$\text{RSS} = \Sigma(Y_i - \hat{Y}_i)^2$$
$$= \Sigma e_i^2$$

is lower than for any other set of b's. OLS possesses optimal statistical properties if errors are normal i.i.d.

We can write the sample regression equation in matrix form:

$$\mathbf{Y} = \mathbf{XB} + \mathbf{e} \qquad [3.7]$$

For n cases and $K - 1$ X variables,

Y is an $n \times 1$ column vector of Y values;

X is an $n \times K$ matrix of X values (with initial column of 1's);

B is a $K \times 1$ column vector of estimated coefficients;

e is an $n \times 1$ column vector of sample residuals.

Computers perform OLS calculations using matrix algebra. Since these calculations are impractical by hand, we will bypass their details and focus instead on understanding the results.

A Three-Variable Example

In Chapter 2 we regressed 1981 water use on household income (Table 2.1):

$$\hat{Y}_i = 1201 + 47.5X_i \qquad [3.8]$$

Income explains about 17% ($R_a^2 = .17$) of the variation in postshortage water use. Predicted water use increases by 47.5 cubic feet with each \$1000 increase in income; this relationship is statistically significant ($t = 10.221$, $P < .0005$).

Table 3.1 shows the effects of including a second X variable, preshortage water use (X_2). (Hereafter income appears as X_1 rather than just X.) Table 3.1 has the same layout as Table 2.1 but with an additional line listing the coefficient, standard

Table 3.1 Regression of postshortage (1981) water use on income and preshortage (1980) water use.

Source	SS	df	MS	
Model	671025350	2	335512675	Number of obs = 496
Residual	422213359	493	856416.551	$F(2, 493)$ = 391.76
				Prob > F = 0.0000
Total	1.0932e + 09	495	2208563.05	R-square = 0.6138
				Adj R-square = 0.6122
				Root MSE = 925.43

Variable	Coefficient	Std. Error	t	Prob > $\|t\|$	Mean
water81					2298.387
income	20.54504	3.38341	6.072	0.000	23.07661
water80	.5931267	.0250482	23.679	0.000	2732.056
_cons	203.8217	94.36129	2.160	0.031	1

error, t-statistic, and P-value for the second X variable. The regression equation is

$$\hat{Y}_i = 203.8 + 20.5X_{i1} + .59X_{i2} \qquad [3.9]$$

The coefficient on household income remains positive and significant ($t = 6.072$, $P < .0005$) but drops from 47.5 to 20.5. Adjusted R^2 more than triples: income and preshortage water use together explain about 61% of the variation in postshortage water use.

The large increase in R_a^2 (from .17 to .61) indicates that preshortage water use contributes substantially to prediction of postshortage water use. This is not surprising; neither is the fact that the coefficient on preshortage use is positive ($b_2 = .59$) and statistically significant ($t = 23.679$, $P < .0005$). Households consuming lots of water before the shortage tended to remain high consumers. Such households still had the big lawns, water-using appliances, and domestic needs that they did a year earlier. Similarly, low-use households tended to remain relatively low.

The coefficient on income declined by 57% (from $b_1 = 47.5$ to $b_1 = 20.5$), indicating that we earlier overestimated income's effect on postshortage water use. The relation between income and postshortage water use seen in Chapter 2 results partly from an omitted third variable: both income and postshortage use correlate with preshortage water use. For example, a major contributor to water use is maintaining a lawn. The proportion of people with green lawns undoubtedly increases with income, but some low-income households have bigger or greener lawns than some high-income households. Lawn watering and similar extravagances, rather than income itself, drive up water use. Preshortage water use partly reflects how much lawn watering a household did. When we statistically *adjust for preshortage water use*, by including it in the regression analysis, income appears less important.

More interesting than the numerical change in the coefficient on income is the fact that this coefficient now has a different meaning. Bivariate analysis established that higher-income households had higher postshortage water use. This tells us little about conservation, because higher-income households had higher preshortage water use too. Table 3.1 reveals something new: even adjusting for preshortage water use, higher-income households consumed more water. Some observers had thought that affluent households would be more responsive to public appeals for water conservation. They apparently were not.[1]

If higher-income households were not more altruistic during the water shortage, what about households with higher levels of education? And what happens if we adjust for other demographic variables, like age or the number of people living in a household? Such questions can be addressed by including more X variables in the regression, which will be done later in this chapter. First we will look more closely at what it means to "adjust for" the effects of other variables.

Partial Effects

A simple experiment might compare two vaccines. One group of subjects receives vaccine 1, and another group receives vaccine 2. The experimenter thus physically controls the X variable (type of vaccine). This control typically involves *random assignment*: randomly determining which subjects get which vaccine. Let X_1 stand for "type of vaccine" and $X_2, X_3, X_4 \ldots$ represent other subject characteristics (age, gender, health history, and so on. Random assignment theoretically ensures no correlation between X_1 and other subject characteristics $X_2, X_3, \ldots$.

If, due to random assignment, there is no correlation between X_1 and all other possible X variables, then whatever relation we detect between X_1 and Y (illness) cannot be spurious. A spurious relation could occur only if some other X variable is related to both X_1 and Y. Furthermore, the time ordering in experiments is generally clear: first administer vaccine; then observe subjects' health. Thus experimental design addresses two of the three criteria for causality mentioned in Chapter 2. Analysis of experimental data can concentrate on comparatively simple questions of how X_1 and Y covary—for example, whether there are significant differences in the illness rates of the two vaccine groups.

Experiments provide strong evidence about causality, but they are impractical for many lines of research. We would not want to experimentally expose people to two types of toxic waste, for example, and then compare their illness rates as we do in testing vaccines. This may be an equally urgent research topic, but it requires a nonexperimental approach.[2] We might be able to collect data about people who already had the misfortune of toxic-waste exposure. Their exposure (X_1) would likely be correlated with any number of other X variables, though—where they lived, socioeconomic status, age, lifestyle, other environmental conditions, and so forth. Since X_1 could be correlated with X_2, X_3, and so on, we cannot rule out the possibility that whatever X_1-Y relation we observe actually is spurious.

In nonexperimental research we try to remove spurious effects by statistical adjustment. That is, we estimate the effect of X_1 on Y while mathematically adjusting for X_1's correlations with $X_2, X_3, \ldots$. Such analysis supports causal interpretations, but less conclusively than do true experiments. We can adjust for other relevant variables only if we know their values and know to include them in the analysis.

To explore the meaning of statistical adjustment, we return to the regression of postshortage water use (Y) on income (X_1) and preshortage water use (X_2):

$$\hat{Y}_i = 203.8 + 20.5X_{i1} + .59X_{i2}$$

If preshortage water use stays constant, predicted postshortage use increases by 20.5 cubic feet with each one-unit ($1000) increase in household income. If household income stays constant, predicted postshortage water use increases by 0.59 cubic feet with each one-cubic-foot increase in preshortage use. In nonexperimental research we cannot actually hold the X variables constant, but the data allow us to estimate their *partial effects.* Adjusting for preshortage use, $b_1 = 20.5$ is our estimate of the effect of income; adjusting for income, $b_2 = .59$ is our estimate of the effect of preshortage use.

Matrix algebra ([3.13]) performs the actual adjustments. Here is one way to understand the results. Regressing Y on X_2 (preshortage use) yields

$$Y_i = 537.9 + .64X_{i2} + e_{i,Y|X_2} \qquad [3.10]$$

Regressing X_1 (income) on X_2 yields

$$X_{i1} = 16.26 + .0025X_{i2} + e_{i,X_1|X_2} \qquad [3.11]$$

Residuals in [3.10] and [3.11] constitute two new variables, written $e_{Y|X_2}$ and $e_{X_1|X_2}$. These represent variations in postshortage water use and in income, respectively, *after we have subtracted the effects of preshortage water use.* Now regressing $e_{Y|X_2}$

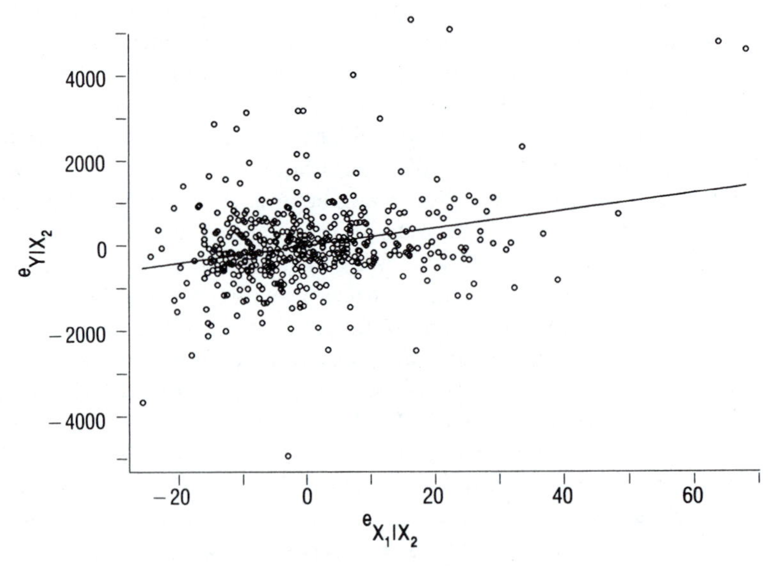

Figure 3.1 Partial regression leverage plot: postshortage water use (Y) versus income (X_1), adjusting for preshortage water use (X_2).

on $e_{X_1|X_2}$ yields the following:

$$\hat{e}_{i,Y|X_2} = 0 + 20.5e_{i,X_1|X_2} \quad [3.12]$$

The coefficient on $e_{X_1|X_2}$ in [3.12] ($b_1 = 20.5$) is the same as the partial coefficient on X_1 in [3.9]. Thus b_1 describes the relation between Y and X_1, adjusting for relations both variables have with X_2.

Figure 3.1 graphs the regression of Equation [3.12]. The slope in Figure 3.1 equals the partial regression coefficient $b_1 = 20.5$. The scatter of dots around this line shows variations in the data independent of X_2. Graphs like Figure 3.1 are called *partial regression leverage plots*. Chapter 4 views some of their applications.

We can derive the partial coefficient on X_2 in the same roundabout way. First we regress both Y and X_2 on X_1:

$$Y_i = 1201.1 + 47.5X_{i1} + e_{i,Y|X_1}$$

$$X_{i2} = 1681.4 + 45.5X_{i1} + e_{i,X_2|X_1}$$

Residuals from these regressions, $e_{Y|X_1}$ and $e_{X_2|X_1}$, represent variations in post- and preshortage water use that are unrelated to income. Regression of $e_{Y|X_1}$ on $e_{X_2|X_1}$ yields

$$\hat{e}_{i,Y|X_1} = 0 + .59e_{i,X_2|X_1}$$

The coefficient in this two-variable regression, $b_2 = .59$, is the same as the partial coefficient on X_2 in [3.9]. Figure 3.2 shows the corresponding leverage plot.

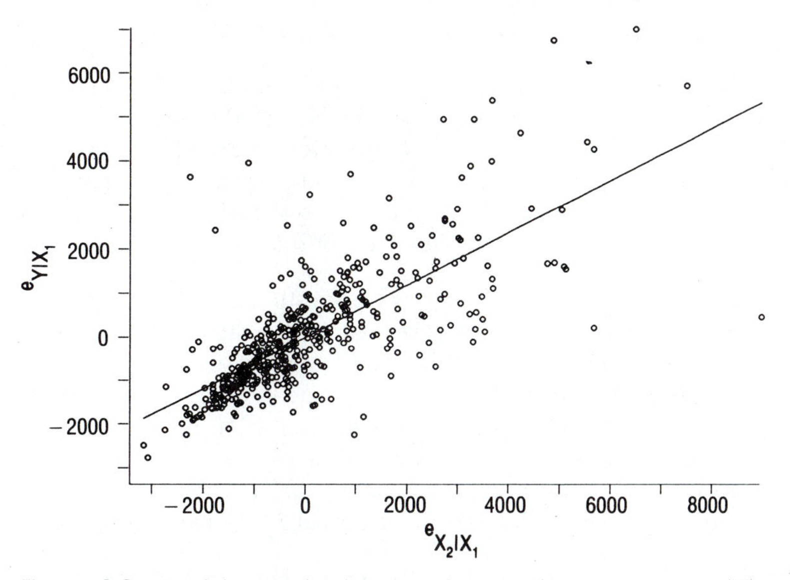

Figure 3.2 Partial regression leverage plot: postshortage water use (Y) versus preshortage water use (X_2), adjusting for income (X_1).

More generally, a leverage plot for variables Y and X_k depicts a regression in which:

The "Y" variable equals the residual from the regression of Y on *all X variables except* X_k.
The "X" variable equals the residual from the regression of X_k on *all X variables except* X_k.

The line in a leverage plot always has an intercept of zero. Its slope equals b_k, the partial coefficient on X_k from the regression of Y on all X variables including X_k.

Partial regression coefficients are not usually obtained in this manner, but they could be. (See below for the usual approach.) Leverage plots provide a less mathematical account of what it means to "statistically adjust for" other variables.

The vector **B** of OLS estimated regression coefficients for [3.7] is obtained most easily by matrix algebra:

$$\mathbf{B} = (\mathbf{X}'\mathbf{X})^{-1}\mathbf{X}'\mathbf{Y} \qquad [3.13]$$

Variable Selection

Occasionally we know the correct model: which X variables to include and how (linearly or otherwise) they relate to Y. More often, however, theory or previous research provides only vague guidance. Then we must look to the data for clues. This section discusses some aspects of variable selection.

Several changes occur as we include more X variables in a regression.

1. Prediction improves. R^2 (but not necessarily R_a^2) increases, and s_e (residual standard deviation) decreases. Is the improvement substantial?
2. Coefficients describe how the additional variables affect $\hat{Y}$. Are these coefficients significantly different from zero and large enough for substantive importance?
3. Spurious coefficients may shrink. Do the added variables substantially alter our conclusions regarding the effects of other X variables?

Affirmative answers to any of these questions support keeping the added variable(s) in the model. Negative answers indicate that the variables contribute little and should be left out unless theoretically important.

One goal of variable selection is to achieve in the final model a balance of simplicity and fit, called *parsimony*. Adjusted R^2 (Equation [2.21]) reflects parsimony, because it combines a measure of fit (R^2) with a measure of the difference in complexity between data (n, sample size) and model (K, number of parameters estimated). This difference, $n - K$, is called the *residual degrees of freedom*:

$$\text{df}_\text{R} = n - K \qquad [3.14]$$

For Table 3.1, $n = 496$ and $K = 3$, so $df_R = 496 - 3 = 493$. The df_R has many applications, including testing hypotheses and adjusting R^2.[3] When X variables are added to a model, R^2 will always increase, but R_a^2 may stay the same or decrease, if the improvement in fit is small relative to the increase in complexity.[4]

In selecting X variables for a regression, we risk two kinds of mistakes:

1. *Including an irrelevant variable.* X_k is irrelevant if the true value of β_k is zero or, more pragmatically, small enough to have no substantive importance. If irrelevant variables are included, coefficient estimates and predictions have greater sample-to-sample variation. Furthermore, we have made our model unnecessarily complex.
2. *Omitting a relevant variable.* X_k is relevant if the true value of β_k is nonzero (and large enough to have substantive importance), and X_k is correlated with other X variables. If relevant variables are omitted, all aspects of the regression (coefficients, standard errors, predictions, and so on) become untrustworthy. Our model is unrealistically simple.

Selecting variables for a regression involves a tradeoff. Which risks are worse depends on strengths of relations and purposes of the research.[5]

The *mean squared error* (MSE) (not to be confused with the sample mean squared error, or residual variance, often printed by regression programs) measures variation of estimates around a parameter. For regression coefficients:

$$\begin{aligned} \text{MSE} &= E[b_k - \beta_k]^2 \\ &= \text{Var}[b_k] + \text{bias}^2 \end{aligned}$$

(see Appendix 1). To have the best chance for our sample estimates to be near the population parameters, we want MSE as low as possible.

Including more X variables generally increases $\text{Var}[b_k]$, though at the same time it may reduce bias. It is possible for increases in $\text{Var}[b_k]$ to offset reductions in bias, so that MSE worsens when we add variables—even if those variables are relevant.

We cannot know whether this applies to the data at hand, but it provides a further reason, besides parsimony, for choosing a model without too many X variables.

Sample data can mislead us about which variables are relevant. Even when $\beta_k \neq 0$, b_k may be nonsignificant. Conversely, even when $\beta_k = 0$, our sample estimate b_k might be far from zero. Theoretically, about 5% of the time b_k could be "significantly different from zero" at the .05 level despite $\beta_k = 0$.

By theoretical and statistical criteria, preshortage water use belongs in the model for predicting postshortage water use (Table 3.1). We next consider some further X variables.

A Seven-Variable Example

Table 3.2 shows the regression of postshortage water use (Y) on six X variables:

X_1—household income, in thousands of dollars
X_2—preshortage water use, in cubic feet
X_3—education of household head, in years
X_4—retirement, coded 1 if household head is retired and 0 otherwise
X_5—number of people living in household at time of water shortage (summer 1981)
X_6—change in the number of people, summer 1981 minus summer 1980

The regression equation is approximately

$$\hat{Y}_i = 242 + 21X_{i1} + .49X_{i2} - 42X_{i3} + 189X_{i4} + 248X_{i5} + 96X_{i6} \qquad [3.15]$$

These six X variables together explain about 67% of the variation in postshortage water use ($R_a^2 = .6734$).

The variables were chosen from a set of nine background characteristics thought to predict household water use. Three variables (house value, occupation, and own/rent) were dropped because their partial effects were near zero. One variable with a near-zero coefficient, change in number of people, was kept in the model. Despite weak statistical evidence, on common-sense grounds it seems likely to affect water use in the population.[6]

Table 3.2 Regression of postshortage water use on income, preshortage water use, education, retirement, number of people resident, and increase in people resident

Source	SS	df	MS
Model	740477522	6	123412920
Residual	352761188	489	721393.022
Total	1.0932e + 09	495	2208563.05

Number of obs = 496
$F(6, 489)$ = 171.08
Prob > F = 0.0000
R-square = 0.6773
Adj R-square = 0.6734
Root MSE = 849.35

Variable	Coefficient	Std. Error	t	Prob > $\lvert t \rvert$	Mean
water81					2298.387
income	20.96699	3.463719	6.053	0.000	23.07661
water80	.49194	.0263478	18.671	0.000	2732.056
educat	−41.86552	13.22031	−3.167	0.002	14.00403
retire	189.1843	95.02142	1.991	0.047	.2943548
peop81	248.197	28.7248	8.641	0.000	3.072581
cpeop	96.4536	80.51903	1.198	0.232	−.0383065
_cons	242.2204	206.8638	1.171	0.242	1

Coefficients in Equation [3.15] are *fifth-order partials*, meaning that each adjusts for five other predictors. (Since [3.9] includes two X variables, the coefficients are *first-order partials*. The single slope in [3.8] is a *zero-order partial*.) We interpret the fifth-order coefficient on education, $b_3 = -42$, as follows:

> Predicted postshortage water use *declines* by 42 cubic feet with each additional year of household head's education, if income, preshortage use, retirement, number of people, and change in number of people remain constant.

Other things being equal, well-educated households tend to conserve more water. The negative effect of education on water use (the more education, the less water used) contrasts with the positive effect of income (the more income, the more water used).

The other coefficients in Equation [3.15] are interpreted similarly:

Income ($b_1 = 21$): Predicted postshortage water use increases by 21 cubic feet with each thousand-dollar increase in household income, other things being equal. This coefficient is virtually unchanged from Equation [3.9], although we now adjust for four more variables.

Preshortage water use ($b_2 = .49$): Predicted postshortage water use increases by .49 cubic feet with each one-cubic-foot increase in preshortage use, other things being equal. This effect is somewhat weaker than that in [3.9]; evidently part of X_2's effect there results from correlations with household size and other variables.

Retirement ($b_4 = 189$): Other things being equal, retired households tend to use 189 cubic feet more water than nonretired households.

Number of people ($b_5 = 248$): Other things being equal, predicted water use goes up by 248 cubic feet with each one additional person living in a household.

Change in number of people ($b_6 = 96$): Predicted postshortage water use increases by an additional 96 cubic feet if one more person lived in the household in 1981 than in 1980, other things being equal. This effect is not significantly different from zero, however.

We might summarize these findings as follows: For any given level of preshortage water use, postshortage use tends to be lower, and hence *water conservation greater*, in households with low income, highly educated and nonretired heads, and few people living there. Conversely, postshortage use will be higher, and hence *less water conserved*, in households with high income, poorly educated and retired heads, and many people.

With data on n cases, we could theoretically estimate coefficients on as many as $n - 1$ predictors. Since such a model would have n parameters ($n - 1$ regression coefficients, plus one Y-intercept), it would be just as complex as the original data—and fit those data perfectly ($R^2 = 1$). We ordinarily seek models that are much simpler than the data, however. Equation [3.15] is reasonably parsimonious: with only six X variables, it explains approximately two-thirds of the variation in water use among 496 households.

Standardized Regression Coefficients

We can express any variable as a *standard score*, measured in standard deviations from its mean:

$$Z_i = \frac{X_i - \bar{X}}{s_X} \qquad [3.16]$$

For example, household #1 in our data has an income of $30,000. Since the sample mean is 23.1 and the standard deviation is 13.1, household #1's income as a standard score is (30 − 23.1)/13.1 = .53. This tells us that household #1 is .53 standard deviations above average in income. Household #23 has an income of $7000, or (7 − 23.1)/13.1 = −1.23 in standard-score form. Household #23 is 1.23 standard deviations below average.

Standardized regression coefficients, or "beta weights," are coefficients we would obtain from a regression with all variables in standard-score form. Standardized coefficients can be found directly from unstandardized coefficients:

$$b_k^* = b_k \frac{s_k}{s_Y} \qquad [3.17]$$

where b_k is the unstandardized coefficient on X_k, s_k is the standard deviation of X_k, and s_Y is the standard deviation of Y.

A standardized version of [3.15] is

$$\hat{Z}_{iY} = .18Z_{i1} + .58Z_{i2} - .09Z_{i3} + .06Z_{i4} + .28Z_{i5} + .03Z_{i6} \qquad [3.18]$$

The standardized regression coefficient or "beta weight" on income, $b_1^* = .18$, tells us that predicted water use increases by .18 standard deviations with each 1-standard-deviation increase in income, other things being equal.

In standardized regression equations like [3.18]:

1. The Y-intercept always equals zero.
2. All variables are expressed as standard scores, measured in standard deviations from their means.
3. Coefficients indicate by how many standard deviations $\hat{Y}$ changes, per 1-standard-deviation increase in X (other things being equal).

Like correlation coefficients, standardized regression coefficients theoretically can range from −1 to +1, with ±1 indicating a perfect linear relationship and 0 indicating no linear relationship. Consequently they seem to provide an easy way to determine whether one variable's effect is "larger" than another's. Such use of standardized coefficients implies a specialized definition of effect size, however: standard deviations change per standard deviation. Effects measured in the natural units of Y and X (as by unstandardized coefficients) usually have clearer practical interpretation.

Standardized regression coefficients are ill suited for comparisons across different samples or subsamples, because they depend partly on the X variables' variance. If $|b_k^*|$ is larger in sample 1 than in sample 2, this could indicate that either:

1. $|b_k|$ is larger in sample 1; *or*
2. X_k has more variance in sample 1.

Similar problems complicate cross-sample comparisons using other standardized statistics, like R^2 and correlation.

t-Tests and Confidence Intervals for Individual Coefficients

Multiple regression *t*-tests and confidence intervals closely resemble their bivariate counterparts. The estimated standard error of b_k, the regression coefficient on X_k, is

$$\mathrm{SE}_{b_k} = \frac{s_e}{\sqrt{\mathrm{RSS}_k}} \qquad [3.19]$$

where s_e is the residual standard deviation ([2.11]). RSS_k, the residual sum of squares from regression of X_k on all the other X variables, measures independent variation in X_k. (In bivariate regression all variation in X is independent, so $\mathrm{RSS}_k = \mathrm{TSS}_X$; compare [3.19] with [2.22].) High standard errors indicate much sample-to-sample variation and hence a higher likelihood that any one sample estimate will be far from the true parameter.

Estimated standard errors of regression coefficients equal square roots of the major diagonal elements of the *variance-covariance matrix* of estimated coefficients:

$$\mathbf{S} = s_e^2(\mathbf{X}'\mathbf{X})^{-1} \qquad [3.20]$$

Off-diagonal elements are covariances between estimated coefficients, which are useful in detecting multicollinearity.

Coefficients in [3.15] estimate unknown β parameters of the model:

$$E[Y_i] = \beta_0 + \beta_1 X_{i1} + \beta_2 X_{i2} + \beta_3 X_{i3} + \beta_4 X_{i4} + \beta_5 X_{i5} + \beta_6 X_{i6} \qquad [3.21]$$

Also, *t*-statistics test hypotheses regarding individual β values:

$$t = \frac{b_k - \beta_k}{\mathrm{SE}_{b_k}} \qquad [3.22]$$

Equation [3.22] measures distance (in estimated standard errors) between sample statistic (b_k) and population parameter (β_k). We substitute a null-hypothesis β_k into [3.21]. Given a true null hypothesis and normal i.i.d. errors, [3.22] follows a t-distribution with $\text{df}_\text{R} = n - K$ degrees of freedom. For the "usual" null hypothesis H_0: $\beta_k = 0$, [3.22] simplifies to

$$t = b_k/\text{SE}_{b_k} \qquad [3.23]$$

The output in Table 3.2 includes this information:

$\hat{Y} = 242 + 21X_1$	$+ .49X_2$	$- 42X_3$	$+ 189X_4$	$+ 248X_5$	$+ 96X_6$
SE_{b_k}: 3.5	.026	13.2	95	28.7	80.5
t: 6.0	18.7	−3.2	2.0	8.6	1.2
P-value: <.0005	<.0005	.002	.047	<.0005	.232

Each t-statistic equals a coefficient divided by its standard error, following [3.23]. P-values could be obtained by looking t-statistics up in Table A4.1, with $496 - 7 = 489$ degrees of freedom. In practice, software calculates more precise P-values.

All of the X variables except X_6 (change in number of people) have coefficients significant at $\alpha = .05$. Thus we can reject the null hypothesis of no relationship in the population (H_0: $\beta_k = 0$) for five of the six X variables. By rejecting the null hypotheses, we conclude in favor of alternative hypotheses H_1: $\beta_k \neq 0$: there exists "some" linear relationship in the population.

A relation can be statistically significant yet too weak to be important. Strength comes into sharper focus with confidence intervals. We construct confidence intervals by adding and subtracting t times the estimated standard error:

$$b_k \pm t(\text{SE}_{b_k}) \qquad [3.24]$$

We choose a t value for the desired level of confidence from the theoretical t-distribution with $\text{df}_\text{R} = n - K$ degrees of freedom. For example, given $\text{df}_\text{R} = 489$, we refer to the t_∞ distribution (bottom row in Table A4.1), finding these values:

90% confidence interval: $t = 1.645$

95% confidence interval: $t = 1.96$

99% confidence interval: $t = 2.576$

Forming a 95% confidence interval for the coefficient on income from Table 3.2, we have

$$b_1 \pm t(\text{SE}_{b_1})$$

$$20.96699 \pm 1.96(3.463719)$$

$$\rightarrow 14.178 \leq \beta_1 \leq 27.756$$

If we drew many random samples and constructed intervals in this manner for each one, 95% of those intervals should include the true value of β_1. Informally, we might say that, based on this sample, we are "95% confident" that the true parameter β_1 lies between 14.178 and 27.756.[7]

Similarly, we can find a 95% confidence interval for the coefficient on education:

$$b_3 \pm t(\mathrm{SE}_{b_3})$$

$$-41.86552 \pm 1.96(13.22031)$$

$$\rightarrow -67.777 \leq \beta_3 \leq -15.954$$

This interval suggests that the population mean water use declines between 67.777 and 15.954 cubic feet with each additional year of education, other things being equal.

Chapter 2 described intervals around predicted Y:

$$\hat{Y}_i \pm t(\mathrm{SE}_{\hat{Y}_i}) \qquad [3.25]$$

again using t values with $\mathrm{df_R} = n - K$. Such intervals have two applications:

1. Confidence intervals: estimating the population mean of Y, given a certain combination of X values.
2. Prediction intervals: predicting the value of Y for an individual case, given a certain combination of X values.

These applications require different standard errors. Computer programs estimate either type; be sure to select the standard error appropriate for your research question.

Let $\mathbf{C}$ be a row vector of specific X values:

$$\mathbf{C} = [1 \quad X_{i1} \quad X_{i2} \quad \cdots \quad X_{i,K-1}]$$

To form *confidence intervals* for the population mean of Y given $\mathbf{C}$, estimate the standard error of $\hat{Y}$ as

$$\mathrm{SE}_{\hat{Y}_i} = s_e(\mathbf{C}(\mathbf{X}'\mathbf{X})^{-1}\mathbf{C}')^{1/2} \qquad [3.26]$$

To form *prediction intervals* for individual predicted values of Y given $\mathbf{C}$, estimate the standard error of $\hat{Y}$ as

$$\mathrm{SE}_{\hat{Y}_i} = s_e(1 + \mathbf{C}(\mathbf{X}'\mathbf{X})^{-1}\mathbf{C}')^{1/2} \qquad [3.27]$$

Equations [3.26] and [3.27] generalize [2.29] and [2.30].

Values of [3.26] and [3.27] depend on the specific combination of X values, so they may be different for every case. Standard errors are smallest, and hence confidence intervals narrowest, when all X variables equal their means.

F-Tests for Sets of Coefficients

A regression with $K - 1$ X variables requires K parameter estimates: one on each X variable, plus a Y-intercept. The t-statistics test hypotheses regarding individual parameters. The F-statistics can test hypotheses regarding *sets* of parameters. They do this by comparing *nested models*: two models, one a subset of the other. We test whether a complex model, with K parameters, significantly improves upon a simpler model with H fewer parameters ($0 < H < K$):

$$F^{H}_{n-K} = \frac{(\text{RSS}\{K - H\} - \text{RSS}\{K\})/H}{(\text{RSS}\{K\})/(n - K)} \qquad [3.28]$$

RSS$\{K\}$ denotes the residual sum of squares for the complex (K parameters) model, and RSS$\{K - H\}$ is the residual sum of squares for a model with $K - H$ parameters. We compare F-statistics calculated from [3.28] to a theoretical F-distribution (Table A4.2) with $\text{df}_1 = H$ and $\text{df}_2 = n - K$ degrees of freedom.

For example, both education and income reflect socioeconomic status. Suppose we wish to test the hypothesis that socioeconomic status has no linear effect on water use; that is, we wish to test the null hypothesis $H_0: \beta_1 = \beta_3 = 0$. This amounts to a test of nested models: a complex model including education and income versus a simpler model without them, but otherwise the same. Table 3.2 gave the more complex regression with six X variables ($K = 7$, RSS$\{7\}$ = 352,761,188). Table 3.3 gives the simpler regression, omitting income and education ($H = 2$, RSS$\{5\}$ = 380,520,363).

Table 3.3 Regression of postshortage water use omitting income and education

Source	SS	df	MS	
Model	712718346	4	178179587	Number of obs = 496
Residual	380520363	491	774990.557	$F(4, 491)$ = 229.91 Prob > F = 0.0000 R-square = 0.6519
Total	1.0932e + 09	495	2208563.05	Adj R-square = 0.6491 Root MSE = 880.34

Variable	Coefficient	Std. Error	t	Prob > |t|	Mean
water81					2298.387
water80	.519741	.026774	19.412	0.000	2732.056
peop81	265.2894	29.63234	8.953	0.000	3.072581
cpeop	134.4626	83.1959	1.616	0.107	−.0383065
retire	67.27992	94.28846	0.714	0.476	.2943548
_cons	48.64897	107.0549	0.454	0.650	1

Applying [3.28], we have

$$F^3_{489} = \frac{(\text{RSS}\{5\} - \text{RSS}\{7\})/2}{(\text{RSS}\{7\})/(496 - 7)}$$

$$= \frac{(380{,}520{,}363 - 352{,}761{,}188)/2}{352{,}761{,}188/489}$$

$$= 19.24$$

With $\text{df}_1 = 2$ and $\text{df}_2 = 489$ (approximately 2 and ∞) degrees of freedom, this F-statistic leads to a P-value below .001. We may reject H_0: $\beta_1 = \beta_3 = 0$ and accept the alternative H_1: not $(\beta_1 = \beta_3 = 0)$.

This result is not surprising, because individual t-tests (Table 3.2) lead to rejection of both H_0: $\beta_1 = 0$ and H_0: $\beta_3 = 0$. Sometimes, however, when t-tests *fail* to reject individual null hypotheses, an F-test nonetheless rejects the combined null hypothesis regarding those same parameters. The next section describes one way (multicollinearity) this occurs.

We can apply F-tests to the null hypothesis that coefficients on *all* of the X variables in a model equal zero. This tests the full model against a model with no X variables and with Y estimated by $\bar{Y}$. (Recall that $\bar{Y}$ is the least-squares estimate of β_0, for a model with no X variables.) For such tests, $H = K - 1$. Since RSS$\{K\}$ is just the usual residual sum of squares, RSS, Equation [3.28] becomes

$$F^{K-1}_{n-K} = \frac{(\text{RSS}\{1\} - \text{RSS}\{K\})/(K - 1)}{\text{RSS}\{K\}/(n - K)}$$

$$= \frac{(\text{TSS}_Y - \text{RSS})/(K - 1)}{\text{RSS}/(n - K)}$$

$$= \frac{\text{ESS}/(K - 1)}{\text{RSS}/(n - K)} \qquad [3.29]$$

This last expression appeared earlier as Equation [2.28]. It is a special case of [3.28].

Most regression programs automatically print F-tests based on [3.29]. For example, one appears at the upper right in Table 3.3: $F(4, 491) = 229.91$, Prob $> F =$ .0000 (meaning that the probability of a greater F is so low that it rounds off to zero, to four decimal places). We could calculate this statistic ourselves using the sums of squares at the upper left:

$$F = \frac{712{,}718{,}346/4}{380{,}520{,}363/491}$$

$$= 229.91$$

With $\text{df}_1 = 4$ and $\text{df}_2 = 491$ degrees of freedom, Table A4.2 indicates that $P < .001$. We reject the null hypothesis that all population coefficients are zero.

Multicollinearity

Collinearity refers to linear relationships between two X variables. *Multicollinearity*, a more general term, encompasses linear relationships between two or more X variables. Multiple regression is impossible in the presence of *perfect* collinearity or multicollinearity. If X_1 and X_2 have no independent variation, we cannot estimate the effects of X_1 adjusting for X_2 or vice versa. One of the variables must be dropped. This is no loss, since a perfect relationship implies perfect redundancy. Perfect multicollinearity is rarely a practical problem, however. Strong (not perfect) multicollinearity, which permits estimation but makes it less precise, is more common.

We have seen that a partial regression coefficient b_1, in a regression with $K - 1$ X variables, equals the slope obtained by regressing two constructed variables:

variation in Y that is unrelated to $X_2 \cdots X_{K-1}$ ($e_{Y|X_2 \cdots X_{K-1}}$) on
variation in X_1 that is unrelated to $X_2 \cdots X_{K-1}$ ($e_{X_1|X_2 \cdots X_{K-1}}$)

Thus b_1 describes the relationship between Y and X_1, adjusting for their relations with variables $X_2 \cdots X_{K-1}$. If X_1 varies closely with other X variables, $e_{X_1|X_2 \dots X_{K-1}}$ will be nearly constant (zero), and the data provide little information for making the adjustment. In that case we cannot reliably estimate b_1.

Unreliable parameter estimates mean large standard errors, or much sample-to-sample variation. Large standard errors in turn produce lower t-statistics and wider confidence intervals. Even a sizable b_k value may not be statistically significant, since we cannot reliably distinguish it from zero.

Although multicollinearity prevents precise estimation of individual coefficients, we may nonetheless test whether these coefficients *all* equal zero, using an F-statistic. If X_1, X_2, and X_3 exhibit multicollinearity, an F-test may reject

$$H_0: \beta_1 = \beta_2 = \beta_3 = 0$$

even though t-tests fail to reject the individual null hypotheses

$$H_0: \beta_1 = 0$$

$$H_0: \beta_2 = 0$$

$$H_0: \beta_3 = 0$$

(An example appears at the top of Table 8.8 in Chapter 8.) Then we conclude that at least one of the multicollinear variables affects Y, but we cannot be more specific.

Search Strategies

In selecting X variables, a common temptation is to search the data for strong (or significant) predictors of Y. Although useful for some purposes, this approach is hazardous. When we search the data, hypothesis tests do not provide their usual protection against overinterpreting sample results (Type I error).[8] For an extreme example, suppose we have sample data on many variables that, in the population, are completely unrelated. By chance about 5% of the *sample* relationships should be statistically significant at $\alpha = .05$. A search of the data will find such relationships, but it is a mistake (*multiple comparison fallacy*) to single them out as "significant." Variable selection based on searching tends to exploit chance patterns in the sample at hand, leading to conclusions that do not apply to other samples or to the population.

Search procedures are nonetheless popular, and they can be automated in several ways. Computers can examine *all possible subsets* of X variables and can rank models with any given number of variables from best to worst fitting. This thorough procedure requires much computation. For example, with 6 X variables there are $2^6 - 1 = 63$ possible subsets; with 12 X variables we would need $2^{12} - 1 = 4095$ separate regressions. Shortcuts may lessen the computational work.

Stepwise regression refers to an easier class of automated search procedures. It has two main variants.

1. *Forward inclusion*: Starting with no X variables in the model, first include the one that has the largest simple correlation with Y. Thereafter, at each step, add the X variable that produces the largest further increase in R^2.
2. *Backward elimination*: Starting with all X variables in the model, seek the one whose deletion will least reduce R^2. Thereafter, at each step, remove the X variable that produces the least further decrease in R^2.

These two variants are often combined. For example, we could start with the full model and use backward elimination but at each step consider whether to bring back any previously dropped variable.

Stepwise procedures can employ stopping rules based on "significance" levels, number of X variables, or other criteria. For example, they might add or subtract variables until the equation contains only "significant" terms; all left-out variables would have "nonsignificant" effects if brought in on the next step. This may be accomplished with forward inclusion by specifying an *F-to-enter*: a minimum F value that is required before an X variable will be added to the equation. When no variable meets this minimum, the process stops. Similarly, backward elimination may employ an *F-to-stay* stopping rule: elimination stops when all remaining terms have F-statistics above a minimum level.

"Significant" and "nonsignificant" are in quotes above because they are not what they seem. All-possible-subsets and stepwise regression share the risks of any search strategy, including searching by hand: sample-specific results and inaccurate P-values. The P-values for t- and F-tests printed by most stepwise regression packages are unrealistically low. This also applies to stopping rules based on F-statistics or their P-values. Adjustments to obtain more accurate P-values are difficult and seldom attempted in practice. Instead, analysts aware of this problem read the printed P-values as descriptive information, recognizing that the real probability of Type I error may be substantially higher. Although a printout reports $P = .01$, the actual probability might be .05 or .10, for instance. The likely discrepancy between printed and true probabilities increases with the ratio of variables searched to variables kept.

Automated search procedures lack the human analyst's ability to consider whether decisions make substantive sense. Stepwise methods also trip over certain technical problems:

1. When there are strong relationships among several predictors (multicollinearity), stepwise procedures tend to exclude one or more of these variables from the regression. We then mistakenly conclude that the excluded variable is unimportant, and we overstate the importance of variables that are included.
2. Offsetting effects: if X_1 and X_2 are positively related to each other but their effects on Y have opposite signs, or if X_1 and X_2 are negatively related but their effects on Y have the same sign, stepwise procedures (which consider variables only one at a time) may exclude one or both variables. We then understate the importance of both variables.

Where possible, an intelligent mix of substantive reasoning and statistical exploration should guide variable selection. But keep in mind that any searching, automatic or otherwise, leads to deceptively low P-values.

Interaction Effects

In simple models, effects of each X variable do not depend on the values of other X variables. For example, Equation [3.15] states that predicted water use increases by 21 cubic feet with each $1000 of household income, regardless of the level of preshortage water use, education, number of people, and so forth. We call it an *interaction* if an X variable's effect *does* depend on the values of other X variables.

Linear models can incorporate interactions in two ways:

1. When the left-hand-side variable is a nonlinear transformation of the variable of substantive interest (for example, log Y), all effects in the model are implicitly interactions.
2. *Interaction terms*, which typically are products of other variables, can be included as right-hand-side variables. Consider a model with two predictors, X and W:

$$\hat{Y}_i = b_0 + b_1 X_i + b_2 W_i$$

If X increases by one unit while W remains constant, $\hat{Y}$ changes by $b_1 \times 1$. If X changes by any amount ΔX, $\hat{Y}$ changes by

$$\Delta \hat{Y} = b_1 \times \Delta X$$

The effect of X on Y is the same, regardless of the value of W. Likewise, the effect of W on Y is the same, regardless of the value of X.

To model an interaction between X and W, include the interaction term X times W:

$$\hat{Y}_i = b_0 + b_1 X_i + b_2 W_i + b_3 X_i W_i$$

This equation implies that, if X increases by one unit while W remains constant, $\hat{Y}$ changes by $b_1 + b_3W$. For any change ΔX, $\hat{Y}$ changes by

$$\Delta\hat{Y} = (b_1 + b_3W)\Delta X$$

Therefore $(b_1 + b_3W)$ is the real "coefficient" on X.[9] Obviously the size of this effect depends on the value of W. A t-test of estimated b_3 establishes whether the interaction is statistically significant; H_0: $\beta_3 = 0$ implies no interaction (that is, $\beta_1 + \beta_3W$ is no different from β_1).

We interpret b_3 as follows: *with every 1-unit increase in W, the coefficient on X changes by* b_3. For example:

If W is:	*The coefficient on X is:*
$W = 0$	$b_1 + b_3W = b_1$
$W = 1$	$b_1 + b_3W = b_1 + b_3$
$W = 2$	$b_1 + b_3W = b_1 + 2b_3$
$W = 3$	$b_1 + b_3W = b_1 + 3b_3$

If b_1 and b_3 have opposite signs, then at large-enough W values (when $|b_3W| > |b_1|$) the relation between Y and X changes sign.[10] That is, the $X-Y$ relationship could be positive at low values of W and negative at high values, or vice versa.

Conditional effect plots (Chapter 5) help us to understand interaction effects. If W is a measurement variable, we might graph the $X-Y$ relationship at the first and third quartiles of W, or at other values of interest.

Intercept Dummy Variables

Dummy variables are (0, 1) dichotomies. For example, retirement (X_4) in Equation [3.15] is a dummy variable, coded 1 if household head has retired and 0 otherwise. Dummy variables permit analysis with categorical predictors.

To see the implications of a dummy variable regression, write out the regression equation substituting 0 and 1. Among nonretired households ($X_4 = 0$), Equation [3.15] becomes

$$\begin{aligned} \hat{Y}_i &= 242 + 21X_{i1} + .49X_{i2} - 42X_{i3} + 189(0) + 248X_{i5} + 96X_{i6} \\ &= 242 + 21X_{i1} + .49X_{i2} - 42X_{i3} + 248X_{i5} + 96X_{i6} \end{aligned} \qquad [3.30]$$

Among retired households ($X_4 = 1$), Equation [3.15] becomes

$$\begin{aligned} \hat{Y}_i &= 242 + 21X_{i1} + .49X_{i2} - 42X_{i3} + 189(1) + 248X_{i5} + 96X_{i6} \\ &= 431 + 21X_{i1} + .49X_{i2} - 42X_{i3} + 248X_{i5} + 96X_{i6} \end{aligned} \qquad [3.31]$$

The Y-intercepts in [3.30] and [3.31] differ by the coefficient on X_4: $b_4 = 189$. Since this is the only difference between the two equations, X_4 here serves as an *intercept dummy variable.*

Does wintertime road salting contaminate rural water supplies? Data on 52 New England wells were collected to address this question, and these data provide further examples of dummy variable regression. The variables are:

Y—*chloride*: natural logarithm of the chloride concentration (log milligrams per liter) in the well's water—an indicator of contamination by road salt;

X_1—*type*: dummy variable coded 0 for shallow wells and 1 for deeper wells drilled into bedrock; and

X_2—*distance*: natural logarithm of the distance (log feet) between the well and the nearest salted road.

In raw form, chloride concentration and distance from road have positively skewed distributions with severe outliers (a few badly polluted or remote wells). Logarithms reduce these problems.

Starting simply, we find that the mean log chloride concentration is lower in deep wells:

Type	$\bar{Y}$	s_Y	n
Shallow wells ($X_1 = 0$)	3.78	1.73	10
Deep wells ($X_1 = 1$)	3.07	1.26	42
Both	3.21	1.37	52

Tested by a two-sample t-test (or ANOVA—next section), the difference between means, $\bar{Y}_1 - \bar{Y}_0 = -.71$, is not statistically significant ($P = .15$).

Dummy variable regression performs essentially the same test. Regressing chloride on well type yields

$$\hat{Y}_i = 3.78 - .71X_{i1} \qquad [3.32]$$

For shallow wells, [3.32] simplifies to

$$\hat{Y}_i = 3.78 - .71(0) = 3.78$$

For deep wells, [3.32] yields

$$\hat{Y}_i = 3.78 - .71(1) = 3.07$$

These predicted values equal the shallow and deep-well means, and the coefficient on well type ($-.71$) equals their difference. A t-statistic for b_1 tests whether the difference of means is significant.[11]

Regressing Y on one dummy X variable, as in Figure 3.3, produces equations with form

$$\hat{Y}_i = b_0 + b_1X_{i1}$$

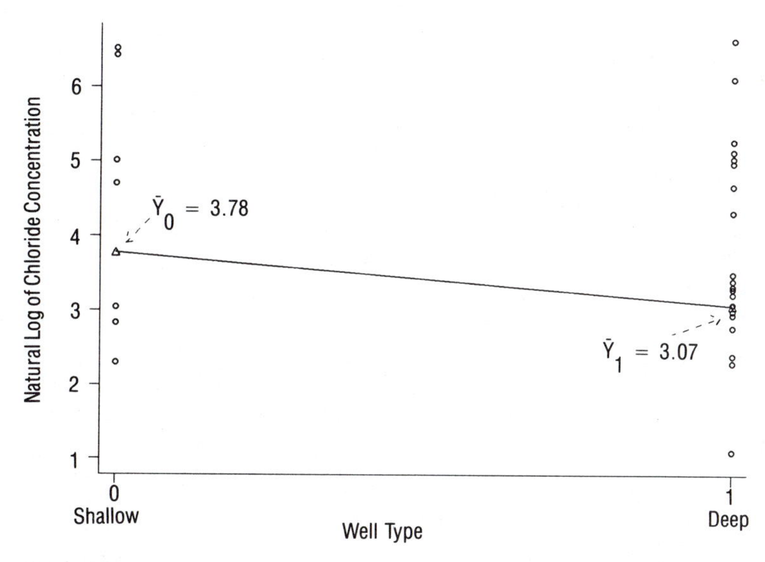

Figure 3.3 Regression of log chloride concentration on a dummy variable for well type.

1. The mean of Y when $X_1 = 0$ equals the Y-intercept: $\bar{Y}_0 = b_0$.
2. The mean of Y when $X_1 = 1$ equals the Y-intercept plus the coefficient on X_1: $\bar{Y}_1 = b_0 + b_1$.
3. The *difference in mean* Y between $X_1 = 0$ and $X_1 = 1$ equals the coefficient on X_1: $\bar{Y}_1 - \bar{Y}_0 = b_1$.

A t-test of H_0: $\beta_1 = 0$ therefore establishes whether the two means significantly differ. The t-statistic equals that obtained from a two-sample t-test for a difference of means.

Since chloride indicates salt contamination, we might expect higher concentrations in wells near roads. Including distance from road (X_2) as a second predictor, we have

$$\hat{Y}_i = 4.21 - .70X_{i1} - .09X_{i2} \qquad [3.33]$$

In [3.33], type becomes an intercept dummy variable. For shallow wells, [3.33] implies

$$\begin{aligned}\hat{Y}_i &= 4.21 - .70(0) - .09X_{i2} \\ &= 4.21 - .09X_{i2}\end{aligned}$$

which is the higher of the two lines in Figure 3.4. For deep wells we obtain the lower line:

$$\begin{aligned}\hat{Y}_i &= 4.21 - .70(1) - .09X_{i2} \\ &= 3.51 - .09X_{i2}\end{aligned}$$

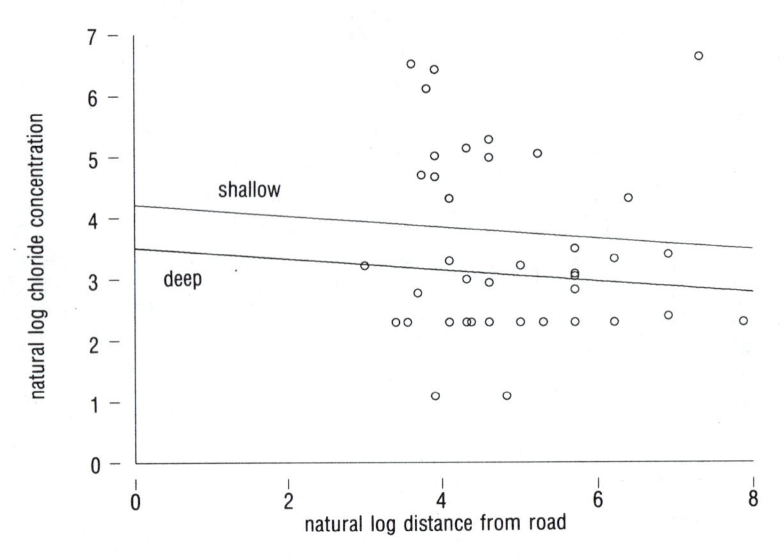

Figure 3.4 Regression of log chloride concentration on log distance from road and an *intercept* dummy variable for well type.

These two lines differ only in intercept. They suggest that chloride decreases slightly (not significantly) with distance, at the same rate for either well type. For any given distance, shallow wells tend to be more contaminated.

Regressing Y on an intercept dummy variable (X_1) in addition to one or more measurement variables (X_2) yields equations with form

$$\hat{Y}_i = b_0 + b_1 X_{i1} + b_2 X_{i2}$$

1. The Y-intercept equals b_0 when $X_1 = 0$.
2. The Y-intercept equals $b_0 + b_1$ when $X_1 = 1$.
3. b_1 equals the *difference in Y-intercepts* between X_1 categories 0 and 1.

A t-test of H_0: $\beta_1 = 0$ therefore determines whether the two intercepts differ significantly.

Slope Dummy Variables

Intercept dummy variables test for differences in intercepts. To test for a difference in slopes, form an interaction term called a *slope dummy variable* by multiplying dummy times measurement variable. If X_1 is a dummy and X_2 a measurement variable, we create the new slope dummy variable X_1X_2 and include it with X_2 in a regression. For example, regressing log chloride (Y) on log distance from road (X_2) and a slope dummy formed from distance and well type (X_1X_2), we get

$$\hat{Y}_i = 3.67 - .03X_{i2} - .08X_{i1}X_{i2} \qquad [3.34]$$

For shallow ($X_1 = 0$) wells, [3.34] becomes

$$\hat{Y}_i = 3.67 - .03X_{i2} - .08(0)X_{i2}$$
$$= 3.67 - .03X_{i2}$$

For deep ($X_1 = 1$) wells, [3.34] becomes

$$\hat{Y}_i = 3.67 - .03X_{i2} - .08(1)X_{i2}$$
$$= 3.67 - .11X_{i2}$$

The deep-wells line is steeper by an amount equal to the (nonsignificant) coefficient on the slope dummy variable, $-.08$ (Figure 3.5).

Regressing Y on both a measurement variable X_2 and a slope dummy variable X_1X_2 yields equations with form

$$\hat{Y}_i = b_0 + b_2X_{i2} + b_3X_{i1}X_{i2}$$

1. The slope relating X_2 to Y equals b_2 when $X_1 = 0$.
2. The slope relating X_2 to Y equals $b_2 + b_3$ when $X_1 = 1$.
3. b_3, the coefficient on X_1X_2, equals the *difference in slopes* between X_1 categories 0 and 1.

A t-test of H_0: $\beta_3 = 0$ determines whether the two slopes differ significantly. Different slopes imply that, for one of the two categories of X_1, the Y–X_2 relation is stronger or has a different sign.

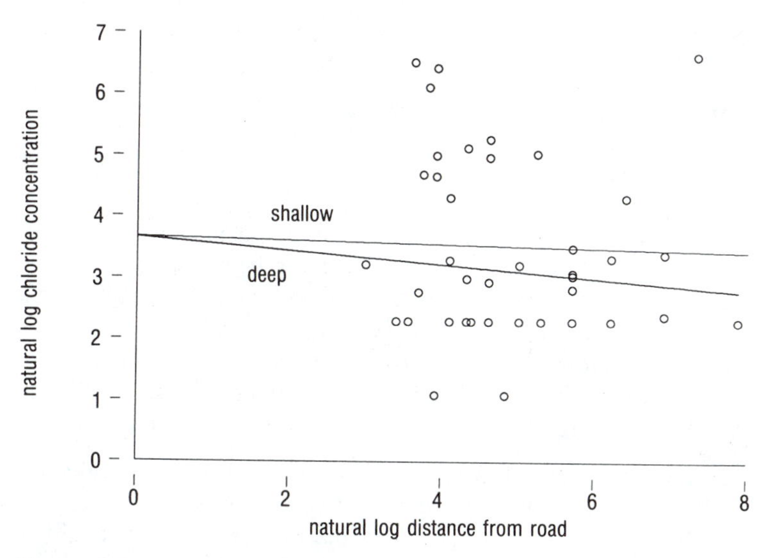

Figure 3.5 Regression of log chloride concentration on log distance from road and a *slope* dummy variable for well type.

Table 3.4 Regression of log chloride concentration on log distance from road, with intercept and slope dummy variables for well type.[1]

Source	SS	df	MS	
Model	18.4831272	3	6.1610424	Number of obs = 52 $F(3, 48)$ = 3.81 Prob > F = 0.0157
Residual	77.5390714	48	1.61539732	R-square = 0.1925
Total	96.0221986	51	1.88278821	Adj R-square = 0.1420 Root MSE = 1.271

Variable	Coefficient	Std. Error	t	Prob > $\|t\|$	Mean
chlor					3.205046
deep	−6.717366	2.094713	−3.207	0.002	.8076923
road	−1.109424	.3844204	−2.886	0.006	4.852686
deeproad	1.255847	.4268777	2.942	0.005	3.934269
_cons	9.073459	1.879384	4.828	0.000	1

[1] The dummy variable for well type, coded 0 = shallow and 1 = deep, is called "deep" in this output. Naming dummy variables for their 1-category often simplifies interpretation. For example, the negative effect of "deep" on chloride in Table 3.4 indicates that deep wells have less chloride (hence shallow wells have more). Had we instead named this variable "type," the meaning of a negative (or positive) coefficient would be less obvious.

If we suspect that two categories might differ with respect to intercepts *and* slopes, both kinds of dummy variable could be included. Table 3.4 shows output from such a regression with the wells data. The result is

$$\hat{Y}_i = 9.07 - 6.72X_{i1} - 1.11X_{i2} + 1.26X_{i1}X_{i2} \qquad [3.35]$$

Substituting $X_1 = 0$, for shallow wells [3.35] becomes

$$\hat{Y}_i = 9.07 - 1.11X_{i2} \qquad [3.36]$$

Substituting $X_1 = 1$, for deep wells [3.35] becomes

$$\hat{Y}_i = 2.35 + .15X_{i2} \qquad [3.37]$$

Figure 3.6 graphs this analysis. Among deep wells we see virtually no relationship between distance and chloride. Deep bedrock wells draw water from far away and are not much affected by nearby pollutants (although, overall, they are almost equally polluted). In shallow wells we see a significant negative relationship. The closer a shallow well is to the road, the higher its chloride concentration tends to be.

In Table 3.4 t-tests indicate that the lines in Figure 3.6 (Equations [3.36] and [3.37]) have significantly different intercepts and slopes:

1. The coefficient on X_1 ("deep" in Table 3.4) is statistically significant ($t = -3.2$, $P = .002$), so the intercepts differ significantly.
2. The coefficient on X_1X_2 ("deeproad") is also statistically significant ($t = 2.9$, $P = .005$), so the slopes differ significantly too.

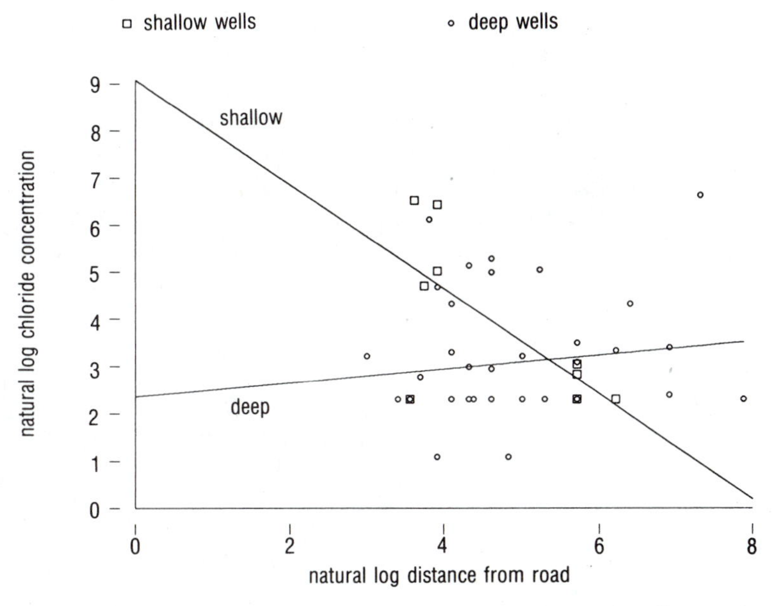

Figure 3.6 Regression of log chloride concentration on log distance from road, with slope and intercept dummy variables for well type.

We could have obtained [3.36] and [3.37] another way: by performing separate regressions of chloride on distance for the 10 shallow wells and then for the 42 deep wells (Figure 3.7). The advantage of the dummy variable approach is that it provides hypothesis tests regarding differences of intercepts or slopes.

An F-statistic can test the null hypothesis that *neither* intercept nor slope is different (that is, test whether the coefficients on intercept and slope dummies both equal zero). This technique works around multicollinearity, a common problem for

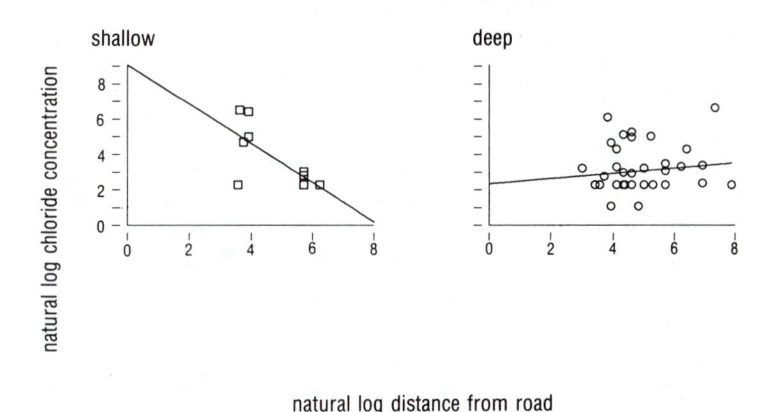

Figure 3.7 Separate regressions for shallow (left) and deep (right) wells; same lines as in Figure 3.6.

regressions including both intercept and slope dummies (or other kinds of interaction terms). Slope dummy variables are defined as a nonlinear (multiplicative) function of two other variables, but linear relations among all three may be strong. Multicollinearity could prevent precise estimation of the difference in intercept adjusting for the difference in slope, and vice versa—so neither is individually distinguishable from zero.

Multicollinearity is not severe in the wells example; t-tests indicate that the individual coefficients are significant (Table 3.4). This lessens the need for an F-test to evaluate whether a model with both dummy variables ([3.35]) significantly improves upon a model with neither:

$$\hat{Y}_i = 3.69 - .10X_{i2} \tag{3.38}$$

For [3.38] the residual sum of squares equals 95.44. For [3.35], RSS = 77.54. The F-statistic, following [3.28], is

$$F = \frac{(95.44 - 77.54)/2}{77.54/48}$$
$$= 5.54$$

Given 2 and 48 degress of freedom, we obtain $P = .007$. We can reject H_0: $\beta_1 = \beta_3 = 0$ and conclude that [3.35] fits significantly better than a model ([3.38]) specifying identical intercepts and slopes for both deep and shallow wells.

Oneway Analysis of Variance

Analysis of variance (ANOVA) refers to a broad class of F-statistic methods for testing the significance of relationships between a measurement Y variable and categorical X variables. The same F-tests can also be obtained with regression. ANOVA is basically a subset of regression methods, providing a simpler approach for some analytical problems. On the other hand, regression gives a clearer view of the underlying model's parameters, predictions, and errors.

This section does not attempt a traditional presentation of ANOVA. Instead, it describes how to recast ANOVA as regression, which can then be analyzed and understood like any other regression.

Table 3.5 contains data from 26 counties in Pennsylvania, New Jersey, and New York. Archer (1987) found a relationship between lung cancer rates and the underlying bedrock type: counties over Reading Prong granite had more cancer. Since these granites emit radon, a potent carcinogen, it seems plausible that radon from the Reading Prong bedrock causes higher cancer rates.

We will begin by testing the relationship between bedrock area and cancer rate. Bedrock area has three categories: Reading Prong, fringe (on Reading Prong borders), and control (nearby but not on Reading Prong). We can encode three categories as two dummy variables: X_1 for Reading Prong (1 if yes, 0 if no) and

Table 3.5 Cancer, bedrock, and radon in 26 counties[1]

County	Lung Cancer[2] Y	Bedrock Area[3]	Reading Dummy X_1	Fringe Dummy X_2	Mean House Radon[4]	Low Radon Dummy X_3	Mid Radon Dummy X_4
Orange, NY	6.0	Reading Prong	1	0	Low	1	0
Putnam, NY	10.5	Reading Prong	1	0	Mid	0	1
Sussex, NY	6.7	Reading Prong	1	0	Mid	0	1
Warren, NJ	6.0	Reading Prong	1	0	High	0	0
Morris, NJ	6.1	Reading Prong	1	0	Low	1	0
Hunterdon, NJ	6.7	Reading Prong	1	0	High	0	0
Berks, PA	5.2	Fringe	0	1	High	0	0
Lehigh, PA	5.6	Fringe	0	1	High	0	0
Northampton, PA	5.8	Fringe	0	1	High	0	0
Pike, PA	4.5	Fringe	0	1	Low	1	0
Dutchess, NY	5.5	Fringe	0	1	Mid	0	1
Sullivan, NY	5.4	Fringe	0	1	Low	1	0
Ulster, NY	6.3	Fringe	0	1	Low	1	0
Columbia, NY	6.3	Control	0	0	Mid	0	1
Delaware, NY	4.3	Control	0	0	Mid	0	1
Greene, NY	4.0	Control	0	0	Mid	0	1
Otswego, NY	5.9	Control	0	0	Mid	0	1
Tioga, NY	4.7	Control	0	0	Mid	0	1
Carbon, PA	4.8	Control	0	0	Mid	0	1
Lebanon, PA	5.8	Control	0	0	High	0	0
Lackawanna, PA	5.4	Control	0	0	Low	1	0
Luzerne, PA	5.2	Control	0	0	Low	1	0
Schuylkill, PA	3.6	Control	0	0	High	0	0
Susquehanna, PA	4.3	Control	0	0	Low	1	0
Wayne, PA	3.5	Control	0	0	Low	1	0
Wyoming, PA	6.9	Control	0	0	Mid	0	1

[1] Data adapted from Archer (1987) and Cohen (1988).
[2] White female lung cancer rate per 100,000 per year, 1950–1969.
[3] Reading Prong areas overlie granite bedrock that has been associated with high indoor radon concentrations. Fringe areas border the Reading Prong, and control areas lie outside it.
[4] Cohen (1988) reports mean radon concentrations in pCi/L, for living areas of hundreds of individual houses within each county. Categories used here are: low, 0–1.5; mid, 1.6–2.4; and high, over 2.5.

X_2 for fringe areas (1 if yes, 0 if no):

If County Is in:	X_1	X_2
Reading Prong	1	0
Fringe	0	1
Control	0	0

For control-area counties, X_1 (Reading Prong) and X_2 (fringe area) both equal zero. Any categorical variable with K categories can be encoded as $K - 1$ dummy variables; a Kth dummy would be redundant. Furthermore, due to multicollinearity we *cannot* estimate models with both a Y-intercept and coefficients on K dummies for a K-category variable.[12]

Table 3.6 shows the regression of cancer rate on Reading and fringe dummy variables (top) and an analysis of variance for cancer rate by bedrock area (bottom). Note the point-for-point correspondences. Both analyses rest on the same underlying model:

$$E[Y_i] = \beta_0 + \beta_1 X_{i1} + \beta_2 X_{i2}$$

They require the same assumptions and lead to identical predictions, residuals, sums of squares, R^2, overall F-statistics, and P-values. *Oneway ANOVA (ANOVA with one K-category X variable) is equivalent to regression on K − 1 dummy variables.*

Table 3.6 Relation between cancer rate and bedrock area (corresponding results in *italics*)

DUMMY VARIABLE REGRESSION

. regress cancer X1 X2 (see Table 3.5 for definitions)
(obs = 26)

Source	SS	df	MS
Model	*16.9087889*	*2*	*8.45439445*
Residual	*30.3373647*	*23*	*1.31901586*
Total	*47.2461536*	*25*	*1.88984614*

Number of obs = *26*
$F(2, 23)$ = *6.41*
Prob > F = *0.0061*
R-square = *0.3579*
Adj R-square = *0.3021*
Root MSE = *1.1485*

Variable	Coefficient	Std. Error	t	Prob > |t|	Mean
cancer					5.576923
X1	2.023077	.5668322	3.569	0.002	.2307692
X2	.4945055	.5384177	0.918	0.368	.2692308
_cons	4.976923	.3185322	15.625	0.000	1

. test X1 X2
(1) X1 = 0.0
(2) X2 = 0.0
$F(2, 23)$ = *6.41*
Prob > F = *0.0061*

ONEWAY ANOVA

. anova cancer area

Number of obs = *26* R-square = *0.3579*
Root MSE = *1.14848* Adj R-square = *0.3021*

Source	Partial SS	df	MS	F	Prob > F
Model	*16.9087889*	*2*	*8.45439445*	*6.41*	*0.0061*
area	*16.9087889*	*2*	*8.45439445*	*6.41*	*0.0061*
Residual	*30.3373647*	*23*	*1.31901586*		
Total	*47.2461536*	*25*	*1.88984614*		

Below the regression output in Table 3.6 is an F-test ([3.28]) of the null hypothesis that coefficients on both X_1 and X_2 equal zero:

$$H_0: \beta_1 = \beta_2 = 0$$

This corresponds both to the regression's overall F-test (upper right) and to ANOVA's F-test for the effect of bedrock area. All three tests indicate a significant ($P = .0061$) bedrock/cancer relationship.

Italics emphasize corresponding regression and ANOVA results in Table 3.6. Notice that regression duplicates all the ANOVA results, but it also provides further information (coefficients, standard errors, and t-tests) not given by ANOVA.

When all X variables are dummies for a single categorical variable, as at top in Table 3.6, we interpret regression coefficients as follows:

> Each coefficient is the difference between mean Y for that category and the mean Y for the omitted category.

Mean Y for the omitted category equals the intercept. Therefore, mean cancer rates within each area are:

Reading Prong: $\bar{Y} = b_0 + b_1 = 4.98 + 2.02 = 7$

Fringe: $\bar{Y} = b_0 + b_2 = 4.98 + .49 = 5.47$

Control: $\bar{Y} = b_0 = 4.98$

We could find the same means directly from Table 3.5.

Twoway Analysis of Variance

Twoway ANOVA involves two categorical X variables. For example, we could test the effect upon cancer rates of bedrock type and radon level from Table 3.5. Cohen (1988) published mean radon measurements taken from hundreds of households scattered among these 26 counties. Table 3.5 encodes radon level as one three-category variable or, alternatively, as two dummy variables (X_3 and X_4).[13] Cohen found no relationship between cancer rates and radon and stated that this finding destroyed Archer's interpretation regarding cancer and bedrock type. Archer (1988) responded by noting several reasons to distrust Cohen's radon data. For example, in five counties Cohen's means are based on fewer than 10 houses.

Without taking sides in the dispute, we can use these data to explore further aspects of ANOVA. *If* radon from bedrock causes variation in county cancer rates, *and if* the variables are well measured, then we might expect the following:

1. The cancer/bedrock relationship weakens when adjusted for radon level (since radon is the omitted third variable that really explains the bedrock/cancer correlation).
2. Radon level significantly predicts cancer rate.

Table 3.7 bears out neither expectation.

Table 3.7 Relation among cancer rate, bedrock area, and radon (corresponding results in *italics*)

DUMMY VARIABLE REGRESSION

. regress cancer X1 X2 X3 X4
(obs = 26)

(see Table 3.5 for definitions)

Source	SS	df	MS
Model	*22.1236128*	*4*	*5.53090321*
Residual	*25.1225407*	*21*	*1.19631146*
Total	*47.2461536*	*25*	*1.88984614*

Number of obs = *26*
$F(4, 21)$ = *4.62*
Prob > F = *0.0078*
R-square = *0.4683*
Adj R-square = *0.3670*
Root MSE = *1.0938*

Variable	Coefficient	Std. Error	t	Prob > |t|	Mean
cancer					5.576923
X1	2.211891	.5510183	4.014	0.001	.2307692
X2	.8669809	.5492147	1.579	0.129	.2692308
X3	−.1166753	.5560187	−0.210	0.836	.3461538
X4	.9058845	.5759487	1.573	0.131	.3846154
_cons	4.525039	.5280793	8.569	0.000	1

. test X1 X2
(1) X1 = 0.0
(2) X2 = 0.0
$F(2, 21)$ = *8.06*
Prob > F = *0.0025*

. test X3 X4
(1) X3 = 0.0
(2) X4 = 0.0
$F(2, 21)$ = *2.18*
Prob > F = *0.1380*

TWOWAY ANOVA

. anova cancer area radon

Number of obs = *26*
Root MSE = *1.09376*

R-square = *0.4683*
Adj R-square = *0.3670*

Source	Partial SS	df	MS	F	Prob > F
Model	*22.1236128*	*4*	*5.53090321*	*4.62*	*0.0078*
area	19.2846335	2	9.64231675	*8.06*	*0.0025*
radon	5.21482394	2	2.60741197	*2.18*	*0.1380*
Residual	*25.1225407*	*21*	*1.19631146*		
Total	*47.2461536*	*25*	*1.88984614*		

Like Table 3.6, Table 3.7 gives regression output at top, followed by regression F-tests and then the corresponding ANOVA. ANOVA's F-test for the effect of area is equivalent to regression's F-test for whether coefficients on both X_1 and X_2 equal zero:

$$H_0: \beta_1 = \beta_2 = 0$$

We reject this hypothesis ($F = 8.06$, $P = .0025$). Similarly, ANOVA's F-test for the effect of radon is equivalent to regression's F-test for whether coefficients on X_3

and X_4 are both zero:

$$H_0: \beta_3 = \beta_4 = 0$$

We fail to reject this second hypothesis ($F = 2.18$, $P = .1380$).

When all X variables are dummies, we interpret their coefficients as follows:

1. The Y-intercept is the predicted Y when all dummies equal zero (cases are in the omitted category of all variables).
2. Coefficients on each dummy are differences between the Y-intercept and predicted values of cases for which that dummy equals 1 but all other dummies equal 0.

Area	Radon	X_1	X_2	X_3	X_4	Predicted Cancer Rate
Control	high	0	0	0	0	$\bar{Y} = b_0 = 4.52$
Reading	high	1	0	0	0	$\bar{Y} = b_0 + b_1 = 4.52 + 2.21 = 6.73$
Fringe	high	0	1	0	0	$Y = b_0 + b_2 = 4.52 + .87 = 5.39$
Control	low	0	0	1	0	$\bar{Y} = b_0 + b_3 = 4.52 - .12 = 4.40$
Control	mid	0	0	0	1	$\bar{Y} = b_0 + b_4 = 4.52 + .91 = 5.43$

We can use t-statistics to test the differences between each of these values and β_0. Predictions for other combinations of categories can be obtained by writing out the regression equation and substituting appropriate zeros and ones as X values.

Twoway (and higher) ANOVAs often include interaction effects. ANOVA-type interactions can be represented in regression as products of dummy variables. For the cancer-rates example, we multiply each of the two bedrock-area dummies (X_1 and X_2) times each of the two radon-level dummies (X_3 and X_4), creating four interaction terms: X_1X_3, X_1X_4, X_2X_3, and X_2X_4. Then we regress Y on the original dummy variables and interactions.

In Table 3.8, ANOVA's F-test for the interaction of radon level and bedrock type corresponds to a regression F-test of the hypothesis that coefficients on all four interaction terms equal zero:

$$H_0: \beta_{13} = \beta_{14} = \beta_{23} = \beta_{24} = 0$$

The data provide little reason to reject this null hypothesis ($F = .78$, $P = .5513$).[14]

If an interaction is not significant, as in Table 3.8, we may drop it and focus on a simpler model (for example, Table 3.7). On the other hand, if an interaction is significant, we need to carefully work out its implications. This could be done by examining $\bar{Y}$ within each combination of X-variable categories; graphs may help. (For Table 3.8 we would be comparing nine different means: mean cancer rates for each combination of bedrock and radon level.) Alternatively, we could work through the regression equation, interpreting individual coefficients as differences between means.

When ANOVA includes an interaction effect, substantive interpretation of the main (noninteraction) effects becomes difficult. ANOVA main effects are then not equivalent to dummy variable regression. Instead, they correspond to a regression

Table 3.8 Relation among cancer rate, bedrock area, and radon, with interaction (corresponding results in *italics*)

DUMMY VARIABLE REGRESSION WITH INTERACTIONS

. regress cancer X1 X2 X3 X4 X1X3 X1X4 X2X3 X2X4
(obs = 26)

Source	SS	df	MS
Model	*26.0351991*	*8*	*3.25439989*
Residual	*21.2109545*	*17*	*1.2477032*
Total	*47.2461536*	*25*	*1.88984614*

Number of obs = *26*
$F(8, 17)$ = *2.61*
Prob > F = *0.0460*
R-square = *0.5511*
Adj R-square = *0.3398*
Root MSE = *1.117*

Variable	Coefficient	Std. Error	t	Prob > $\|t\|$	Mean
cancer					5.576923
X1	1.65	1.117006	1.477	0.158	.2307692
X2	.8333333	1.019683	0.817	0.425	.2692308
X3	−.1	.9673559	−0.103	0.919	.3461538
X4	.5714286	.8955975	0.638	0.532	.3846154
X1X3	−.1999999	1.477661	−0.135	0.894	.0769231
X1X4	1.678571	1.431712	1.172	0.257	.0769231
X2X3	−.0333332	1.329503	−0.025	0.980	.1153846
X2X4	−.6047619	1.570254	−0.385	0.705	.0384615
_cons	4.7	.7898428	5.951	0.000	1

. test X1X3 X1X4 X2X3 X2X4
(1) X1X3 = 0.0
(2) X1X4 = 0.0
(3) X2X3 = 0.0
(4) X2X4 = 0.0
$F(4, 17)$ = *0.78*
Prob > F = *0.5513*

Note: When the model includes interactions, ANOVA and dummy variable F-tests are equivalent only for the highest-order interaction (here area*radon). ANOVA main effect (or low-order interaction) F-tests are equivalent to those for effect-coded regression.

TWOWAY ANOVA WITH INTERACTION

. anova cancer area radon area*radon

Number of obs = *26* R-square = *0.5511*
Root MSE = *1.11701*   Adj R-square = *0.3398*

Source	Partial SS	df	MS	F	Prob > F
Model	*26.0351991*	*8*	*3.25439989*	*2.61*	*0.0460*
area	17.4002224	2	8.7001112	6.97	0.0061
radon	4.35810394	2	2.17905197	1.75	0.2043
area*radon	3.91158629	4	.977896573	*0.78*	*0.5513*
Residual	*21.2109545*	*17*	*1.2477032*		
Total	*47.2461536*	*25*	*1.88984614*		

model in which the X variables have *effect coding*. Effect coding assigns values of 1 for the present category, 0 for all other categories except the last (omitted) category, and -1 for the omitted category. Dichotomous variables are effect-coded as $(-1, 1)$ instead of dummy-coded as $(0, 1)$. Table 3.9 shows effect coding for bedrock areas from Table 3.5. Under effect coding, the variable denoting Reading Prong (V_1) is 1 for counties in this area, 0 for all other counties except the omitted category (control), and -1 for counties in the omitted category. Interaction terms can be formed by multiplying effect-coded variables, just as they are with dummy variables.

An effect-code version of Table 3.8 appears as Table 3.10. We regress cancer rate on V_1, V_2, V_3, V_4, and the four interaction terms V_1V_3, V_1V_4, V_2V_3, and V_2V_4. This obtains the same R^2, overall F-statistic, predicted values, and residuals as before. The F-test for the four interaction-term coefficients equals ANOVA's area*radon F in Table 3.8. An F-test for the two area coefficients (on V_1 and V_2) would equal ANOVA's area F in Table 3.8, and the F for the two radon coefficients (V_3 and V_4) would equal ANOVA's radon F. Thus effect coding allows us to duplicate all ANOVA F-tests using regression. Dummy coding, on the other hand, duplicates ANOVA F-tests only for the highest-order interactions.[15]

Table 3.9 Effect coding of bedrock area from Table 3.5

County	Bedrock Area	Reading Dummy X_1	Reading Effect V_1	Fringe Dummy X_2	Fringe Effect V_2
Orange, NY	Reading Prong	1	1	0	0
Putnam, NY	Reading Prong	1	1	0	0
Sussex, NY	Reading Prong	1	1	0	0
Warren, NJ	Reading Prong	1	1	0	0
Morris, NJ	Reading Prong	1	1	0	0
Hunterdon, NJ	Reading Prong	1	1	0	0
Berks, PA	Fringe	0	0	1	1
Lehigh, PA	Fringe	0	0	1	1
Northampton, PA	Fringe	0	0	1	1
Pike, PA	Fringe	0	0	1	1
Dutchess, NY	Fringe	0	0	1	1
Sullivan, NY	Fringe	0	0	1	1
Ulster, NY	Fringe	0	0	1	1
Columbia, NY	Control	0	−1	0	−1
Delaware, NY	Control	0	−1	0	−1
Greene, NY	Control	0	−1	0	−1
Otswego, NY	Control	0	−1	0	−1
Tioga, NY	Control	0	−1	0	−1
Carbon, PA	Control	0	−1	0	−1
Lebanon, PA	Control	0	−1	0	−1
Lackawanna, PA	Control	0	−1	0	−1
Luzerne, PA	Control	0	−1	0	−1
Schuylkill, PA	Control	0	−1	0	−1
Susquehanna, PA	Control	0	−1	0	−1
Wayne, PA	Control	0	−1	0	−1
Wyoming, PA	Control	0	−1	0	−1

Table 3.10 Relation among cancer rate, bedrock area, and radon, with interaction (corresponding results in *italics*)

EFFECT-CODED VARIABLE REGRESSION WITH INTERACTIONS

. regress cancer V1 V2 V3 V4 V1V3 V1V4 V2V3 V2V4
(obs = 26)

Source	SS	df	MS
Model	*26.0351991*	*8*	*3.25439989*
Residual	*21.2109545*	*17*	*1.2477032*
Total	*47.2461536*	*25*	*1.88984614*

Number of obs = *26*
$F(8, 17)$ = *2.61*
Prob > F = *0.0460*
R-square = *0.5511*
Adj R-square = *0.3398*
Root MSE = *1.117*

Variable	Coefficient	Std. Error	t	Prob > |t|	Mean
cancer					5.576923
V1	1.221693	.3631098	3.365	0.004	−.2692308
V2	−.3005291	.3735647	−0.804	0.432	−.2307692
V3	−.4283069	.3355503	−1.276	0.219	.0769231
V4	.678836	.3720891	1.824	0.086	.1153846
V1V3	−.5216931	.5012262	−1.041	0.313	−.0769231
V1V4	.921164	.5263877	1.750	0.098	−.1923077
V2V3	.3505291	.4856171	0.722	0.480	−.0769231
V2V4	−.6566138	.5950653	−1.103	0.285	−.2692308
_cons	5.778307	.2500637	23.107	0.000	1

. test V1V3 V1V4 V2V3 V2V4
(1) V1V3 = 0.0
(2) V1V4 = 0.0
(3) V2V3 = 0.0
(4) V2V4 = 0.0
$F(4, 17)$ = *0.78*
Prob > F = *0.5513*

TWOWAY ANOVA WITH INTERACTION

. anova cancer area radon area*radon

Number of obs = *26*
Root MSE = *1.11701*
R-square = *0.5511*
Adj R-square = *0.3398*

Source	Partial SS	df	MS	F	Prob > F
Model	*26.0351991*	*8*	*3.25439989*	*2.61*	*0.0460*
area	17.4002224	2	8.7001112	6.97	0.0061
radon	4.35810394	2	2.17905197	1.75	0.2043
area*radon	3.91158629	4	.977896573	*0.78*	*0.5513*
Residual	*21.2109545*	*17*	*1.2477032*		
Total	*47.2461536*	*25*	*1.88984614*		

Often dummy variables permit more substantively meaningful tests than effect coding (or ANOVA) does. This is particularly true for the main effects in models that include interactions. Effect coding or ANOVA F-tests for such main effects may refer to obscure and uninteresting null hypotheses, except in the special case of *balanced designs*, in which each combination of X-variable categories has an equal number of cases.[16]

In principle, ANOVA can accommodate many categorical X variables, just as regression can. Mixtures of categorical and measurement X variables require a related method called *analysis of covariance* (ANCOVA), which corresponds to regression with slope and intercept dummy variables.

Conclusion

Regression methods readily extend to linear models with two or more predictors. The most widely used approach, ordinary least squares (OLS), obtains coefficient estimates that minimize the sum of squared residuals. The popularity of OLS stems from its simplicity and widely demonstrated practical value, as well as from its theoretical advantages over other estimators under ideal conditions (normal i.i.d. errors).

Sample regression coefficients provide estimates of corresponding population parameters. Hypotheses about individual parameters may be tested with t-statistics. F-statistics test hypotheses about nested models, or sets of parameters. F-tests are especially useful when several X variables strongly correlate with one another or measure the same general concept.

Dummy variables, or $(0, 1)$ dichotomies, are useful in several different ways. A single dummy X variable makes regression equivalent to a difference of means test. Other possibilities include intercept dummy variables and slope dummy variables, which test for differences of intercepts and slopes. Appropriately chosen dummy or effect variables permit ANOVA-type analyses within a regression framework. All these applications employ dummies as X variables only. Chapter 7 considers models for dummy Y variables.

Data analysis is rarely as simple in practice as it appears in books. Like other statistical techniques, regression rests on certain assumptions and may produce unrealistic results if those assumptions are false. Furthermore it is not always obvious how to translate a research question into a regression model. The following chapters illustrate practical methods for diagnosing and solving problems.

Exercises

Some red spruce forests in the Appalachian Mountains show signs of decline, with many dead or dying trees. Environmental stress may contribute to this decline; deposition of airborne pollutants such as metals or acids tends to be heavier at high elevations, where red spruce predominate. The following table contains data on elevation and the percentage of dead or badly damaged trees, from 64 Appalachian sites (Johnson and Siccama, reported by the Committee on Monitoring and Assessment of Trends in Acid Deposition, 1986). Eight of the sites are in southern states (West Virginia, Virginia, and North Carolina); the remainder are northern (New Hampshire, Vermont, and New York). Exercises 1–11 use these data.

Elevation and percentage dead or damaged red spruce trees

Location	Elevation (Meters)	% Trees Damaged	Location	Elevation (Meters)	% Trees Damaged
1 South	1615	5	33 North	1000	67
2 South	1768	13	34 North	1050	70
3 South	1524	6	35 North	650	17
4 South	1311	21	36 North	700	31
5 South	1128	4	37 North	750	38
6 South	1005	20	38 North	800	44
7 South	1128	17	39 North	850	37
8 South	1052	31	40 North	900	50
9 North	670	10	41 North	950	70
10 North	720	28	42 North	1000	65
11 North	780	5	43 North	1060	58
12 North	850	20	44 North	1120	87
13 North	910	30	45 North	700	20
14 North	950	28	46 North	750	25
15 North	1000	55	47 North	800	20
16 North	1070	72	48 North	850	45
17 North	1150	80	49 North	900	40
18 North	1220	53	50 North	950	30
19 North	650	15	51 North	1000	60
20 North	700	15	52 North	1050	55
21 North	750	6	53 North	1100	50
22 North	800	37	54 North	1150	50
23 North	850	21	55 North	650	15
24 North	900	21	56 North	700	47
25 North	950	17	57 North	750	31
26 North	1000	30	58 North	800	50
27 North	1060	42	59 North	850	22
28 North	1110	80	60 North	900	60
29 North	800	40	61 North	950	35
30 North	850	55	62 North	1000	42
31 North	900	52	63 North	1060	58
32 North	950	60	64 North	1120	24

1. Create a computer file from these data, coding location as a dummy variable (0 = South, 1 = North). Regress percentage damaged (Y) on location (X_1) and elevation (X_2). Write out the regression equation, and interpret the values of the Y-intercept, both regression coefficients, and R_a^2.

2. At $\alpha = .01$, which of these null hypotheses can we reject?

$$H_0: \beta_0 = 0$$

$$H_0: \beta_1 = 0$$

$$H_0: \beta_2 = 0$$

$$H_0: \beta_1 = \beta_2 = 0$$

3. Construct 95% and 99% confidence intervals for:
   a. regression coefficients on X_1 and X_2.
   b. the population mean percentage damaged at northern sites 1500 meters above sea level.
4. Draw a Y-versus-X_2 scatterplot, and show two regression lines derived from your analysis of Exercise 1: one line for $X_1 = 0$ (South) and one for $X_1 = 1$ (North) sites. How does this graph illustrate the concept of an "intercept dummy variable"?
5. The regression of Exercises 1–4 assumes that the relationship between elevation and percentage damaged is the same for southern and northern sites. Do the data support this assumption? Draw separate Y-versus-X_2 scatterplots showing the data for southern and northern sites alone; describe what you see.
6. To allow for the possibility of interaction (the elevation/damage relationship changing with location), we can redo the regression to include both slope and intercept dummy variables. Generate a slope dummy variable X_1X_2, and regress Y on X_1 (location), X_2 (elevation), and X_1X_2 (location times elevation). Does this model fit much better, as measured by R_a^2, than the simpler model of Exercise 1? Are coefficients on both X_1 and X_1X_2 significant, as assessed by an F-test? Write out the regression equation, and explore its implications for southern and northern forests by substituting $X_1 = 0$ and $X_1 = 1$.
7. Run separate regressions of damage on elevation for southern and northern sites. Confirm that the equations from these two regressions match those derived from your slope and intercept dummy variable regression in Exercise 6. What does Exercise 6 tell us that separate North and South regressions do not?
8. Graph the regression of Exercise 6 on separate South and North scatterplots. How do these graphs illustrate the ideas of "slope" and "intercept" dummy variables?
9. Construct a residuals versus predicted values (e-versus-$\hat{Y}$) plot for your regression in Exercise 6. Compare this plot with Figures 2.10 and 2.11 in Chapter 2. Do you see any signs of trouble, or is your plot basically "all clear"?
10. Regress percentage trees damaged on a dummy variable for location (South = 0, North = 1), as you did in Exercise 1. Then perform a oneway ANOVA: percentage trees damaged by location. Point out correspondences and differences between the regression and ANOVA results.
11. Repeat the slope and intercept dummy variable regression of Exercise 6. Calculate three F-statistics, testing the coefficients on location, elevation, and

location*elevation. Then perform ANCOVA: percentage trees damaged by location (categorical), elevation (measurement), and location*measurement (interaction). Point out correspondences and differences between the regression and ANCOVA results.

The next table presents data from a study of beryllium exposure among employees of a Utah surface mine (Rom, Lockey, Bang, Dewitt, and Johns, 1983). Researchers expected that beryllium exposure would be associated with higher rates of blastogenic lymphocyte transformation (LT ratio), which in turn is a precursor of chronic beryllium disease. Exercises 12–14 refer to these data. (We return to analyze them by different methods in Exericse 11 of Chapter 6.)

Lymphocyte transformation (LT ratio) among 15 mine workers

Worker	Age	Years of Employment	Exposure Level	LT Ratio
1	42	13	1	5.4
2	46	14	2	7.3
3	43	4	4	3
4	25	3	3	2
5	26	3	4	5.4
6	55	3	4	5
7	23	9	4	3.7
8	24	4	4	5
9	38	3	3	2.8
10	24	3	4	2.2
11	28	3	4	2.5
12	38	3	4	3.1
13	26	3	4	2.5
14	28	9	4	.8
15	26	4	2	1.2

12. Regress LT ratio on age, employment, and exposure. Describe your results.
13. Use an F-statistic to test whether the coefficients on employment and exposure are *both* zero.
14. Construct leverage plots showing the partial effects of age, employment, and exposure on LT ratio. Visually check whether the slopes of lines in these plots equal your regression coefficients from Exercise 12.
15. In an article on salamander evolution, Ruben and Boucot (1989) note a positive correlation between lung volume and head width, due to the mechanisms of salamander breathing. Consequently, lunged salamanders with relatively wider heads tend to absorb a smaller proportion of their oxygen cutaneously (through their skin, necessary if lungs are insufficient). Many salamanders have no lungs, however, and among them head size appears unrelated to cutaneous oxygen uptake.

    Fit a model that expresses these ideas, by regressing cutaneous oxygen uptake on relative head width and both intercept and slope dummy variables for type (lungless = 0, lunged = 1). Draw a graph illustrating your analysis.

Salamander Species	Type	% Cutaneous Oxygen Uptake	Head Width/Snout-to-Vent Length
G. porphyriticus	lungless	81	.13
P. ruber	lungless	84	.13
P. glutinosus	lungless	76	.16
D. quadramaculatus	lungless	86	.19
A. tiginum	lunged	45	.23
A. talpoideum	lunged	49	.23
A. opacum	lunged	64	.17
A. macrodactylum	lunged	65	.16
A. maculatum	lunged	69	.16
R. olympicus	lunged	74	.14

Notes

1. Equation [3.9] has 1981 water use (Y) on the left-hand side and 1980 water use (X_2) on the right-hand side:

   $$Y_i = 203.8 + 20.5X_{i1} + .59X_{i2} + e_i$$

   We could instead have specified a model in which the left-hand-side variable is *water savings*, or reduction in water use (1980 minus 1981):

   $$X_{i2} - Y_i = -203.8 - 20.5X_{i1} + .41X_{i2} - e_i$$

   This is an example of *change-score analysis*. The change-score regression equation may be obtained either by computing the change score ($X_2 - Y$) and regressing it on X_1 and X_2 or, equivalently, just by rearranging the terms in the original regression. Change-score regression clarifies the relation between income and conservation: income had a negative effect on water savings. Adjusted for preshortage water use, higher-income households *conserved less water* than did lower-income households.

   In general, if

   $$Y_i = b_0 + b_1X_{i1} + b_2X_{i2} + e_i$$

   then

   $$X_{i2} - Y_i = -b_0 - b_1X_{i1} + (1 - b_2)X_{i2} - e_i$$

   Like X_1, additional right-hand-side variables will have coefficients with opposite signs but the same standard errors and P-values, in either change-score or original versions.
2. The science of *epidemiology* employs multivariate statistics to investigate the causes of illness in nonexperimental settings.

3. Two other applications are the residual variance (sometimes called "mean squared error"):

$$s_e^2 = \text{RSS}/\text{df}_\text{R}$$

and its square root, the residual standard deviation:

$$s_e = \sqrt{\text{RSS}/\text{df}_\text{R}}$$

R_a^2 can be defined from the ratio of residual variance to Y's variance:

$$R_a^2 = 1 - \frac{s_e^2}{s_Y^2}$$

4. Another statistic to guide variable selection is Mallows' C_p:

$$C_p = \frac{\text{RSS}_p}{s_e^2} + 2p - n$$

where s_e^2 is the residual variance from a "full" regression with all relevant X variables. Let K represent the number of parameter estimates in this full regression. RSS_p is the residual sum of squares from a regression with some subset ($p \leq K$) of these parameters.

Calculate C_p for subsets of X variables. For the full model, $K = p = C_p$. The model is adequate if C_p approximately equals p. If C_p is much larger than p, we infer that one or more important variables (from the full model) have been omitted. Plots of C_p versus p help in selecting the best subset of predictors. We seek the subset that combines low p and low C_p with $p \approx C_p$; tradeoffs may be necessary.

This approach assumes that all relevant variables (plus perhaps some extraneous ones) were included in the full model. See Rawlings (1988) or Draper and Smith (1981) for examples and further discussion.

5. See Johnston (1972) for a mathematical treatment of this problem. Rawlings (1988) provides a good general discussion of issues and techniques for variable selection.
6. Chapters 4 and 5 discuss statistical problems with the regression of Table 3.2. When these problems are corrected, change in the number of people appears more important (and retirement less so).
7. This is *not* to imply that there is a 95% probability that β_1 lies within the interval. Either it does or it does not.
8. A Type I error is rejection of the null hypothesis when in fact it is true.
9. By similar reasoning, $b_2 + b_3X$ is the real "coefficient" on W.
10. This statement describes the simplest case, nonnegative W. If W has some negative values, then $b_1 + b_3W$ could change signs even when b_1 and b_3 have the same sign.

11. Here we have $b_1 = -.71$ and $SE_{b_1} = .48$, so $t = -.71/.48 = -1.48$ (Equation [2.25]) with $52 - 2 = 50$ degrees of freedom ($P = .15$). A two-sample t-test (see an introductory text for details) reaches the same conclusion: ($\bar{Y}_1 - \bar{Y}_0 = -.71$, $SE_{\bar{Y}_1 - \bar{Y}_0}$, $t = -.71/48 = -1.48$, df $= 50$, $P = .15$.
12. We can include K dummy variables in a "no-constant" regression. Then each dummy's coefficient equals the mean Y (or, with measurement X variables in the model, the Y-intercept) for that category. For these coefficients t-statistics test whether each mean (or intercept) differs *from zero.*
13. Cohen's data involve actual measurements rather than categories; I split these into three categories for illustration purposes in Table 3.5. Although categorizing measurement data is generally a bad practice (because it throws away information), in this example it does not change the substantive conclusions.
14. Interaction terms formed by multiplying other variables often introduce multicollinearity, as noted earlier with respect to slope dummy variables. Comparing the regressions in Tables 3.7 and 3.8, notice that including the interaction terms almost doubles estimated standard errors on $X_1 - X_4$.
15. When no interactions are included, as in Tables 3.6 and 3.7, the "highest-order interactions" are simply the main effects. In that case, regression F-tests on dummy or effect-coded data are the same, and both equal their ANOVA counterparts.
16. See Kleinbaum and Kupper (1978) for a more detailed presentation of correspondences between ANOVA and regression.

4 Regression Criticism

Like any research method, regression analysis requires certain assumptions. They guide decisions at every step, from model specification to interpretation of results. Assumptions simplify reality, focusing research attention onto manageable tasks and answerable questions. Although it is a necessary element of science, simplification carries obvious risks. *Regression criticism* involves studying the data to answer the questions "Have we simplified too much?" and "Are our assumptions plausible, or should we change our approach?"

This chapter begins by outlining the most commonly made assumptions regarding ordinary least squares (OLS). Theoretical advantages of OLS over other methods follow from these assumptions. If the assumptions do not hold, the theoretical advantages erode—though OLS still may have practical attractions. The remainder of the chapter illustrates techniques for regression criticism, a systematic effort to detect bad assumptions and other statistical problems.

Assumptions of Ordinary Least Squares

OLS estimates parameters of linear models:

$$E[Y_i] = \beta_0 + \beta_1 X_{i1} + \beta_2 X_{i2} + \cdots + \beta_{K-1} X_{i,K-1} \quad [4.1]$$

where $E[Y_i]$ denotes the expected value (population mean) of Y given the ith set of $X_1 \cdots X_{K-1}$ values. Actual Y values equal expected Y plus random error:

$$\begin{aligned} Y_i &= E[Y_i] + \varepsilon_i \\ &= \beta_0 + \beta_1 X_{i1} + \beta_2 X_{i2} + \cdots + \beta_{K-1} X_{i,K-1} + \varepsilon_i \end{aligned} \quad [4.2]$$

Errors represent "everything else," besides the linear effects of $X_1 \cdots X_{K-1}$, that explains variation in Y.

Under certain conditions, OLS is BLUE: the *b*est (most efficient) *l*inear *u*nbiased *e*stimator. Those necessary conditions, expressed as a series of assumptions, are numbered 1–4 below.

1. **Fixed *X***: we could theoretically obtain many random samples, each with the same X values but different Y_i, due to different ε_i values.[1]
2. **Errors have zero mean**:

$$E[\varepsilon_i] = 0 \qquad \text{for all } i$$

Assumptions 1 and 2 together ensure independence of errors and X variables, which is sufficient for *unbiased* estimation of all β_k parameters. OLS estimates of $\beta_1 \cdots \beta_{K-1}$ (but not β_0) remain unbiased under the weaker assumption that errors all have the same mean, not necessarily zero:

$$E[\varepsilon_i] = \mu \qquad \text{for all } i$$

Unbiasedness means that, over the long run, sample estimates (b_k) center on the true parameter values (β_k). An unbiased estimator may still often be wildly wrong. Some unbiased estimators are more *efficient* than others, or have less sample-to-sample variation. Efficiency improves the chances that any one sample estimate will be near the true parameter.

b_k is an *unbiased* estimator of β_k if the expected value of b_k (mean b_k over all possible size-n random samples) equals β_k:

$$E[b_k] = \beta_k$$

b_k is more *efficient* than a_k if both are unbiased estimators of β_k and $\text{Var}[b_k] < \text{Var}[a_k]$, meaning:

$$E[b_k - \beta_k]^2 < E[a_k - \beta_k]^2$$

3. **Errors have constant variance** (*homoscedasticity*):

$$\text{Var}[\varepsilon_i] = \sigma^2 \qquad \text{for all } i$$

4. **Errors are uncorrelated with each other** (*no autocorrelation*):

$$\text{Cov}[\varepsilon_i, \varepsilon_j] = 0 \qquad \text{for all } i \neq j$$

Var[ ] and Cov[ ] denote population variance and covariance, respectively; see Appendix 1.

Under assumptions 3 and 4 (in addition to 1 and 2):

a. Standard errors estimated by [3.20] are unbiased.
b. OLS is more efficient than any other linear unbiased estimator (BLUE).

This last result, called the *Gauss-Markov Theorem*, is an important point in favor of OLS. It rests on assumptions 1–4, however. The plausibility of these assumptions must be a matter of degree; in actual research they are seldom, if ever, literally true. Furthermore, even if they are true and OLS is BLUE:

a. Certain *biased* estimators can have a lower mean squared error than OLS, making them preferable under some conditions. One example, *ridge regression*, is useful with highly correlated X variables.
b. Certain *nonlinear robust* estimators (Chapter 6) are unbiased and more efficient than OLS with nonnormal error distributions. ("Nonlinear" here refers to the estimation of parameters, not the functional form of the model.)

Consequently the Gauss-Markov Theorem, by itself, is not a compelling reason to use OLS.

Assuming the same variance (σ^2) for all errors, and zero covariance between errors, the errors' *covariance matrix* has the form:

$$\begin{bmatrix} \sigma^2 & 0 & 0 & \cdots & 0 \\ 0 & \sigma^2 & 0 & \cdots & 0 \\ 0 & 0 & \sigma^2 & \cdots & 0 \\ \cdot & \cdot & \cdot & \cdots & \cdot \\ \cdot & \cdot & \cdot & \cdots & \cdot \\ \cdot & \cdot & \cdot & \cdots & \cdot \\ 0 & 0 & 0 & \cdots & \sigma^2 \end{bmatrix} = \sigma^2\mathbf{I}$$

where $\mathbf{I}$ is an $n \times n$ identity matrix. Thus assumptions 3 and 4 can be expressed compactly:

The errors' covariance matrix equals $\sigma^2\mathbf{I}$.

OLS estimates remain *consistent*, but not necessarily unbiased or efficient, under less restrictive conditions. Given a correct model ([4.1]):

a. Under assumptions 1–2 (3–4 not required), if Y has bounded range, then coefficient estimates are consistent.[2]

b. Under assumptions 2–4 (1 not required), plus the assumption that errors are uncorrelated with X variables:

$$\text{Cov}[X_{ik}, \varepsilon_i] = 0 \qquad \text{for all } i, k$$

coefficient and standard error estimates are consistent.

Consistency implies that estimates based on large enough samples are likely to be close to the true parameters.

b_k is a *consistent* estimator of β_k if the probability that they are very close approaches 1 as n approaches infinity:

$$\lim_{n \to \infty} P(|b_k - \beta_k| < c) = 1$$

for any small constant c.

5. **Errors are normally distributed:**[3]

$$\varepsilon_i \sim N(0, \sigma^2) \qquad \text{for all } i$$

Normal (Gaussian) error distributions are unnecessary for OLS to be BLUE or even consistent, but:

a. They justify use of the theoretical t and F distributions in confidence intervals and hypothesis tests, especially with small samples.
b. Given normal errors, OLS is more efficient than any other unbiased estimator, linear or not.

Nonnormal error distributions can drastically reduce the efficiency of OLS.

Assumptions 1–5 could be summarized as follows:

> We assume the linear model is correct, with normal, independent, and identically distributed (normal i.i.d.) errors.

A "correct model" means that the expected value of Y is a linear function of the model's X variables ([4.1]). No other X variables that affect $E[Y_i]$, and correlate with $X_1, X_2, \ldots, X_{K-1}$, have been left out of the model. And all of the X variables included do affect $E[Y_i]$.

We can use sample data to check some assumptions, including linearity, the choice of which X variables to include, homoscedasticity, no autocorrelation, and normality. However, we cannot check certain other assumptions with sample data. For example, we assume (through 1 and 2) no correlation between errors (ε_i) and X variables. But even if they really do correlate, we will see zero correlation between the sample residuals (e_i) and any X variable. OLS ensures that, in the sample,

Table 4.1 Some common statistical problems and their consequences for OLS

	Undesirable Statistical Consequences			
Problem	**Biased *b***	**Biased SE**	**Invalid *t* & *F* Tests**	**High Var[*b*] (Inefficient)**[1]
Nonlinear relationship	yes	yes	yes	—
Omit relevant *X*	yes	yes	yes	—[2]
Include irrelevant *X*	no	no	no	yes
X measured with error	yes	yes	yes	—
Heteroscedasticity	no	yes	yes	yes
Autocorrelation	no	yes	yes	yes
X correlated with ε	yes	yes	yes	—
Nonnormal ε distribution	no	no	yes	yes
Multicollinearity	no	no	no	yes

[1] Inefficiency (Var[b]) noted only for unbiased estimators. For biased estimators a more general criterion, mean squared error (MSE = Var[b] + bias^2), reflects variation around the true parameter.
[2] Omission of relevant variables sometimes improves MSE, because Var[b] shrinks more than bias^2 grows.

$\text{Cov}[X_{ik}, e_i] = 0$ always, even if in the population $\text{Cov}[X_{ik}, \varepsilon_i] \neq 0$. Similarly, we assume that errors have zero mean, but sample OLS residuals always have zero mean, whether the errors do or not. And we cannot use the data to test whether *all* relevant X variables have been included in the model—infinite possibilities may exist.

Table 4.1 summarizes how some common problems affect OLS coefficient and standard error estimates. Subsequent sections examine ways to detect such problems and deal with them if they arise. We will follow through the example of Chapter 3 (see Table 3.2):

$$\hat{Y}_i = 242 + 21X_{i1} + .49X_{i2} - 42X_{i3} + 189X_{i4} + 248X_{i5} + 96X_{i6} \quad [4.3]$$

where

Y is postshortage (1981) water use;
X_1 is household income;
X_2 is preshortage (1980) water use;
X_3 is education of household head;
X_4 is a dummy variable for retirement;
X_5 is number of people in household at time of water shortage; and
X_6 is increase in number of people, summer 1981 minus summer 1980.

Correlation and Scatterplot Matrices

Multivariate analysis builds upon bivariate and univariate analysis. Chapter 2 showed how problems or unexpected results in bivariate analysis can often be

Table 4.2 Correlation matrix for 1981 household water use and predictors

(obs = 496)	income	water80	educat	retire	peop81	cpeop	water81
income	1.0000	0.3371	0.3463	−0.3806	0.3113	0.0911	0.4178
water80	0.3371	1.0000	0.0982	−0.2919	0.5251	−0.0312	0.7648
educat	0.3463	0.0982	1.0000	−0.1742	0.0587	0.0055	0.0404
retire	−0.3806	−0.2919	−0.1742	1.0000	−0.3757	−0.0585	−0.2731
peop81	0.3113	0.5251	0.0587	−0.3757	1.0000	0.1443	0.6183
cpeop	0.0911	−0.0312	0.0055	−0.0585	0.1443	1.0000	0.0661
water81	0.4178	0.7648	0.0404	−0.2731	0.6183	0.0661	1.0000

traced back to features of the univariate distributions. We should therefore examine univariate distributions before venturing into bivariate analysis. Simiarly, we should examine bivariate distributions before venturing into multivariate analysis.

The sheer number of bivariate combinations may be daunting. The seven variables in [4.3] yield 21 different two-variable relationships ($(7^2 - 7)/2$). Carefully studying each relationship would take much effort.

One shortcut is to examine a *correlation matrix*, like that in Table 4.2. Correlations indicate the sign and strength of linear relationships. In theory they range from ± 1 (perfect linear relationship) to 0 (no linear relationship). Table 4.2 shows positive correlations as strong as +.7648, and negative correlations as strong as −.3806. Note that the correlation of any variable with itself is perfect ($r_{11} = 1$), and the correlation of X_1 with X_2 is the same as the correlation of X_2 with X_1 ($r_{12} = r_{21}$). Consequently many of the correlations in Table 4.2 are redundant; a simpler display could include only those coefficients below the major diagonal.

Correlations convey limited information; they may conceal problems with curvilinearity, outliers, heteroscedasticity, or distributional shape. Scatterplots reveal more details and can be scanned almost as quickly as correlation matrices. Figure 4.1 shows a *scatterplot matrix* corresponding to the lower-diagonal part of the correlation matrix of Table 4.2. (As with correlations, a full 7 × 7 matrix is unneeded.) Scatterplots of the Y variable (summer 1981 water use) against each of the six X variables appear in the bottom row of Figure 4.1. Other plots show relations among the X variables.[4]

Heteroscedasticity (nonconstant variance) and outliers appear in several plots, notably those involving 1980 water use, 1981 water use, and income. Any curvilinearity in these relationships is too mild to detect in this scatterplot matrix. In general, Figure 4.1 visually confirms the correlations of Table 4.2. The strongest relations are those between 1980 and 1981 water use ($r = .7648$) and between number of people and 1981 water use ($r = .6183$). Education and 1981 water use have a negligible correlation ($r = .0404$).

Scatterplot matrices provide a quick overview of zero-order relationships, but from such a distance that only major problems will show up. For example, curvilinearity is striking in Figure 4.2, a scatterplot matrix of data on energy consumption, population growth, fertilizer use, and birth rates in 122 countries. It would be a mistake to try fitting a linear model to these data. Until we have dealt with curvilinearity, as described in Chapter 5, other statistical problems (heteroscedasticity, nonnormality, and so on) are of secondary concern.

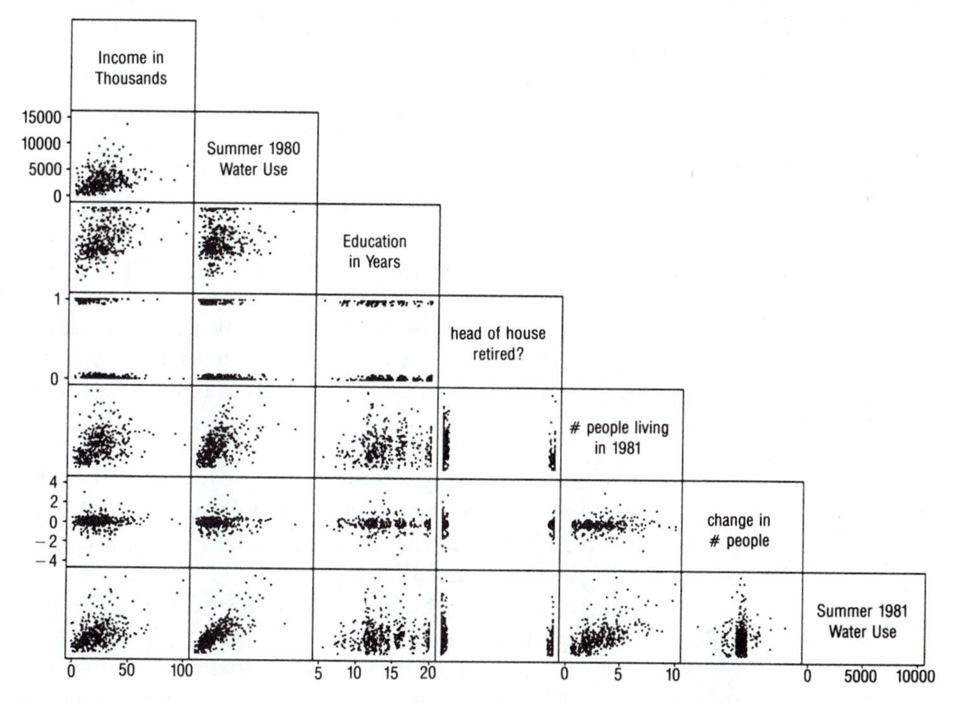

Figure 4.1 Scatterplot matrix corresponding to Table 4.2 (household water use and predictors).

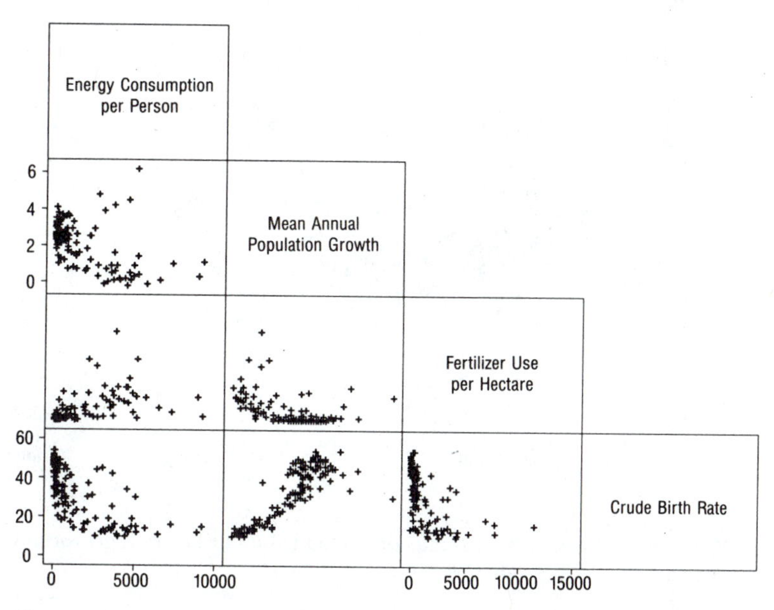

Figure 4.2 Scatterplot matrix for data from 122 countries.

Residual Versus Predicted Y Plots

Residual versus predicted values (e-versus-$\hat{Y}$) plots, introduced with bivariate regression in Chapter 2, provide a starting point for criticism in multiple regression. They may uncover problems such as curvilinearity, heteroscedasticity, nonnormality, or outliers (see Figure 2.11). More specialized graphs bring each of these problems into sharper focus.

Figure 4.3 shows an e-versus-$\hat{Y}$ plot based on [4.3]. Several problems appear:

1. The scatter fans out to the right, indicating heteroscedasticity. Residuals vary more (predictions are less accurate) when predicted water use is high.
2. The residuals distribution has heavy tails, numerous outliers, and mild positive skew.

Heteroscedasticity (violating assumption 3) leads to inefficiency and biased standard error estimates. Nonnormal errors (violating assumption 5) compound inefficiency and undermine the rationale for t- and F-tests. These problems cast doubt on the P-values printed in Table 3.2.

We also might wonder whether some of the outliers visible in Figure 4.3 unduly influence the regression results. In fact, the two high outliers at the upper left are not especially influential. However, two other households near the upper right (households #127 and #134) do affect the coefficient on income, and the outlier

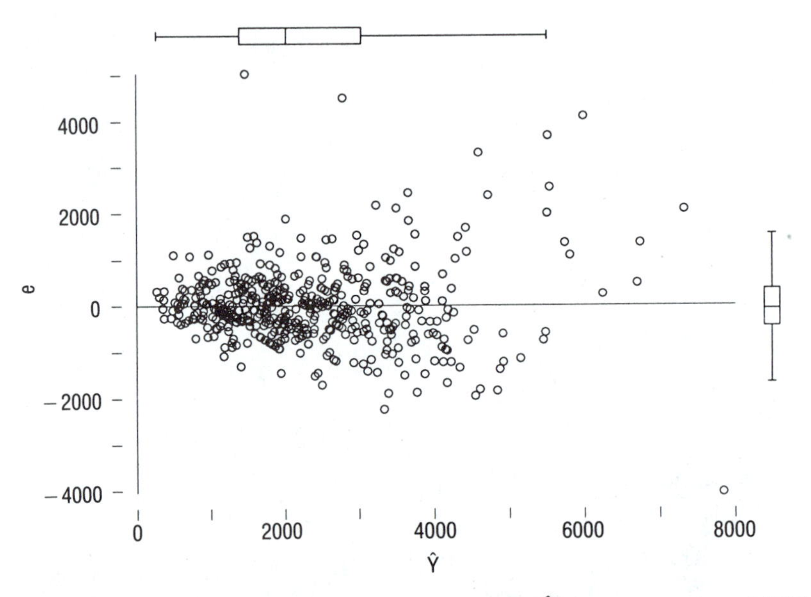

Figure 4.3 Residuals (e) versus predicted values ($\hat{Y}$) from regression of 1981 household water use on seven predictors.

at the lower right (household #101) affects the coefficient on preshortage water use. An e-versus-$\hat{Y}$ plot like Figure 4.3 cannot diagnose these specific problems, but it does suggest that we need a detailed influence analysis.

Heteroscedasticity is not always so obvious as in Figure 4.3. To make it clearer, we can graph the absolute values of residuals against $\hat{Y}$. *Homoscedasticity implies that the median (or mean) absolute residual should be about the same for all values of* $\hat{Y}$. Median absolute residuals that systematically change with $\hat{Y}$ indicate heteroscedasticity.

Figure 4.4 graphs $|e|$ versus $\hat{Y}$ for the water-use regression. A line shows that the median $|e|$, and hence the variance of e, increases with $\hat{Y}$. The line illustrates a technique called *band regression*. For this example:

1. The range of $\hat{Y}$ values is divided into six equal-width vertical bands. (Number of bands is arbitrary.)
2. Median $(\hat{Y}, |e|)$ points within each band are connected by line segments.

Band regression can be applied to any scatterplot, not just $|e|$-versus-$\hat{Y}$ plots. Because it is based on medians rather than means or squared deviations, band regression resists the pull of outliers. The number of bands and other details can be varied as needed.

We could also employ OLS to determine whether there is a linear trend to the mean absolute residuals in Figure 4.4 (there is). Although quick and easy, OLS is not very robust, since any statistical problems that trouble the original regression will likely affect the regression of $|e|$ on $\hat{Y}$ as well.

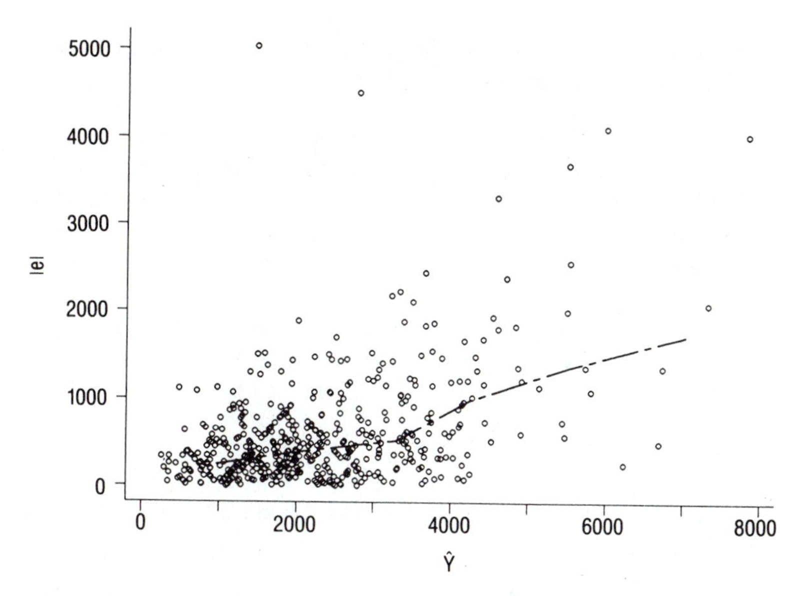

Figure 4.4 Absolute residuals $|e|$ versus $\hat{Y}$, with band regression line indicating heteroscedasticity (household water-use regression).

Autocorrelation

Autocorrelation refers to correlation between values of the same variable across different cases. For example, errors for case i might correlate with errors for case $i - 1$. Autocorrelated errors (violating assumption 4) reduce the efficiency of OLS and bias its estimated standard errors.

Autocorrelation reflects connections among cases. It often occurs with time series or with geographically linked cases like the U.S. states. On the other hand, when cases are independently and randomly selected from a large population, as with a typical sample survey, autocorrelation is less likely.

Tests for autocorrelation of error terms (ε) examine the sample residuals (e). Often this is done by calculating a *Durbin-Watson* statistic, d. If successive residuals are uncorrelated, $d = 2$. Positive autocorrelation leads to $d < 2$; negative autocorrelation produces $d > 2$. Table A4.4 tabulates critical values for the sampling distribution of d, under the null hypothesis of no positive correlation between ε_i and ε_{i-1}. Critical values depend on sample size (n) and the number of X variables in the regression ($K - 1$).

■ The Durbin-Watson test statistic is

$$d = \frac{\sum_{i=2}^{n} (e_i - e_{i-1})^2}{\sum_{i=1}^{n} e_i^2} \qquad [4.4a]$$

where e_i is the ith-case sample residual. Compare d with critical values (Table A4.4) for n and $K - 1$ degrees of freedom. ■

The Durbin-Watson test is not appropriate for *autoregressive* models, in which a previous value of Y appears on the right-hand side of the regression equation. We therefore cannot apply it to the water-use example, which has 1981 water use as Y and 1980 water use as one of the X variables.

To illustrate the Durbin-Watson test, we turn first to the mine workers data from Exercise 12 in Chapter 3. Regression of LT ratio on age, employment, and exposure yields $d = 1.76$. For each combination of n and $K - 1$, Table A4.4 gives two critical values: d_L (lower) and d_U (upper). Considering only the possibility of positive autocorrelation ($0 < d < 2$), the decision rule is:

Reject H_0 (at level α) if $d < d_L$.
Do not reject H_0 (at level α) if $d > d_U$.
The *test is inconclusive* if $d_L < d < d_U$; we do not know whether H_0 should be rejected.[5]

Given 15 cases and 3 X variables, the $\alpha = .05$ critical values of d are $d_L = .82$ and $d_U = 1.75$. Since $d > d_U$ ($1.76 > 1.75$), we fail to reject the null hypothesis of no positive autocorrelation. That is, autocorrelation is not significant at $\alpha = .05$.

Figure 4.5 shows another example, this one using time series data. The graph plots Concord's average daily water use over 137 consecutive months, from January 1970 through May 1981. The conservation campaign began in late summer 1980 (vertical line) and may partly account for the subsequent decline in water use.

To quantify the campaign's impact, we might regress average daily water use (Y) on average daily temperature (X_1), monthly precipitation (X_2), and a dummy variable representing the conservation campaign (X_3). Results are:

$$\begin{array}{rlll} Y_i = 3.83 + & .013X_{i1} - & .047X_{i2} - & .247X_{i3} + e_i \\ \text{SE}_b: & .002 & .021 & .113 \\ t = b/\text{SE}_b: & 6.5 & -2.2 & -2.2 \end{array} \qquad [4.5]$$

With each one-degree rise in temperature (X_1), average daily water use increased by .013 million gallons (13,000 gallons). With each additional inch of monthly precipitation (X_2), average daily water use declined by .047 million gallons. Adjusting for temperature and precipitation, the conservation campaign reduces average water use by .247 million gallons per day. All three effects are statistically significant. However, the standard errors and t-tests depend on assumption 4: no autocorrelation.

This assumption is implausible. Errors represent everything else (besides temperature, precipitation, and the conservation campaign) that influenced Concord water consumption. Many such influences, such as the general U.S. trend toward owning more labor-saving, water-wasting appliances, persist over long periods of time. Errors for month i therefore almost certainly correlate with errors for previous

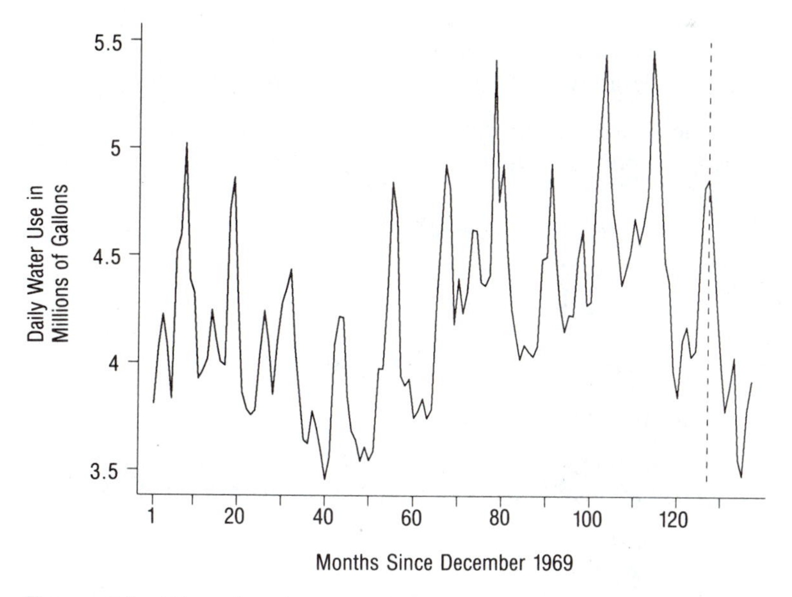

Figure 4.5 Time plot of average daily water use by the city of Concord, 1970–1981 (dashed line shows start of conservation campaign).

months. A Durbin-Watson test confirms this suspicion: $d = .535$; given 137 cases (approximately 100) and 3 predictors, we can reject the null hypothesis even at $\alpha = .01$ because $d < d_L$ ($.535 < 1.48$). Significant autocorrelation invalidates the standard errors and t-tests of Equation [4.5].

The Durbin-Watson test has several limitations, which can most easily be discussed using a time series notation. Let e_t represent the residual at time t; e_{t-1} is the previous value, and e_{t+1} is the next. The time series consists of $t = 1, 2, 3, \ldots, T$ observations. Thus t corresponds to the case index i, and T to the sample size n.

In this notation the Durbin-Watson statistic becomes

$$d = \frac{\sum_{t=2}^{T} (e_t - e_{t-1})^2}{\sum_{t=1}^{T} e_t^2} \qquad [4.4b]$$

It tests for *first-order* autocorrelation: correlation between current and previous values of e (e_t and e_{t-1}). Durbin-Watson statistics tell nothing about higher-order autocorrelations, such as between e_t and e_{t-2} (*second-order*) or between e_t and e_{t-12} (*12th-order*). High-order autocorrelations characterize time series with trends or repeating cycles, like those visible in Figure 4.5.

Autocorrelation coefficients may be correlations of any order. For residuals from [4.5], the first-, second-, and third-order autocorrelations are:

$$r_1 = .7319$$

$$r_2 = .4759$$

$$r_3 = .3498$$

Residuals strongly correlate with residuals from the previous month ($r_1 = .7319$) and moderately correlate with residuals from three months earlier ($r_3 = .3498$).

■ The *first-order autocorrelation coefficient*, r_1, equals the correlation between present and previous value of e:

$$r_1 = \frac{\sum_{t=1}^{T-1} (e_t - \bar{e})(e_{t+1} - \bar{e})}{\sum_{t=1}^{T} (e_t - \bar{e})^2} \qquad [4.6]$$

Also, r_1 is related to the Durbin-Watson d-statistic:

$$d \approx 2(1 - r_1) \qquad [4.7]$$

The *mth-order autocorrelation coefficient*, r_m (also called "autocorrelation at lag m"), equals the correlation between present and mth previous value of e:

$$r_m = \frac{\sum_{t=1}^{T-m} (e_t - \bar{e})(e_{t+m} - \bar{e})}{\sum_{t=1}^{T} (e_t - \bar{e})^2} \qquad [4.8]$$

■

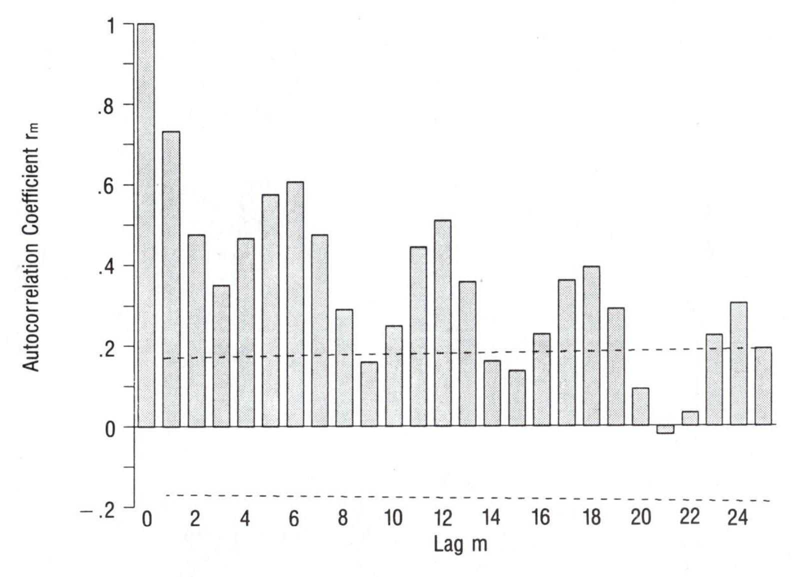

Figure 4.6 Correlogram showing autocorrelation of Concord water-use residuals, at monthly lags 0–25.

Correlograms like Figure 4.6 display autocorrelation coefficients at many lags. Vertical bars show r_m; Figure 4.6 includes lags from $m = 0$ ($r_0 = 1$) to $m = 25$ ($r_{25} = .19$). Dashed lines mark off approximate 95% confidence bands ($0 \pm 2/\sqrt{T - m}$). r_m values that fall outside these bands are significantly different from zero. Note that nearly all of these autocorrelations (of residuals from [4.5]) are positive and most are individually significant. Distinct peaks occur at six-month intervals, reflecting the raw data's seasonal cycles. Autocorrelation here obviously involves more than a relationship between e_t and e_{t-1}. In some data a nonsignificant Durbin-Watson test (which examines only r_1) could occur despite autocorrelation at other lags.

Correlograms provide detailed guidance for fitting models to autocorrelated data, as described in books on time series analysis.[6] Less formally, time plots of residuals also help in understanding the pattern of the autocorrelation. Figure 4.7 shows these residuals to have wild month-to-month fluctuations, but with autocorrelation evident in long sequences of mostly negative and mostly positive values.

Smoothing clarifies the underlying patterns of wildly fluctuating series. There are several basic types of smoother. Considering only those of *span 3*, which smooth present values together with the previous and next values, we have the following:

Running means. The smoothed series e^* is formed by replacing values of e (except the first and last, e_1 and e_T) with the mean of the current, previous, and next values:

$$e_t^* = \frac{e_{t-1}}{3} + \frac{e_t}{3} + \frac{e_{t+1}}{3}$$

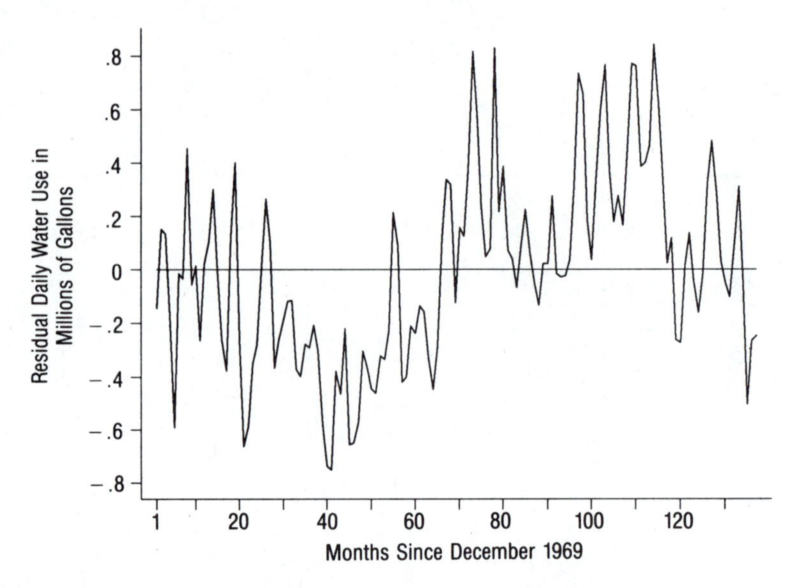

Figure 4.7 Time plot of Concord water-use residuals.

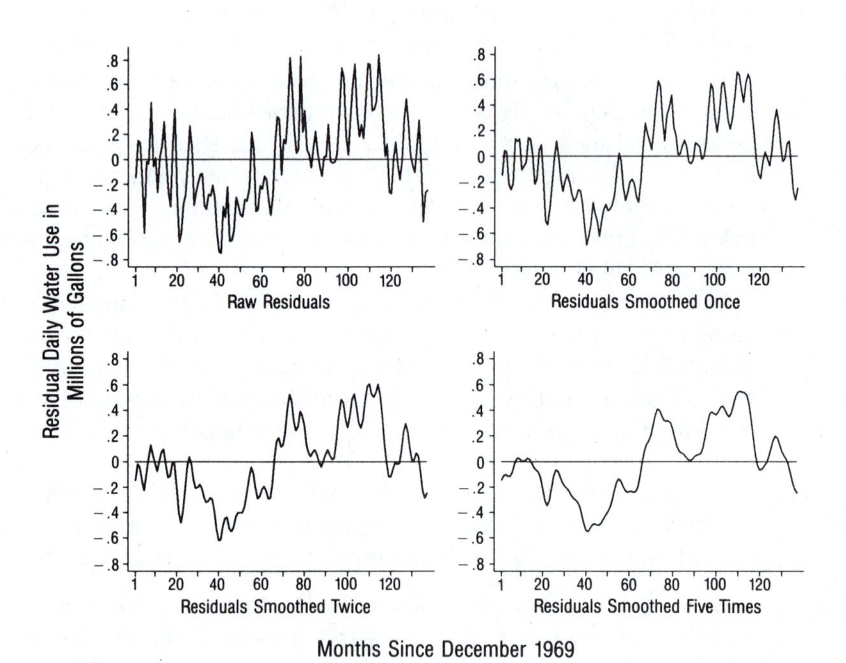

Figure 4.8 Time plots of Concord water-use residuals, showing effects of repeated smoothing by running means of span 3.

Hanning. A milder transformation than running means is obtained by:

$$e_t^* = \frac{e_{t-1}}{4} + \frac{e_t}{2} + \frac{e_{t+1}}{4}$$

Running medians:

$$e_t^* = \text{median}\{e_{t-1}, e_t, e_{t+1}\}$$

Smoothers can be combined or applied repeatedly to achieve the desired effect. Spans longer than 3 increase the degree of smoothing.[7]

Figure 4.8 shows time plots of the residuals smoothed once, twice, and five times by running means of span 3. Smoothing brings out an obvious pattern. Residuals tend to be negative for the first half of the decade, to be positive for most of the second half, and to vary around zero at the end. That is, daily water use was generally less than predicted (on the basis of temperature, precipitation, and conservation) in months 1–70. Daily water use was higher than predicted in months 70–120. This change may reflect national trends toward higher water consumption, mentioned earlier. From month 120 on, residuals fluctuate around zero—[4.5] no longer systematically over- or underpredicts actual water consumption.

If we find significant autocorrelation among residuals, we have several alternatives:

1. Hypothesis tests and confidence intervals are untrustworthy, but regression might still be employed for sample description. Time plots of residuals add a further dimension to this description and may help identify the nature of autocorrelation.
2. It is possible to estimate standard errors that are consistent (not necessarily unbiased) under heteroscedastic or autocorrelated residuals of a very general sort (White, 1980; Hansen, 1982). The necessary calculations require special programs.[8]
3. Autocorrelated residuals result when variables are omitted that affect adjacent cases in similar ways. If we can identify such variables and include them in the regression, autocorrelation may disappear.
4. We can use formal methods for time series analysis (Note 6). For example, working with *first differences* ($Y_t - Y_{t-1}$ instead of Y_t; $X_t - X_{t-1}$ instead of X_t, and so on) can sometimes eliminate autocorrelation.

All tests for autocorrelation refer to the data *as presently sorted.* With time series, the appropriate order is obvious, but other kinds of data, like the 15 mine workers, may possess no inherent order. We could list these workers alphabetically, or by Social Security number, age, LT ratio, or residual. Each ordering results in a different Durbin-Watson statistic or correlogram. Indeed, sorting any sample from lowest to highest residual guarantees positive autocorrelation, even when the cases themselves are completely independent. Conversely, alphabetical sorting of U.S. states, for example, could hide spatial autocorrelation; an appropriate test requires geographical sorting.[9] Autocorrelation tests make sense only to the extent that cases have inherent order; this is problematic when data are not in a time or spatial series.

Nonnormality

If errors are not normally distributed:

1. We lose the justification for applying t and F distributions, especially with small samples.
2. Since OLS tends to track outliers, heavy-tailed error distributions can cause great sample-to-sample variation.

To check the normality assumption, examine sample residuals. Can we believe that our residuals came from normal populations?

Chapter 1 described criteria for approximate normality in sample data. Mean-median comparison provides a simple check for symmetry. The mean OLS residual always equals zero; a nonzero median indicates skew. Residuals from [4.3] have median Md $= -69.5$. Since $\bar{X} >$ Md ($0 > -69.5$), sample residuals exhibit mild positive skew. Given an approximately symmetrical distribution, comparison of standard deviation with IQR/1.35 provides a simple check for normality. If we

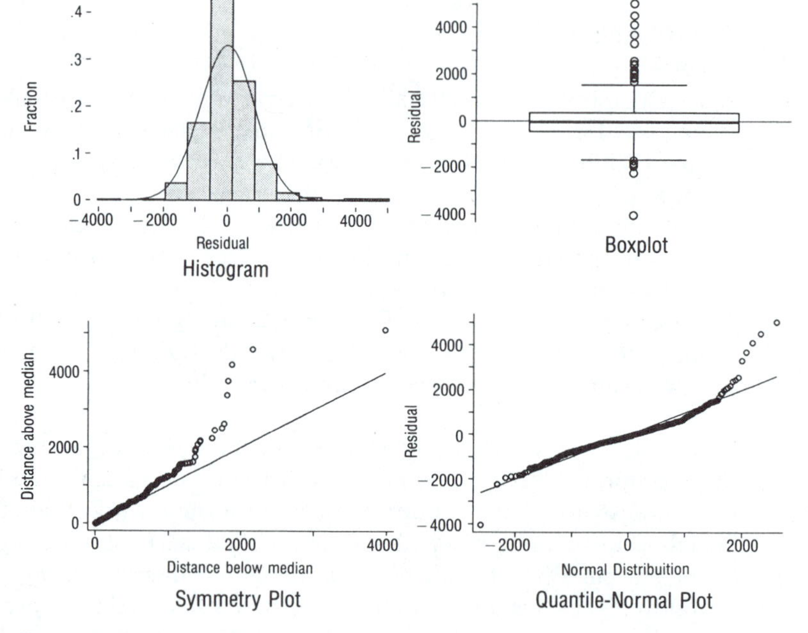

Figure 4.9 Four views showing nonnormality of residuals from regression of 1981 household water use on six predictors.

accept residuals from [4.3] as "approximately symmetrical" (because their skew is mild), $s_e > \text{IQR}/1.35$ ($855 > 603$) indicates heavier-than-normal tails.

Figure 4.9 shows four graphical views of these residuals: histogram with normal curve, boxplot, symmetry plot, and quantile-normal plot. The graphs confirm that the distribution is heavy tailed, with numerous high outliers and mild positive skew. The obvious nonnormality of this large ($n = 496$) sample argues strongly against assuming a normal population.

If e has a substantially nonnormal distribution, transformations of one or more variables may help. Nonnormal e distributions often result from skewed Y and/or X distributions; making the variable distributions more symmetrical (by power transformation) may also improve the distribution of e. Furthermore, transformations sometimes reduce several statistical problems at the same time. Chapter 5 examines how transformations affect the water-use analysis.

Robust methods provide alternatives to OLS for dealing with nonnormal errors. If other problems, such as heteroscedasticity or curvilinearity, are present too, robust regression can be used in combination with power transformations. Chapter 6 illustrates such analysis.

Influence Analysis

A case is *influential* if its deletion substantially changes the regression results. In bivariate regression we can usually spot influential cases by examining a scatterplot, but in multiple regression influence is harder to see. Case i might influence coefficients yet not appear as an outlier in ordinary scatterplots. Conversely, not all outliers need be influential. Influence results from a particular *combination* of values on all variables in the regression, not necessarily from unusual values on one or two of these variables.

The regression coefficient on X_k is b_k. Let $b_{k(i)}$ represent the same coefficient when the ith case is deleted. Deleting the ith case therefore changes the coefficient on X_k by $b_k - b_{k(i)}$. We can express this change in standard errors:

$$\text{DFBETAS}_{ik} = \frac{b_k - b_{k(i)}}{s_{e(i)}/\sqrt{\text{RSS}_k}} \qquad [4.9]$$

DFBETAS_{ik} measures the influence of the ith case on the kth regression coefficient (Belsley, Kuh, and Welsch, 1980). Here $s_{e(i)}$ represents the residual standard deviation of a regression with the ith case deleted, and RSS_k equals the residual sum of squares from regressing X_k on all the other X variables in the analysis (*not* deleting case i). The denominator in [4.9] modifies the usual estimate of the standard error of b_k, $\text{SE}_{b_k} = s_e/\sqrt{\text{RSS}_k}$.[10]

Interpreting DFBETAS:

If $\text{DFBETAS}_{ik} > 0$, case i pulls b_k up.
If $\text{DFBETAS}_{ik} < 0$, case i pulls b_k down.

The larger the absolute value of DFBETAS_{ik}, the more influence case i exerts on b_k. How large is "large"? Belsley et al. note three types of criteria for assessing DFBETAS and other influence statistics:

1. *External scaling.* On normal-theory grounds we might employ an *absolute cutoff*, such as $|\text{DFBETAS}_{ik}| > 2$. As sample size increases, however, it becomes less likely that a single case could move b_k by as much as two standard errors. Alternatively, we could use a *size-adjusted cutoff*. $|\text{DFBETAS}_{ik}| > 2/\sqrt{n}$ should detect roughly the top 5% of influential cases, whether these are influential in absolute terms or not.
2. *Internal scaling.* We can apply univariate outlier-detection procedures to sample DFBETAS_{ik} values. For example, we consider a case a severe influence outlier if $\text{DFBETAS}_{ik} < Q_1 - 3\text{IQR}$ or $\text{DFBETAS}_{ik} > Q_3 + 3\text{IQR}$ (the boxplot definition of *severe outlier*; see Chapter 1).
3. *Gaps.* Boxplots, quantile plots, or histograms may reveal gaps in the distribution of DFBETAS_{ik}, separating a few exceptionally influential cases from the rest.

By any of these criteria, cases identified as influential deserve closer scrutiny.

DFBETAS_{ik} answers the question "By how many standard errors does b_k change, if we drop case i?" For example, the coefficient on income in [4.3] is $b_1 = 20.97$, based on $n = 496$ households. Repeating the regression without household #134 (so $n = 495$), we obtain a lower estimate: $b_{1(134)} = 16.42$. The difference between these two coefficients, $b_1 - b_{1(134)} = 4.55$, reflects the influence of one high-income, high-water-use household.

The standard deviation of the residuals, with household #134 deleted, is $s_{e(134)} = 832.01$. The residual sum of squares from regressing income on the other five X variables is $\text{RSS}_1 = 60{,}120.36$. Therefore

$$\begin{aligned}\text{DFBETAS}_{134,1} &= \frac{20.97 - 16.42}{832.01/\sqrt{60{,}120.36}} \\ &= 1.34\end{aligned}$$

With household #134 included, the coefficient on income is *1.34 standard errors higher* than it otherwise would be. This is our most influential case; for no other household is $|\text{DFBETAS}_{i1}|$ so large.

In contrast, household #71 *lowers* the coefficient on income: $b_{1(71)} = 21.83$, $s_{e(71)} = 849.01$, so

$$\begin{aligned}\text{DFBETAS}_{71,1} &= \frac{20.97 - 21.83}{849.01/\sqrt{60{,}120.36}} \\ &= -.25\end{aligned}$$

This household, possessing the largest negative DFBETAS_{i1} in the sample, reduces the coefficient on income by .25 standard errors.

Table 4.3 shows summary statistics for DFBETAS in the water-use regression. For four of the six regression coefficients, DFBETAS do not exceed $\pm.5$, but for

Table 4.3 Summary statistics for DFBETAS of coefficients in household water-use regression of Equation [4.3]

Variable		Obs	Mean	Std. Dev.	Min	Max
X_1	DBincome	496	.0004404	.0907714	−.248154	1.340371
X_2	DBwatr80	496	−.0002074	.0895012	−1.38789	.6845321
X_3	DBeducat	496	−.0001973	.0434508	−.3601133	.1023189
X_4	DBretire	496	.0000682	.0445025	−.216704	.2785343
X_5	DBpeop81	496	.0000747	.0612796	−.4263415	.4554408
X_6	DBcpeop	496	−.0000535	.0406626	−.23371	.3507265

the coefficient on income they range as high as 1.34 (discussed earlier), and for 1980 water use they go as low as −1.39.

Leverage plots (Chapter 3) aid in influence analysis. Figure 4.10 plots the relation between 1981 water use and income, adjusting for the five other X variables in [4.3].[11] The slope in this plot equals the coefficient on income in [4.3], $b_1 = 20.97$. Two influential cases stand out at the upper right. Without these high-income, high-water-use households, the line would be less steep.

A *proportional leverage plot* displays influence graphically. Figure 4.11 corresponds to Figure 4.10, but with point size proportional to DFBETAS: larger circles denote greater $|\text{DFBETAS}_{i1}|$.[12] Household #134 ($\text{DFBETAS}_{134,1} = 1.34$) appears at upper right. Nearby is the second-most-influential household, #127, with $\text{DFBETAS}_{127,1} = .98$. Their positions and positive DFBETAS indicate that these two households steepen the partial regression line. Deleting both households reduces the coefficient on income from 20.97 to 12.46. Thus two cases—only 2/496

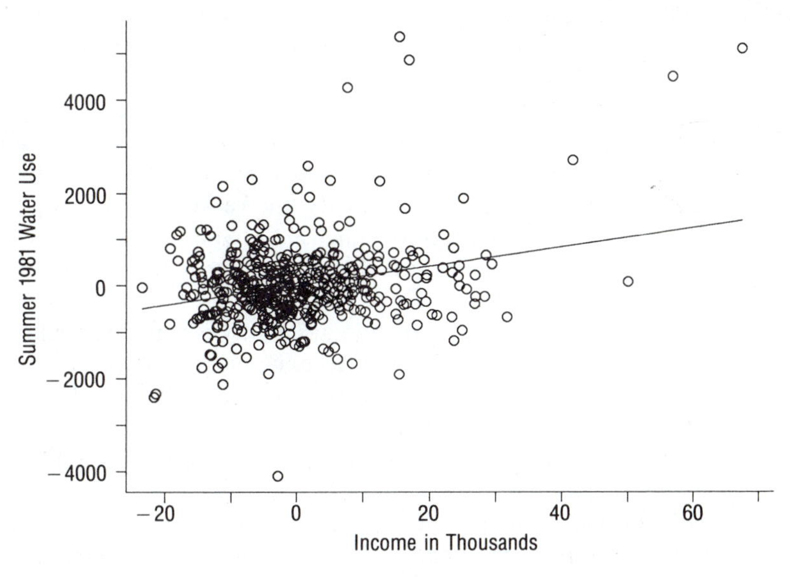

Figure 4.10 Leverage plot of 1981 household water use versus income, adjusting for other predictors.

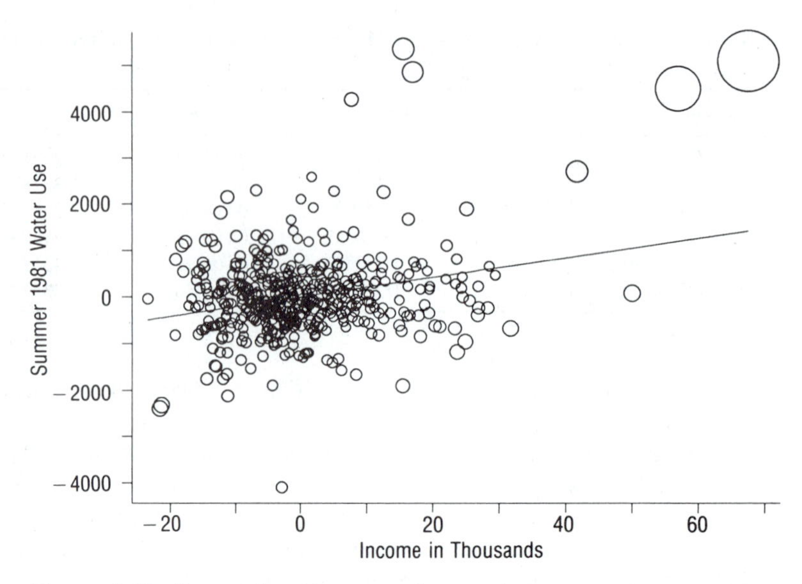

Figure 4.11 Proportional leverage plot (symbols proportional to DFBETAS) of 1981 household water use versus income, adjusting for five other predictors.

or .4% of our data—account for more than 40% of the magnitude of the full-sample regression coefficient!

Figure 4.12 shows a proportional leverage plot of post- versus preshortage water use, adjusting for the other five X variables. One influential case stands out at lower right: household #101, with $\text{DFBETAS}_{101,2} = -1.39$. This household pulls the regression line down; deleting it steepens the line by 1.39 standard errors, from $b_2 = .49$ to $b_{2(101)} = .53$.

Leverage plots sometimes uncover other statistical problems, such as heteroscedasticity and curvilinearity. For example, Figure 4.12 hints that a curve, climbing most steeply at low X/low Y, might describe this relation better than a straight line does. We also see heteroscedasticity: the variability of Y increases with X_2.

Proportional leverage plots for other X variables of [4.3] appear in Figure 4.13. These plots reveal no obvious curvilinearity, heteroscedasticity, or influence problems.

Influential cases bedevil regression and many other statistical methods. Routine use of diagnostic tools like DFBETAS and leverage plots lessens the risk of overlooking influence problems. Plots sometimes show clusters of cases that are jointly influential; purely numerical methods like DFBETAS have more difficulty in detecting clusters. (Chapter 8 describes another graphical method for detecting influential clusters.)

Once detected, influential data points should not necessarily be thrown out. Perhaps their information is valid and essential for accurate conclusions. In between

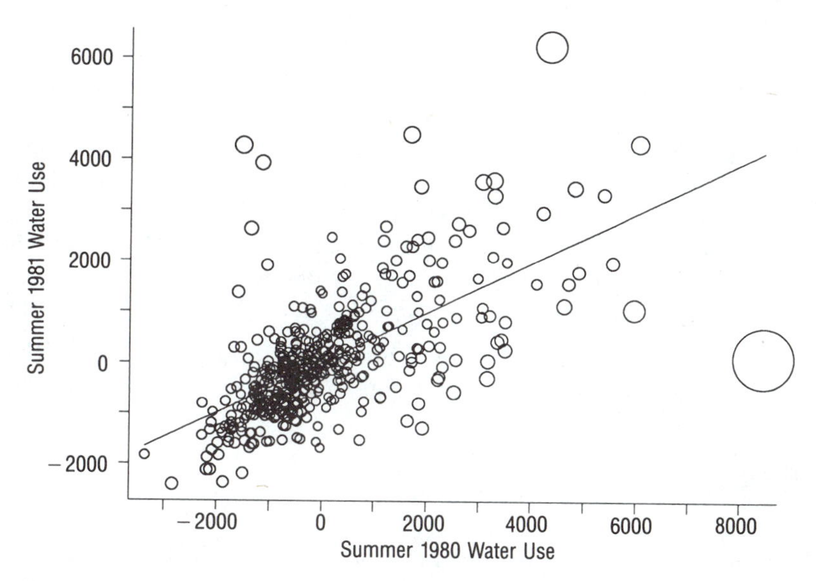

Figure 4.12 Proportional leverage plot of 1981 household water use versus 1980 water use, adjusting for five other predictors.

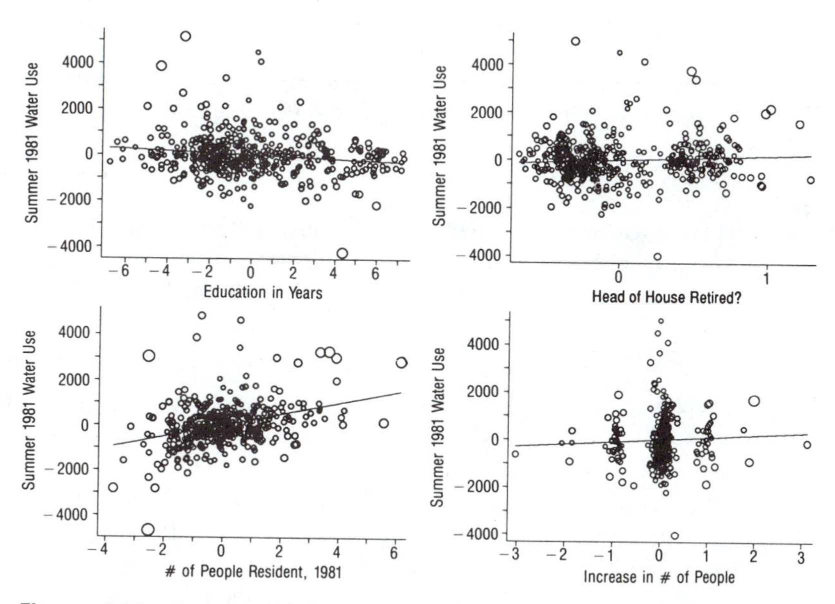

Figure 4.13 Proportional leverage plots of 1981 household water use versus X, each adjusting for five other predictors.

dropping cases and doing nothing, we have several alternatives:

1. Report results both with and without the influential cases, so that readers can draw their own conclusions. This approach is simple and honest.
2. Examine influential cases closely. If they reflect measurement error or belong to a different population, deletion is justified. Alternatively, they may suggest the need to include further X variables.
3. When influential cases come from the heavier tail of a skewed distribution, skew-reducing transformations may also reduce influence.
4. Try robust regression, which is less susceptible to influence than OLS is.

Chapter 5 describes the use of nonlinear transformations, which can solve several statistical problems at once. Robust regression is discussed in Chapter 6.

More Case Statistics

Residuals and DFBETAS are *case statistics*, which have values for every case in the data. A more fundamental case statistic is *leverage*, the *potential* for influence resulting from unusual X values. In bivariate regression, the farther X_i is from $\bar{X}$, the greater the leverage of case i. In multiple regression, leverage reflects how unusual the ith *combination* of X values is. The most common measure of leverage is the *hat statistic*, h_i. *Leverage cases* are those with high h_i.

The *leverage* of case i (h_i) equals the ith diagonal element of the *hat matrix*, **H**:

$$\mathbf{H} = \mathbf{X}(\mathbf{X}'\mathbf{X})^{-1}\mathbf{X}' \qquad [4.10]$$

where **X** is the $n \times K$ matrix of X values, including an initial column of 1's. **H** is a square matrix, $n \times n$.

The hat matrix is so named because it "puts the hat on" Y:

$$\hat{\mathbf{Y}} = \mathbf{HY}$$

where $\hat{\mathbf{Y}}$ and **Y** are $n \times 1$ vectors of predicted and actual Y values, respectively.

Leverage theoretically can range from $1/n$ to 1. Magnitudes depend on sample size; the sample mean is $\bar{h}_i = K/n$. Belsley et al. (1980) suggest that high-leverage cases be identified by the size-adjusted cutoff $h_i > 2K/n$ (approximately the top 5%). Huber (1981) proposed absolute guidelines based on $\max(h_i)$, the sample's highest leverage:

$\max(h_i) \leq 0.2$	safe
$0.2 < \max(h_i) \leq 0.5$	risky
$\max(h_i) > 0.5$	avoid if possible

Leverage above .2 becomes risky because too much of the sample's information about the Y–X relationship comes from a single case. As h_i approaches 1, the ith case almost completely controls the regression.

Figure 4.14 illustrates the distinction between leverage and influence, using manufactured samples of $n = 100$ cases each. At right in both scatterplots is one X-outlier or leverage case ($h_i = .34$, placing it in Huber's "risky" category and well above $2K/n = 4/100 = .04$). The two samples are identical, except for the Y value of that high-leverage case. In the left plot, this high-leverage case has little influence on the slope ($\text{DFBETAS}_i = .06$, much less than $2/\sqrt{n} = .2$). Deleting this case leaves the regression line nearly unchanged, because the leverage case's Y value is consistent with the pattern of the rest of the sample. Such cases have been called "good" or "nondiscordant" outliers.

Changing a leverage case's Y value shifts the regression line, as seen at right in Figure 4.14. Given a much lower Y value, this case (now a "bad" or "discordant" outlier) pulls down the slope. With case i included, the regression slope is about 9 standard errors lower than it would be if i were deleted ($\text{DFBETAS}_i = -9.1$). Influence (DFBETAS) thus depends on both X and Y values.[13] In contrast, leverage, which depends only on X values, is the same in either plot: $h_i = .34$.

Leverage appears in definitions of several other case statistics. The estimated *variance of the ith residual* is

$$\text{Var}[e_i] = s_e^2(1 - h_i) \qquad [4.11]$$

where s_e is the standard deviation of the residuals: $s_e = \sqrt{\text{RSS}/(n - K)}$. The greater the leverage, the smaller the variance of the ith residual—high-leverage points tend to be fitted closely. The estimated *standard deviation of the ith residual* equals the square root of [4.11].

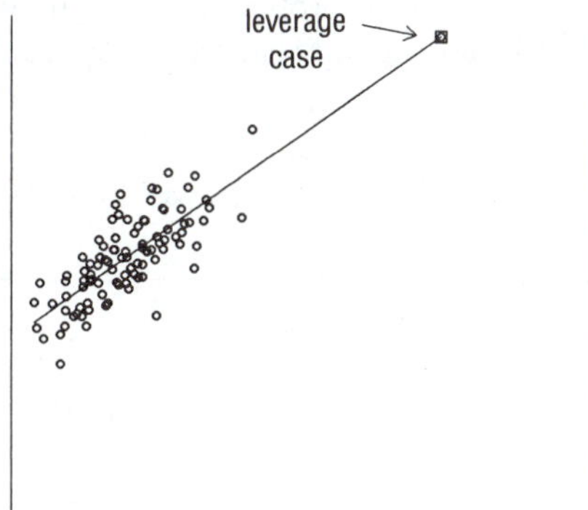

High Leverage, Low Influence

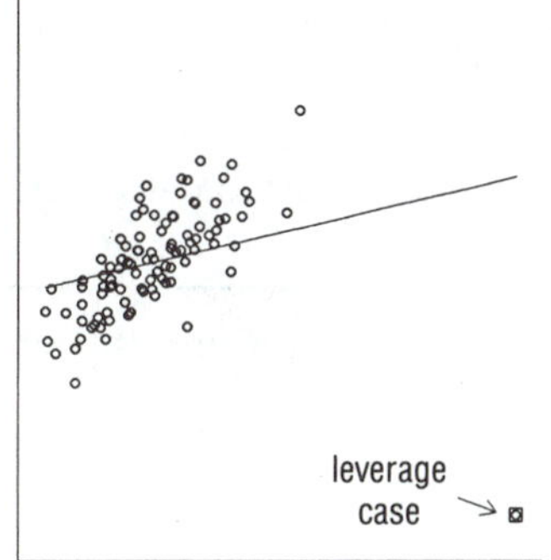

High Leverage, High Influence

Figure 4.14 "Good" (left) and "bad" (right) outliers: "bad" outliers influence the slope (artificial data).

Dividing a residual by its estimated standard deviation produces a *standardized residual*:

$$z_i = \frac{e_i}{s_e\sqrt{1 - h_i}} \qquad [4.12]$$

A related statistic is the *studentized residual*:

$$t_i = \frac{e_i}{s_{e(i)}\sqrt{1 - h_i}} \qquad [4.13]$$

where $s_{e(i)}$ denotes the standard deviation of the residuals with the ith case deleted.[14]

Studentized residuals ([4.13]) amount to a t-test of the null hypothesis that coefficient δ equals zero ($H_0: \delta = 0$) in the model

$$E[Y_i] = \beta_0 + \beta_1 X_{i1} + \beta_2 X_{i2} + \cdots + \beta_{K-1} X_{i,K-1} + \delta I_i$$

where I is a dummy variable coded 1 for case i and 0 for all other cases. This tests whether case i causes a significant shift in the regression intercept and so should be considered an outlier. Although individual t_i follow a t distribution (df $= n - K - 1$), special tables provide P-values for the maximum absolute value of t_i ($\max|t_i|$) based on n tests.[15] Alternatively, applying the Bonferroni inequality, $\max|t_i|$ indicates a significant outlier at level α if (given a t distribution with df $= n - K - 1$) t_i is significant at level α/n.[16] α/n is a low number, and computer programs (rather than ordinary t tables) may be necessary for calculating P-values.[17]

Cook's distance, or *Cook's D* (D_i), measures influence on the model as a whole, rather than on a specific coefficient as DFBETAS does (Cook and Weisberg, 1982). That is, D_i reflects the ith case's influence on all K estimated regression coefficients or, equivalently, on all n predicted values of Y. Interpretation of D_i is less intuitive than DFBETAS_{ik}.[18] Internal scaling and gaps can highlight unusual values of either influence statistic. Two external-scaling criteria for D_i are the following:

1. An absolute cutoff: cases are influential if $D_i > 1$.
2. A size-adjusted cutoff: unusually influential cases will have $D_i > 4/n$.

Cook's D measures the influence of the ith case:

$$D_i = \frac{z_i^2 h_i}{K(1 - h_i)} \qquad [4.14]$$

D_i (influence) increases with:

a. the size of the standardized residual, z_i (Equation [4.12]), and
b. leverage, h_i (Equation [4.10]).

Table 4.4 Case statistics for three influential households (see Figures 4.11 and 4.12)

Household i	Residual e_i	Leverage h_i	Standardized Residual z_i	Studentized Residual t_i	Cook's D_i	Income DFBETAS$_{i1}$	Water DFBETAS$_{i2}$
101	−4037	.08	−4.96[3]	−5.08[3]	.31[1]	0.06	−1.39[1]
127	3316	.06	4.03	4.09	.15[3]	0.98[2]	−0.18
134	3687	.09[3]	4.55	4.65	.29[2]	1.34[1]	0.25

[1] Absolute value of this statistic is largest in sample.
[2] Absolute value of this statistic is second-largest in sample.
[3] Absolute value of this statistic is third-largest in sample.

A case with lower D_i may still substantially influence one regression coefficient; DFBETAS$_{ik}$ is better able to detect such cases. On the other hand, a case's influence may be spread out over several coefficients and hence may be more readily detected by D_i.

Figures 4.11 and 4.12 revealed three influential households, described by case statistics in Table 4.4. All three possess relatively large residuals and leverage well above $2K/n = 14/496 = .03$. Furthermore, these households have the three largest D_i (well above $4/n = .008$, but not above 1) and the three largest |DFBETAS$_{ik}$| values (far above $2/\sqrt{n} = .09$) in the sample. DFBETAS most clearly diagnose the specific problems posed by these influential households: #101 pulls down the coefficient on 1980 water use, whereas #127 and #134 increase the coefficient on income.

The Bonferroni inequality provides a P-value for the largest studentized residual, which belongs to household #85 (not in Table 4.4): $\max|t_i| = 6.19$. For this case to be a significant outlier at α, we require $P(|t| > \max|t_i|) < \alpha/n$. The t-distribution ($df = n - K - 1 = 488$) probability $P(|t_i| > 6.19)$ is very small, much less than our cutoff $\alpha/n = .05/496 = .0001$.[19] Therefore household #85 is a significant outlier at the $\alpha = .05$ level.

Symptoms of Multicollinearity

Multicollinearity, or too-high intercorrelations among X variables, causes trouble but can easily go unnoticed. Define R_k^2 as the coefficient of determination obtained by regressing X_k on all other X variables in the model. R_k^2 equals the proportion of X_k's variance that is shared with the other X variables. The higher R_k^2, the greater the degree of multicollinearity; $R_k^2 = 1$ implies perfect multicollinearity. One way to detect multicollinearity is to regress each X variable on all the other X variables and examine the resulting R_k^2 values.

The *tolerance* of X_k is the proportion of X_k's variance *not* shared with the other X variables:

$$\text{tolerance} = 1 - R_k^2$$

With perfect multicollinearity, tolerance equals zero and regression is impossible. Low tolerance (below .2 or .1) does not prevent regression but makes the results less stable.

Table 4.5 shows R_k^2 and tolerance values for X variables in [4.3]. Income, for example, has a tolerance of .71, meaning that 71% of the variation in income is independent of the other five variables. The lowest tolerance is .64, so there is no problem with multicollinearity here.

The usual formula for the estimated standard error of a regression coefficient ([3.19]) can be rewritten:

$$\mathrm{SE}_{b_k} = \frac{s_e}{\sqrt{(1 - R_k^2)\mathrm{TSS}_k}} \qquad [4.15]$$

where $1 - R_k^2$ is the tolerance of X_k and TSS_k is its total sum of squares ($\mathrm{TSS}_k = \Sigma(X_{ik} - \bar{X}_k)^2$). Equation [4.15] indicates that, other things being equal, the higher the R_k^2 (multicollinearity), the higher the standard error.

Figure 4.15 shows the form of the relationship: SE_{b_k} is proportional to $1/\sqrt{1 - R_k^2}$. Suppose we have a series of regressions, all with identical values of s_e and TSS_k (an idealized, hypothetical situation). Let SE_0 represent the estimated standard error of b_k when X_k is completely independent of the other X variables, $R_k^2 = 0$. Then SE_{b_k} equals:

$1.4(\mathrm{SE}_0)$ when $R_k^2 = .5$

$1.8(\mathrm{SE}_0)$ when $R_k^2 = .7$

$2.2(\mathrm{SE}_0)$ when $R_k^2 = .8$

$3.2(\mathrm{SE}_0)$ when $R_k^2 = .9$

$10(\mathrm{SE}_0)$ when $R_k^2 = .99$

and escalates toward infinity as R_k^2 approaches 1. With real data, SE_{b_k} could increase slower or faster than these calculations indicate.

High standard errors are a principal symptom of multicollinearity. They indicate that coefficient estimates are imprecise—expected to vary widely across different

Table 4.5 Checking for multicollinearity: tolerances of X variables in water-use regression [4.3]

Variable	**Shared Variation R_k^2**	**Independent Variation (Tolerance) $1 - R_k^2$**
X_1 Income	.29	.71
X_2 1980 Water Use	.33	.67
X_3 Education	.13	.87
X_4 Retirement	.22	.78
X_5 # of People	.36	.64
X_6 Increase in People	.04	.96

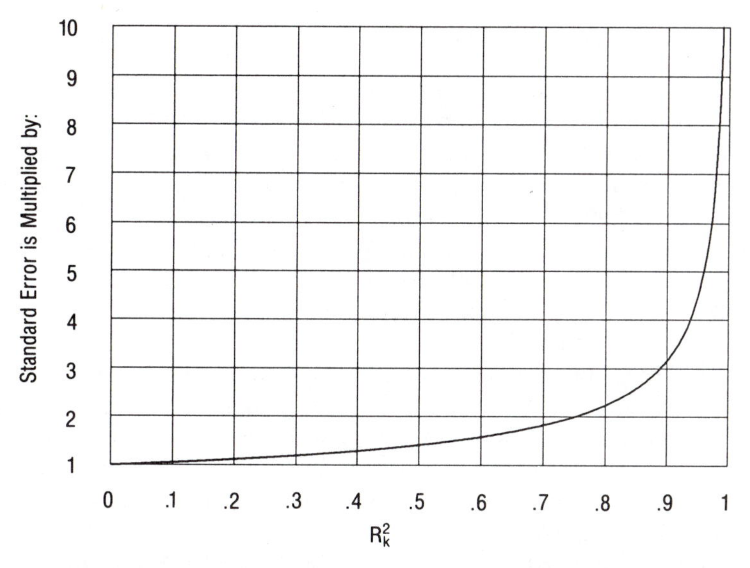

Figure 4.15 Effect of multicollinearity on standard errors (simplified).

samples. Small changes such as adding or deleting one variable or case may also substantially change the estimated coefficients. Sometimes coefficients take unexpected signs. For example, if two collinear variables both correlate positively with Y, one of their partial regression coefficients will often be negative. With high standard errors, even large regression coefficients may not be statistically significant. Multicollinearity can also produce standardized regression coefficients greater than ± 1. Although such symptoms provide warning, they may go unnoticed.

Matrices of correlations between variables, like Table 4.2, provide a fast but unreliable check for multicollinearity. High correlations suggest that multicollinearity may be a problem. Unfortunately, *multi*collinearity can occur even without strong correlations between variable pairs (simple *co*llinearity). A better, equally quick, procedure is to examine the matrix of correlations *between the estimated coefficients.* High correlations (for example, greater than $\pm .9$) between coefficients indicate that multicollinearity is present but do not necessarily identify which variables are involved. Regressions of each X on the other X variables most reliably serve that purpose.

The correlation between b_j and b_k, the estimated regression coefficients on X_j and X_k, is

$$r_{jk} = \frac{s_{jk}}{\sqrt{s_{jj}}\sqrt{s_{kk}}} \qquad [4.16]$$

where s_{jk} is covariance between b_j and b_k: the (j, k)th element of **S**, the variance-covariance matrix of estimated coefficients ([3.20]).

In summary, to guard against multicollinearity problems:

1. Examine a matrix of correlations between coefficients. If high correlations appear, investigate further by regressing each X on all the other X variables.
2. Experiment with adding and deleting the suspect variables. Do standard errors or coefficient estimates change substantially?
3. Employ one of the following coping strategies:
   a. Keep the variables in the equation, but understand that we cannot generalize (beyond the sample) about their *separate* effects. An F-test of their *joint* effects should still be valid (Chapter 3), and the variables may all contribute to the prediction of Y.
   b. Drop one or more of the offending variables, since their information is mostly redundant. With this approach we risk overestimating the importance of the variables kept in the equation, when much of their apparent effect is actually due to the dropped variables.
   c. Combine variables, since there is evidence that they are measuring the same thing (see Chapter 8).
   d. Consider ridge regression, a biased estimation strategy that has greater precision (lower mean squared error) than OLS in the presence of multicollinearity.[20]
   e. Collect more data. Multicollinearity is basically a problem of not enough information. Adding cases generally makes the coefficient estimates more precise.

The coping strategy should be chosen with regard to research goals. For example, (a) through (c) give up on estimating the individual effects of each variable. Theoretically, (e) is the most attractive, although it is often impractical.

Conclusion

Under certain ideal conditions (a true model and normal i.i.d. errors), OLS is superior to other estimators. These conditions rarely occur in real data, so the superiority of OLS cannot be taken for granted. Fortunately, OLS remains useful under a broader range of practical conditions. The purpose of regression criticism is to determine whether these conditions are exceeded. That is, do the data suggest that basic regression assumptions are not even plausible as approximations? Are there reasons (such as influential cases or multicollinearity) to suspect that our findings are sample-specific? Some analytical problems are intractable, but others can be remedied—if we know a problem exists.

Diagnostic tools help detect possible problems:

1. As a preliminary step, examine the scatterplot matrix. Severe curvilinearity, heteroscedasticity, outliers, and collinearity may be noticed before further efforts are wasted. Because they hide curvilinearity and outlier problems, correlation matrices should not be trusted without viewing the corresponding scatterplots.
2. Several problems are most easily diagnosed by examining residuals and other case statistics:

a. *Curvilinearity* appears in leverage plots, and perhaps also in the overall e-versus-$\hat{Y}$ plot.

b. *Heteroscedasticity* appears in leverage plots, e-versus-$\hat{Y}$ plots, and, most clearly, in $|e|$-versus-$\hat{Y}$ plots.

c. First-order *autocorrelation* is tested by the Durbin-Watson statistic. Autocorrelation coefficients are needed for higher-order autocorrelation; correlograms show their pattern graphically. Smoothed e-versus-i (or e-versus-time) plots add graphical information.

d. *Nonnormality* of errors can be partially tested by applying univariate methods such as mean-median comparisons and quantile-normal plots to sample residuals.

e. *Influence* is measured by DFBETAS and Cook's D. Proportional leverage plots display influence graphically.

3. Check for *multicollinearity* by examining the matrix of correlations among the estimated coefficients. If some correlations are high, or other symptoms appear, regress each X on the other X's to identify the source of the problem.

Regression criticism may uncover no problems, allowing more confident conclusions. Often problems are found, however, and we must decide what to do about them. The remaining chapters describe ways to cope with some common difficulties.

Exercises

Cassava or manioc, a plant with starchy roots, is a dietary staple in many tropical countries. The following table contains data on cassava production in the Central African Republic. Although problematic, these data have been employed for a serious purpose: forecasting future food production in a country with rapidly growing population.

Cassava production in the Central African Republic

Year	Cultivated Area in 1000's of Hectares	Cassava Production in 1000's of Metric Tons
1961	204	746
1962	204	746
1963	204	746
1964	204	746
1965	204	746
1969	262	767
1970	262	767
1971	262	767
1974	302	898
1975	286	850
1976	295	850
1977	304	900
1978	315	940
1979	320	970
1980	300	920
1981	287	900
1982	331	676

1. Regress cassava production (Y) on year (X_1) and cultivated area (X_2). Write out and interpret the regression equation and R^2. Is either predictor significant?
2. Calculate residuals and predicted values for your regression of Exercise 1. Construct an e-versus-$\hat{Y}$ plot, and describe any problems you see.
3. Obtain case statistics: studentized residuals, leverage, Cook's D, and DFBETAS on year and area. Use these to discuss possible outliers or influential cases.[21]
4. Conduct a Durbin-Watson test for autocorrelation. Graph residuals or smoothed residuals against year, and describe what these diagnostics reveal.
5. Regress production on year alone and then on area alone. Compare results from these two regressions with your three-variable regression of Exercise 1. Do you see any symptoms of multicollinearity? Also find and discuss:
   a. the matrix of correlations among the coefficient estimates.
   b. R_k^2 for the regression of year on area, and of area on year. (In these two-variable regressions, what is the relationship between either R^2 and the simple correlation between area and year?)
6. What aspects of the original regression appear questionable in light of the problems seen in Exercises 2–5?

Coal miners often encounter methane gas. Methane in coal mines has several important implications: it can cause dangerous explosions; if captured, methane is potentially a cleaner fuel source than the coal itself; and, if simply released into the atmosphere, methane contributes substantially to the "greenhouse gases" that may change the earth's climate. Deeper coal tends to contain more methane. The following table gives depth and methane content of 15 core samples from the Hartshorne coalbed of Oklahoma (from Iannacchione and Puglio, 1979). Exercises 7–9 use these data.

Depth and methane content of Hartshorne coalbed samples

Sample	Depth in Feet	Methane in cm^3/g
1	175	2.5
2	252	5.7
3	318	8.7
4	356	10.8
5	516	11.8
6	553	13.1
7	571	11.8
8	561	11.5
9	556	10.9
10	823	15.5
11	892	16.8
12	1439	17.1
13	1440	16.7
14	489	10.9
15	1296	17.5

7. Regress methane content (Y) on depth (X). Calculate residuals and predicted values, and plot e versus $\hat{Y}$. What is the most obvious problem?

8. Regress methane content (Y) on the logarithm of depth ($\log_{10} X$). (Taking logarithms linearizes the relationship between Y and X.) Calculate a new set of predicted values, and use these to show the regression curve on a scatterplot of methane against depth (not log depth). Describe what the curve implies.
9. Calculate standard errors for predicted Y values, and use them to construct 95% confidence bands for the mean methane content of Hartshorne coal. Add confidence bands to your scatterplot from Exercise 8.

The following table gives data on stream sediment yield, precipitation, and runoff for 19 mountain basins mostly in the Southern Alps of New Zealand (from Hicks, McSaveney, and Chinn, 1990). The authors found that sediment yield was determined largely by runoff and precipitation. Exercise 10 analyzes these data. (We return to these data, which contain one anomalous case, in Chapter 8.)

Runoff, precipitation, and sediment yield of 19 mountain basins

Basin	Mean Annual Runoff (mm)	Mean Annual Precipitation (mm)	Mean Annual Sediment Yield (tons/km²)
1 Ivory	11,300	8,600	14,900
2 Cropp	11,500	10,070	32,600
3 U. Waitangitoana	6,800	7,300	12,500
4 Hokitika	8,870	9,400	18,800
5 Haast	5,970	6,500	14,000
6 L. Hopwood Burn	1,430	2,100	1,640
7 Shotover	1,060	1,600	1,120
8 Arrow	410	1,100	270
9 Manuherikia	130	830	44
10 Karamea	2,860	3,300	360
11 Buller A	1,660	2,400	300
12 Buller B	2,070	2,600	300
13 Inangahua A	1,750	2,500	190
14 Inangahua B	2,300	3,000	800
15 Grey	2,360	3,000	610
16 Butchers Creek	2,320	2,900	300
17 Cleddau	6,510	7,000	14,600
18 Hooker	6,430	6,500	4,600
19 Tsidjiore Nouve	1,480	1,947	3,010

10. To linearize relationships, take base 10 logarithms of yield, runoff, and precipitation. Perform three regressions:
    a. regress $\log_{10}$(yield) on $\log_{10}$(runoff).
    b. regress $\log_{10}$(yield) on $\log_{10}$(precipitation).
    c. regress $\log_{10}$(yield) on both $\log_{10}$(runoff) *and* $\log_{10}$(precipitation).

    Compare the three regressions. What major problem appears in the third? Investigate.
11. Return to the gasoline sales/infant lead regression of Exercise 7 in Chapter 2. Why might we suspect autocorrelation? Conduct a Durbin-Watson test. Do the test results cast doubt on our earlier conclusion that this relationship is statistically significant?

12. The 64 spruce forest sites listed at the end of Chapter 3 are not independent. Sites 9–28 are all on Whiteface Mountain, NY; sites 29–44 are on Mt. Mansfield, VT; and sites 45–64 are on Mt. Washington, NH. Consequently, we should be wary of spatial autocorrelation. Does a Durbin-Watson test ($\alpha = .05$) on residuals from the regression of Exercise 6, Chapter 3, detect this problem? Investigate graphically (for example, with a residual versus case number plot). For which mountain are the model's predictions consistently too low (positive residuals)? Reestimate the model including a dummy variable for this mountain, and perform another Durbin-Watson test. Comment on the results.

The following data are from a study of the relation between children's respiratory illness (percentage of children with symptoms) and air pollution (sulphur dioxide concentration, in micrograms per cubic meter of air), carried out in France's Gardanne coal basin (Charpin, Kleisbauer, Fondarai, Graland, Viala, and Gouezo, 1988). Cases are two-week periods. For Exercises 13–14, assume that autocorrelation is not a problem.

Town		Mean 2-Week SO_2 Levels	Mean Prevalence of Pulmonary Symptoms
1	Biver	69	17
2	Biver	147	24
3	Gardanne	59	22
4	Gardanne	132	31
5	Bouc	33	14
6	Bouc	116	33
7	Meyrueil	67	14
8	Meyrueil	120	25
9	Greasque	58	19
10	Greasque	92	27

13. Does air pollution affect illness? Regress children's pulmonary symptoms on SO_2 level. Evaluate potential influence problems by calculating h_i, D_i, and DFBETAS$_i$, comparing each to the appropriate size-adjusted criterion. On a scatterplot with regression line, identify any unusually influential cases you found.
14. Redo the regression of Exercise 13 without the influential cases. Does this change your conclusions?
15. On common-sense grounds, can you think of reasons to suspect autocorrelation in Exercises 13–14?

Notes

1. Notations, as well as statistical properties, change if X variables are random rather than fixed. For example, with random X the left-hand side of [4.1] becomes $E[Y_i \mid X_i]$: the conditional expectation of Y_i given X-vector $\mathbf{X}_i$.

19. Less than 10^{-8} or .00000001. Most statistical programs will calculate a seemingly more precise P-value, but these are actually just approximations (Note 17).
20. For example, see Rawlings (1988).
21. The high leverage of 1981 reflects the fact that this case has an exceptional *combination* of X_1 and X_2 values: for such a late year, cultivated area is unusually low.

2. Bounded range means that there exists some real $\Phi < \infty$ such that $|Y_i| < \Phi$ for all i. See Achen (1982) for a proof of OLS consistency and for additional arguments that OLS remains useful well beyond the scope of the Gauss-Markov assumptions.
3. The notation $Z \sim N(\mu, \sigma^2)$ signifies that "Z is distributed normally, with mean μ and variance σ^2."
4. Scatterplots work poorly with variables that take on only a few discrete values. For example, in Figure 4.16, the plot at left shows a regression of household head's education on whether he/she is retired. Although regression finds a significant, negative relationship (retirees have lower mean education), we see no evidence of it in the scatterplot. Retirement is a dummy variable with only two values (0 or 1), and education must be a whole number from 6 to 20. Only 25 distinct combinations of education and retirement occur here, so all 496 cases plot at only 25 points.

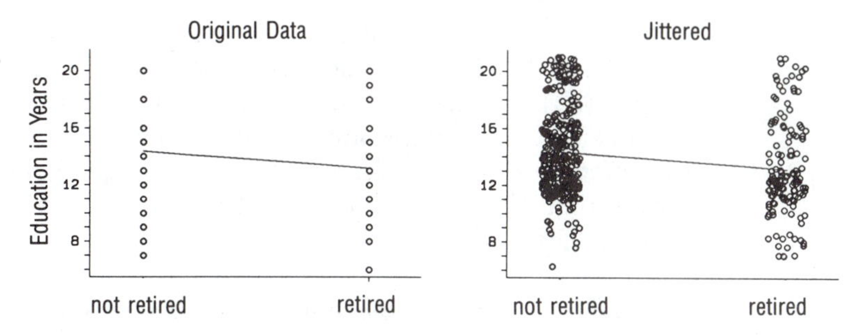

Figure 4.16 Scatterplot and corresponding jittered plot.

Jittering (at right) reduces overplotting by adding spherical random variation to each data point. We can then see that the nonretired distribution has a higher center of gravity; the regression line makes graphical sense. Jittering does not change the actual data—only their appearance in a graph. It often improves readability of scatterplots for discrete data. Jittering is also used in Figures 4.1, 8.13, and 8.17 but not elsewhere in this book.

5. Negatively autocorrelated residuals are much less common. To test for significant *negative* autocorrelation, replace d by $4 - d$ and apply the same decision rules using Table A4.4.
6. Techniques that *are* appropriate for analyzing the Concord water data include a type known as *autoregressive integrated moving average* (ARIMA) models. ARIMA modeling makes heavy use of correlograms to help identify the best models and evaluate their fit. For an introduction to ARIMA modeling, see McCleary, Hay, Meidinger, and McDowall (1980).
7. See Velleman (1982) and also Velleman and Hoaglin (1981) for more detail, including elaborate compound smoothers.
8. For example, the MCOV procedure of the Regression and Time Series (RATS) statistical package.
9. Spatial autocorrelation, which involves at least two dimensions, is more

complicated than the one-dimensional autocorrelation of time series. Durbin-Watson and similar tests are designed for time series and may fail to detect spatial autocorrelation.

10. A related statistic called DFFITS measures the ith case's influence on $\hat{Y}_i$:

$$\text{DFFITS}_i = \frac{\hat{Y}_i - \hat{Y}_{(i)i}}{s_{e(i)}/\sqrt{h_i}}$$

Belsley et al. (1980) suggest the size-adjusted cutoff:

Case i is relatively influential if $|\text{DFFITS}_i| > 2\sqrt{K/n}$

Computational formulas allow calculation of DFBETAS_{jk} and DFFITS_j without actually deleting the ith case and redoing the regression. See Belsley et al. for details.

11. The "Y" variable in this plot consists of residuals from the regression of 1981 water use (Y) on: 1980 water use (X_2), education (X_3), retirement (X_4), number of people (X_5), and change in number of people (X_6). The "X" variable consists of residuals from the regression of household income (X_1) on X_2 through X_6.

12. Proportional leverage plots can be constructed with a computer program (such as Stata) that will draw weighted scatterplots. First, calculate DFBETAS for each case. Next transform them:

$$\text{DFBETAS}^*_{ik} = (99/18)|\text{DFBETAS}_{ik}|(|\text{DFBETAS}_{ik}| + 1)^2 + 1 \qquad \text{if } |\text{DFBETAS}_{ik}| \leq 2$$

$$\text{DFBETAS}^*_{ik} = 100 \qquad \text{if } |\text{DFBETAS}_{ik}| > 2$$

The transformation rescales the absolute values of DFBETAS to range from 1 to 100:

if $\text{DFBETAS}_{ik} = 0$ (deleting case i has no impact on b_k), then $\text{DFBETAS}^*_{ik} = 1$; and

if $|\text{DFBETAS}_{ik}| \geq 2$ (deleting case i shifts b by two standard errors or more), $\text{DFBETAS}^*_{ik} = 100$.

Finally, construct a leverage plot using DFBETAS^*_{ik} values as weights, so that the area of each circle is proportional to DFBETAS^*_{ik}. Cases with $|\text{DFBETAS}_{ik}| \geq 2$ will appear as circles with 100 times the area of cases with $\text{DFBETAS}_{ik} = 0$ and about four times the area of cases with $|\text{DFBETAS}_{ik}| = 1$.

Household #134 has the largest $|\text{DFBETAS}_{i1}|$, 1.34, which transforms to $\text{DFBETAS}^*_{134,1} = 41.36$. Consequently, the largest circle in Figure 4.11 has about 41.36 times the area of the smallest circle.

13. This example, like DFBETAS, refers to influence on estimated regression coefficients. Although "good" outliers by definition do not affect regression coefficients, they still affect standard errors, hypothesis tests, R^2, and other statistics. The only way to assess the full impact of suspicious cases is to redo the analysis with and without them—and closely compare the details, not just regression slopes.

14. The statistic $s_{e(i)}$ is also used in defining DFBETAS ([4.9]). It may be calculated without actually deleting the ith case and redoing the regression:

$$s^2_{e(i)} = \frac{n-K}{n-K-1}s^2_e - \frac{e^2_i}{(n-K-1)(1-h_i)}$$

15. For examples, see Weisberg (1980).

16. The *Bonferroni inequality* states that the probability of at least one of a set of events occurring cannot be more than the sum of their individual probabilities.

The individual probabilities of Type I error, using studentized residuals to test whether each case is an outlier, equal the nominal significance level, α. If we test n cases, then, according to the Bonferroni inequality, the probability that at least one of them appears to be a "significant" outlier (despite a true null hypothesis) cannot be more than $n\alpha$.

We wish to run n tests (calculate studentized residuals for every case) yet have an *overall* Type I error risk of α. This can be done by testing each residual individually at the $\alpha' = \alpha/n$ level. Over n tests, the probability of at least one Type I error at level α' is $n\alpha' = \alpha$. For example, to test whether any case in a sample of $n = 20$ is a significant outlier at $\alpha = .05$, we simply check whether the sample's $\max|t_i|$ is significant at $\alpha' = .05/20 = .0025$.

17. Computer programs only approximate the theoretical P-values, however, and these approximations become less accurate for very small probabilities. Different software or hardware may obtain different results, so the answer produced by your system should be taken with a grain of salt. We may be confident that a P-value is very small (enough to reject H_0) without necessarily believing that we know precisely how small.

18. Cook and Weisberg (1982) offer two geometrical interpretations of D_i:
    a. the Euclidean distance between $\hat{\mathbf{Y}}$ and $\hat{\mathbf{Y}}_{(i)}$ in the n-dimensional observation space; or
    b. the squared distance from $\mathbf{B}$ to $\mathbf{B}_{(i)}$ relative to the fixed geometry of $\mathbf{X'X}$.

In identifying the most influential cases, D_i performs similarly to DFFITS_i (Note 10), which has an interpretation related to point (a) above. The size-adjusted cutoffs

$$|\text{DFFITS}_i| > 2\sqrt{K/n}$$

and

$$D_i > 4/n$$

should flag essentially the same cases as "influential."

$D_i > 1$ corresponds to distances between $\mathbf{B}$ and $\mathbf{B}_{(i)}$ beyond a 50% confidence region.

2. Bounded range means that there exists some real $\Phi < \infty$ such that $|Y_i| < \Phi$ for all i. See Achen (1982) for a proof of OLS consistency and for additional arguments that OLS remains useful well beyond the scope of the Gauss-Markov assumptions.
3. The notation $Z \sim N(\mu, \sigma^2)$ signifies that "Z is distributed normally, with mean μ and variance σ^2."
4. Scatterplots work poorly with variables that take on only a few discrete values. For example, in Figure 4.16, the plot at left shows a regression of household head's education on whether he/she is retired. Although regression finds a significant, negative relationship (retirees have lower mean education), we see no evidence of it in the scatterplot. Retirement is a dummy variable with only two values (0 or 1), and education must be a whole number from 6 to 20. Only 25 distinct combinations of education and retirement occur here, so all 496 cases plot at only 25 points.

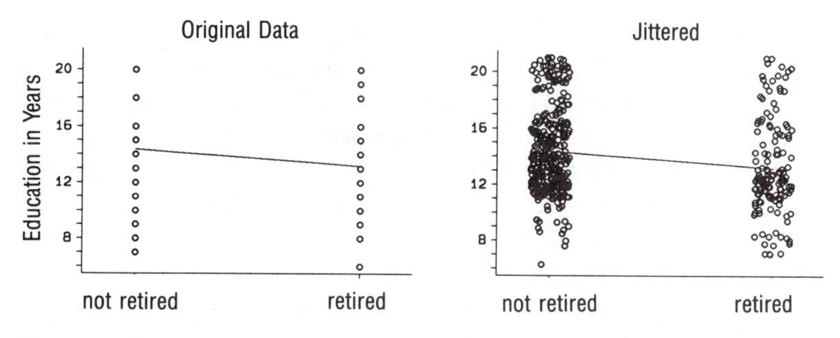

Figure 4.16 Scatterplot and corresponding jittered plot.

Jittering (at right) reduces overplotting by adding spherical random variation to each data point. We can then see that the nonretired distribution has a higher center of gravity; the regression line makes graphical sense. Jittering does not change the actual data—only their appearance in a graph. It often improves readability of scatterplots for discrete data. Jittering is also used in Figures 4.1, 8.13, and 8.17 but not elsewhere in this book.

5. Negatively autocorrelated residuals are much less common. To test for significant *negative* autocorrelation, replace d by $4 - d$ and apply the same decision rules using Table A4.4.
6. Techniques that *are* appropriate for analyzing the Concord water data include a type known as *autoregressive integrated moving average* (ARIMA) models. ARIMA modeling makes heavy use of correlograms to help identify the best models and evaluate their fit. For an introduction to ARIMA modeling, see McCleary, Hay, Meidinger, and McDowall (1980).
7. See Velleman (1982) and also Velleman and Hoaglin (1981) for more detail, including elaborate compound smoothers.
8. For example, the MCOV procedure of the Regression and Time Series (RATS) statistical package.
9. Spatial autocorrelation, which involves at least two dimensions, is more

complicated than the one-dimensional autocorrelation of time series. Durbin-Watson and similar tests are designed for time series and may fail to detect spatial autocorrelation.

10. A related statistic called DFFITS measures the ith case's influence on $\hat{Y}_i$:

$$\text{DFFITS}_i = \frac{\hat{Y}_i - \hat{Y}_{(i)i}}{s_{e(i)}/\sqrt{h_i}}$$

Belsley et al. (1980) suggest the size-adjusted cutoff:

Case i is relatively influential if $|\text{DFFITS}_i| > 2\sqrt{K/n}$

Computational formulas allow calculation of DFBETAS_{jk} and DFFITS_j without actually deleting the ith case and redoing the regression. See Belsley et al. for details.

11. The "Y" variable in this plot consists of residuals from the regression of 1981 water use (Y) on: 1980 water use (X_2), education (X_3), retirement (X_4), number of people (X_5), and change in number of people (X_6). The "X" variable consists of residuals from the regression of household income (X_1) on X_2 through X_6.

12. Proportional leverage plots can be constructed with a computer program (such as Stata) that will draw weighted scatterplots. First, calculate DFBETAS for each case. Next transform them:

$$\text{DFBETAS}^*_{ik} = (99/18)|\text{DFBETAS}_{ik}|(|\text{DFBETAS}_{ik}| + 1)^2 + 1 \quad \text{if} \quad |\text{DFBETAS}_{ik}| \leq 2$$

$$\text{DFBETAS}^*_{ik} = 100 \quad \text{if} \quad |\text{DFBETAS}_{ik}| > 2$$

The transformation rescales the absolute values of DFBETAS to range from 1 to 100:

if $\text{DFBETAS}_{ik} = 0$ (deleting case i has no impact on b_k), then $\text{DFBETAS}^*_{ik} = 1$; and

if $|\text{DFBETAS}_{ik}| \geq 2$ (deleting case i shifts b by two standard errors or more), $\text{DFBETAS}^*_{ik} = 100$.

Finally, construct a leverage plot using DFBETAS^*_{ik} values as weights, so that the area of each circle is proportional to DFBETAS^*_{ik}. Cases with $|\text{DFBETAS}_{ik}| \geq 2$ will appear as circles with 100 times the area of cases with $\text{DFBETAS}_{ik} = 0$ and about four times the area of cases with $|\text{DFBETAS}_{ik}| = 1$.

Household #134 has the largest $|\text{DFBETAS}_{i1}|$, 1.34, which transforms to $\text{DFBETAS}^*_{134,1} = 41.36$. Consequently, the largest circle in Figure 4.11 has about 41.36 times the area of the smallest circle.

13. This example, like DFBETAS, refers to influence on estimated regression coefficients. Although "good" outliers by definition do not affect regression coefficients, they still affect standard errors, hypothesis tests, R^2, and other

statistics. The only way to assess the full impact of suspicious cases is to redo the analysis with and without them—and closely compare the details, not just regression slopes.

14. The statistic $s_{e(i)}$ is also used in defining DFBETAS ([4.9]). It may be calculated without actually deleting the ith case and redoing the regression:

$$s_{e(i)}^2 = \frac{n - K}{n - K - 1} s_e^2 - \frac{e_i^2}{(n - K - 1)(1 - h_i)}$$

15. For examples, see Weisberg (1980).
16. The *Bonferroni inequality* states that the probability of at least one of a set of events occurring cannot be more than the sum of their individual probabilities.

    The individual probabilities of Type I error, using studentized residuals to test whether each case is an outlier, equal the nominal significance level, α. If we test n cases, then, according to the Bonferroni inequality, the probability that at least one of them appears to be a "significant" outlier (despite a true null hypothesis) cannot be more than $n\alpha$.

    We wish to run n tests (calculate studentized residuals for every case) yet have an *overall* Type I error risk of α. This can be done by testing each residual individually at the $\alpha' = \alpha/n$ level. Over n tests, the probability of at least one Type I error at level α' is $n\alpha' = \alpha$. For example, to test whether any case in a sample of $n = 20$ is a significant outlier at $\alpha = .05$, we simply check whether the sample's $\max|t_i|$ is significant at $\alpha' = .05/20 = .0025$.
17. Computer programs only approximate the theoretical P-values, however, and these approximations become less accurate for very small probabilities. Different software or hardware may obtain different results, so the answer produced by your system should be taken with a grain of salt. We may be confident that a P-value is very small (enough to reject H_0) without necessarily believing that we know precisely how small.
18. Cook and Weisberg (1982) offer two geometrical interpretations of D_i:

    a. the Euclidean distance between $\hat{\mathbf{Y}}$ and $\hat{\mathbf{Y}}_{(i)}$ in the n-dimensional observation space; or

    b. the squared distance from $\mathbf{B}$ to $\mathbf{B}_{(i)}$ relative to the fixed geometry of $\mathbf{X'X}$.

    In identifying the most influential cases, D_i performs similarly to DFFITS_i (Note 10), which has an interpretation related to point (a) above. The size-adjusted cutoffs

    $$|\text{DFFITS}_i| > 2\sqrt{K/n}$$

    and

    $$D_i > 4/n$$

    should flag essentially the same cases as "influential."

    $D_i > 1$ corresponds to distances between $\mathbf{B}$ and $\mathbf{B}_{(i)}$ beyond a 50% confidence region.

19. Less than 10^{-8} or .00000001. Most statistical programs will calculate a seemingly more precise P-value, but these are actually just approximations (Note 17).
20. For example, see Rawlings (1988).
21. The high leverage of 1981 reflects the fact that this case has an exceptional *combination* of X_1 and X_2 values: for such a late year, cultivated area is unusually low.

5 Fitting Curves

Chapters 3 and 4 described linear models, which specify straight-line relations between variables. Such models provide simple, useful approximations for many real-world phenomena. Other phenomena, however, cannot realistically be modeled with straight lines. This chapter introduces three ways to fit curves to data: exploratory band regression, curvilinear regression, and nonlinear regression.

Exploratory regression methods attempt to reveal unexpected patterns, so they are ideal for a first look at the data. Unlike other regression techniques, they do not require that we specify a particular model beforehand. Thus exploratory techniques warn against mistakenly fitting a linear model when the relation is curved, a waxing curve when the relation is *S*-shaped, and so forth.

Chapter 2 gave a bivariate example of *curvilinear regression* (Figure 2.17). We specify models with *intrinsically linear* form:

$$Y_i^* = \beta_0 + \beta_1 X_{i1}^* + \beta_2 X_{i2}^* + \cdots + \beta_{K-1} X_{i,K-1}^* + \varepsilon_i \qquad [5.1]$$

but Y^* and $X_1^* \cdots X_{K-1}^*$ could be any functions of Y and $X_1 \cdots X_{K-1}$, for instance, power transformations. If Y^* and/or X_k^* represents a nonlinear transformation, then Equation [5.1] describes a curvilinear relation between Y and X_k. Such curvilinear regression extends linear methods to fit a variety of curves without requiring any new statistical tools.

Models that cannot be expressed in the intrinsically linear form of [5.1] call for *nonlinear regression.* Nonlinear regression programs estimate parameters in a series of steps called *iterations*, each step improving on the previous step's estimates. This chapter ends with a brief, nonmathematical introduction to nonlinear regression.

Exploratory Band Regression

Band regression, a comparatively simple approach with many variants, begins by dividing the scatterplot into a series of vertical bands. We then find central points, perhaps median$\{X, Y\}$, within each band, and we use these points to define a line or curve.[1] Figure 4.4 in Chapter 4 employed band regression as a diagnostic tool. Any systematic trend in median absolute residuals signals heteroscedasticity; band regression can reveal such trends, without requiring us beforehand to specify their shape. More broadly, band regression provides a useful tool for exploring patterns in any two-variable scatter.

Figure 5.1 and Table 5.1 illustrate band regression, using data on the ratio of chromium to iron (Cr/Fe) in sediments sampled from Great Bay, New Hampshire

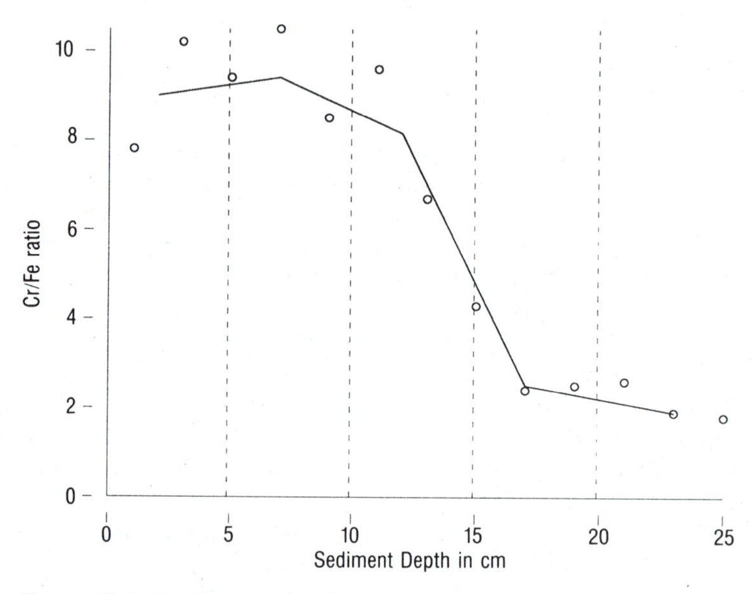

Figure 5.1 Exploratory band regression curve (5 bands) based on cross-medians from Table 5.1

Table 5.1 Cross-medians for exploratory regression with five bands; Ratio of chromium (Cr) to iron (Fe) in Great Bay sediments.[1]

Sample	Depth (cm)	Cr/Fe Ratio	Median{X, Y}
1	1	7.8	{2,9}
2	3	10.2	
3	5	9.4	{7,9.4}
4	7	10.5	
5	9	8.5	
6	11	9.6	{12,8.15}
7	13	6.7	
8	15	4.3	{17,2.5}
9	17	2.4	
10	19	2.5	
11	21	2.6	{23, 1.9}
12	23	1.9	
13	25	1.8	

[1] Data from Hurd (1986).

(Hurd, 1986). Much of the chromium originated from a leather tannery that for 30 years had dumped contaminated water into a river flowing into Great Bay. This dumping ended in the late 1960s, by which time chromium had spread widely through the river and bay. The rate of iron accumulation depends on the overall rate of sedimentation; the ratio of chromium to iron should therefore reflect the rate of pollution, controlling for the natural rate of sedimentation.

For Figure 5.1, the X (sediment depth) range was divided into 5 vertical bands (0–4.99 cm, 5–9.99 cm, and so forth) and median{X, Y} points obtained within each band. Line segments connecting these cross-medians reveal a nonlinear relation between Cr/Fe ratio and depth. Since deeper sediments are older, we could read this graph historically from right to left. Sediments below 17 cm apparently reflect stable background (pretannery) Cr/Fe levels. The increasing contamination at shallower depths (15–11 cm) corresponds to the increasing chromium discharge and buildup during the tannery's years of operation. The tannery's closing may explain the relatively lower Cr/Fe value in top (1-cm) sediments. Upper (0–12-cm) sediments experience much mixing by marine organisms (bioturbation), so measured Cr/Fe ratios vary erratically around the regression curve at these depths.

For basic descriptive purposes, analysis could stop with a graph like Figure 5.1. For other purposes, such as theoretical-model testing or comparisons with other areas, researchers may wish to fit a more formal parametric model, as described in the remainder of this chapter. Band regression provides guidance for parametric model fitting. It shows the general shape of the curve we need to fit, which simplifies the search for a workable model.

Regression with Transformed Variables

Chapter 1 introduced power transformations as a way to pull in outliers and make skewed distributions more symmetrical, perhaps more normal, and hence easier to analyze. Chapter 2 noted that transformations sometimes reduce heteroscedasticity. More importantly, a curvilinear relation can often be linearized by transforming Y and/or X. Thus transformations provide simple ways both to fix statistical problems and to fit curves to data (curvilinear regression).

An overriding reason for using transformations in regression is to linearize relationships. We may have to sacrifice other goals (homoscedasticity, normality, and so on) to achieve linearity. Often, however, the same transformation achieves more than one goal. Partly this works because *linear relationships require similar-shaped distributions.* For instance, a symmetrically distributed Y variable cannot be linearly related (with i.i.d. errors) to a skewed X variable, or vice versa. Also, heteroscedasticity commonly accompanies positively skewed X variables; power transformations with $q < 1$ may reduce both.

Variable transformation adds first and last steps to data analysis:

1. Select appropriate transformations, and generate transformed variables.
2. Analyze the data as usual (using regression, graphs, or whatever), but with the transformed variables.
3. If necessary, reexpress results in natural units for substantive interpretation.

An "appropriate transformation" does not always exist. Some relations cannot be linearized, or a transformation that reduces one problem may worsen others. Experimenting with the ladder of powers (Chapter 1) provides a quick impression of how various powers alter the data at hand.

Transformation may simplify analysis, but it adds an extra step to interpretation. Analytical results refer to the unfamiliar units of transformed variables, like the .3 power of income dollars. For substantive understanding, we want to describe these results in the more familiar terms of the variables' natural units (for example, dollars).

Curvilinear Regression Models

Because they are simple, linear relations make a good starting point—unless prior knowledge or the data suggest that a specific nonlinear form is more suitable. For example, theory might predict that Y increases as X goes from low to middle values, then decreases again as X goes from middle to high. Or a scatterplot might reveal a pattern of initial steep climb, followed by leveling off. In such instances our choice of model is guided by the shape of the desired curve.

Curvilinear regression permits many different curves. Figure 5.2 shows examples of semilog curves: relations with either Y or X (but not both) expressed in logarithms. At left are two curves with form

$$\hat{Y}_i = b_0 + b_1 \log X_i \qquad [5.2]$$

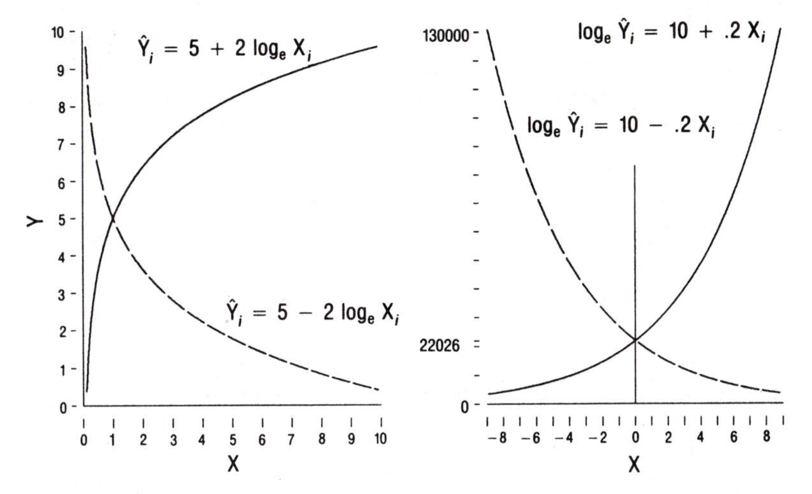

Figure 5.2 Examples of semilog curves.

Obtain coefficient estimates for [5.2] by regressing Y on the logarithm of X (either natural or base 10). If $b_1 > 0$, the curve rises quickly at first, then more slowly. If $b_1 < 0$, the curve falls steeply at first, then more slowly.

At right in Figure 5.2 are two curves with form

$$\log \hat{Y}_i = b_0 + b_1 X_i \qquad [5.3]$$

For $b_1 > 0$, such curves steepen to the right in a pattern called *exponential growth*; for $b_1 < 0$, they exhibit *exponential decay* (see box).

Figure 5.3 depicts several log-log curves:

$$\log \hat{Y}_i = b_0 + b_1 \log X_i \qquad [5.4]$$

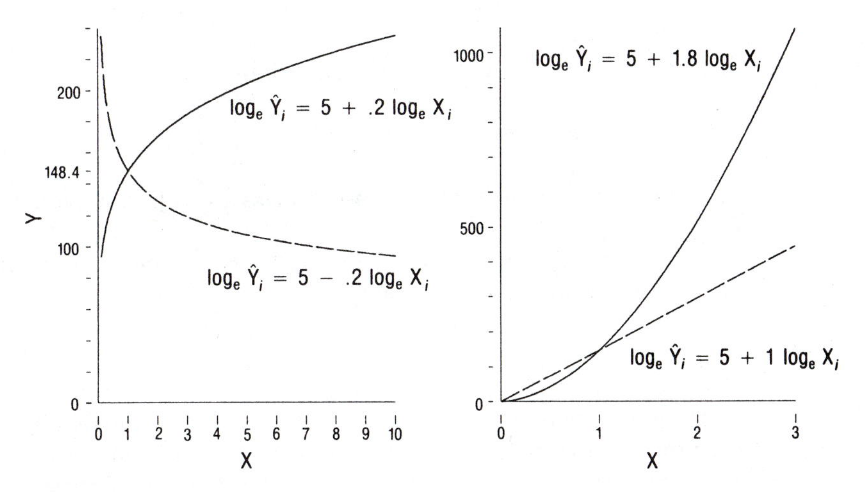

Figure 5.3 Examples of log-log curves.

Since both Y and X are transformed the same way, such relationships are linear if $b_1 = 1$. $b_1 > 1$ implies the steeper-to-right shape of a *waxing exponential* curve. $0 < b_1 < 1$ implies the flatter-to-right shape of a *waning exponential*. Negative b_1 values result in down-to-right versions of these curves.

Logarithms have special mathematical properties. They change multiplication into addition:

$$\log(cX) = \log c + \log X \qquad [5.5]$$

and change raising to a power into multiplication:

$$\log(X^b) = b \log X \qquad [5.6]$$

Similarly, logs transform division into subtraction and taking of roots into division.

These properties can linearize some seemingly nonlinear models. For example, the model

$$Y_i = \alpha e^{\beta X_i} \varepsilon_i \qquad [5.7a]$$

becomes linear if we take natural logarithms of both sides:

$$\log_e Y_i = \log_e \alpha + \beta X_i + \log_e \varepsilon_i \qquad [5.7b]$$

Equation [5.7] thus simplifies to a semilog model like [5.3]; we estimate its parameters by OLS, with b_0 estimating $\log \alpha$.

The exponential model

$$Y_i = \alpha X_i^{\beta} \varepsilon_i \qquad [5.8a]$$

likewise becomes linear if we take logarithms:

$$\log Y_i = \log \alpha + \beta \log X_i + \log \varepsilon_i \qquad [5.8b]$$

Equation [5.8] simplifies to the log-log form of [5.4], which can be estimated using OLS.

Note that [5.7a] and [5.8a] both assume multiplicative errors. If we believe additive errors instead—for example,

$$Y_i = \alpha e^{\beta X_i} + \varepsilon_i$$

then our exponential model cannot be linearized or estimated by OLS. Instead, we need the nonlinear regression methods described later in this chapter.

We obtain other curves by mixing different transformations. Figure 5.4 illustrates log-reciprocal curves:

$$\log \hat{Y}_i = b_0 + b_1 X_i^{-1}$$
$$= b_0 + \frac{b_1}{X_i} \qquad [5.9]$$

The curves in Figure 5.4 approach Y limits of e^{b_0}; $e^{.1} \approx 1.105$ at left, and $e^{.5} \approx 1.649$ at right. (Log-reciprocal curves using base 10 logarithms instead of natural logarithms approach 10^{b_0} instead of e^{b_0}.) The $b_1 > 0$ curve slopes *down*, and the $b_1 < 0$ curve (which is somewhat S shaped) slopes *up*. These directions reflect the fact that the X^{-1} transformation (like any negative-power transformation) *reverses the order* of the original data: high X values become low X^{-1} values. To avoid possible confusion, we could employ $-(X^{-1})$ rather than X^{-1}. This has the same statistical effect but retains the original order.

Figures 5.2–5.4 depict *monotonic* curves: as X increases, $\hat{Y}$ changes in just one direction, either up or down. *Nonmonotonic* curves reverse direction, from ascending to descending or vice versa. *Polynomial regression* provides a simple way to fit nonmonotonic curves. Polynomials involve more than one power of X. Figure 5.5 shows two *second-order polynomials*, of the general form

$$\hat{Y}_i = b_0 + b_1 X_i + b_2 X_i^2 \qquad [5.10]$$

We estimate coefficients in [5.10] by regressing Y on X and X^2. If $b_1 > 0$ and $b_2 < 0$, the curve ascends first and then descends (left in Figure 5.5). The opposite happens with $b_1 < 0$ and $b_2 > 0$, producing a U shape (right in Figure 5.5). Second-order polynomials can change direction once, so they have a single maximum or

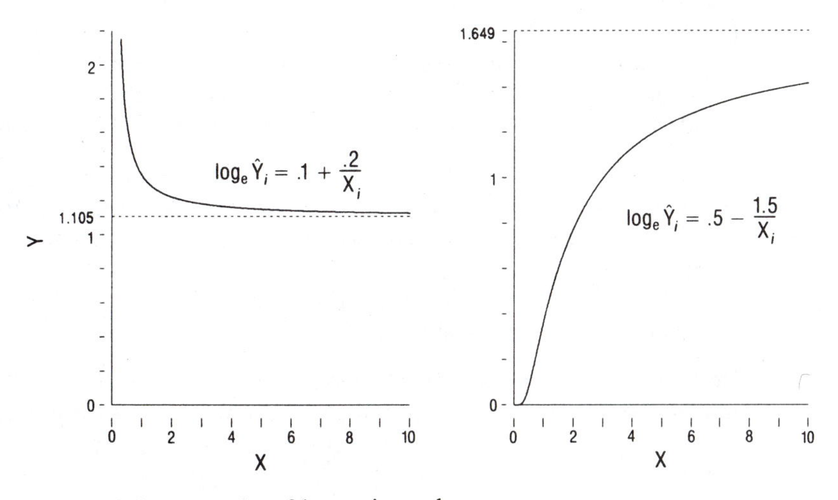

Figure 5.4 Examples of log-reciprocal curves.

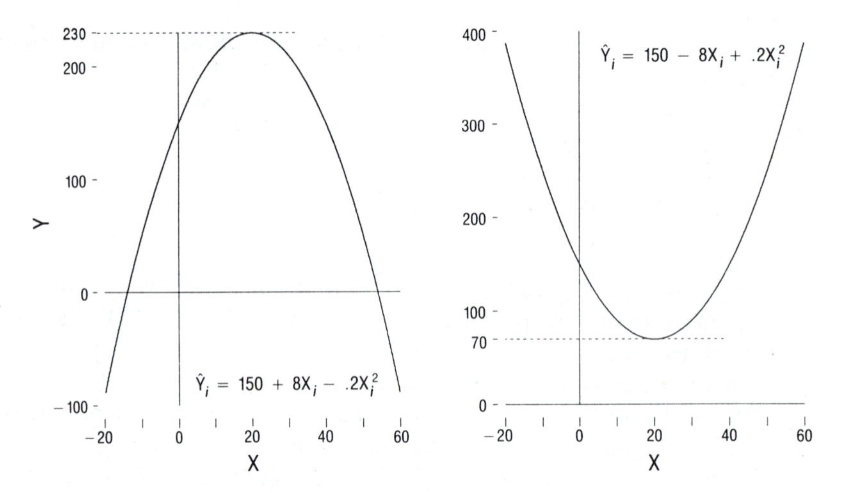

Figure 5.5 Two second-order polynomials.

minimum. Predicted Y reaches its maximum or minimum ($\hat{Y}_{\mathrm{m}}$) when $X = X_{\mathrm{m}}$:

$$X_{\mathrm{m}} = -\frac{b_1}{2b_2} \tag{5.11}$$

We find $\hat{Y}_{\mathrm{m}}$ by substituting X_{m} into [5.10].

Third-order polynomials, which can reverse direction twice, include first, second, and third powers of X:

$$\hat{Y}_i = b_0 + b_1X_i + b_2X_i^2 + b_3X_i^3 \tag{5.12}$$

If $b_1 > 0$, $b_2 < 0$, and $b_3 > 0$, the curve may first rise, then fall, then rise again as X increases (left in Figure 5.6). At right is a fall-rise-fall pattern, produced by $b_1 < 0$, $b_2 > 0$, and $b_3 < 0$.

Higher-order polynomials permit curves with any number of reversals. Although such curves may closely fit the data, they are hard to theoretically justify or understand. Furthermore, polynomial regression encounters two statistical problems:

1. *Influence*—transformation by powers greater than 1 may increase positive skew and create outliers. Large positive or negative X values lead to extremely large X^2 or X^3 values. Polynomial regression curves tend to fit large-X (high-leverage) cases closely.
2. *Multicollinearity*—high correlations among various powers of X can produce unstable coefficient estimates.

Both problems increase the likelihood of sample-specific results. A polynomial model may fit one sample well but provide a weak basis for generalization.

Centering, or subtracting the mean of X, can reduce multicollinearity among powers of X:

$$X^* = X - \bar{X}$$

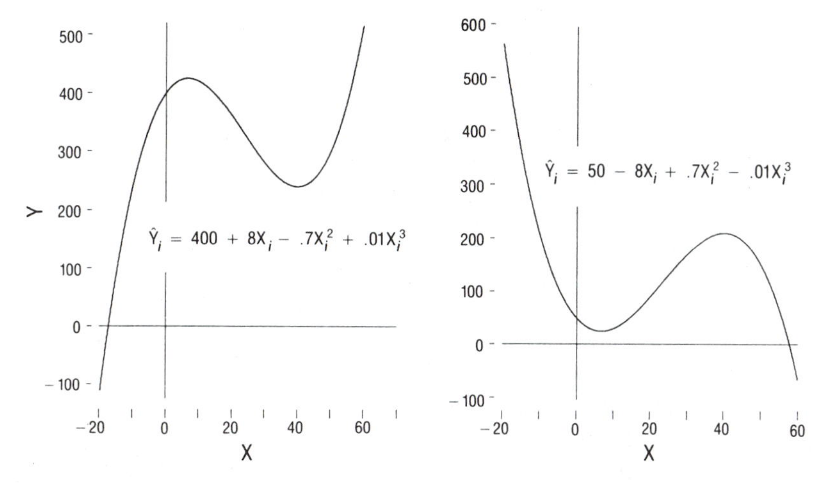

Figure 5.6 Two third-order polynomials.

Then we use $X^*, X^{*2}, X^{*3}, \ldots$ as right-hand-side variables, instead of X, X^2, and X^3. This changes the coefficients on all but the highest-power X term, without affecting predicted values or the shape of the curve. Standard errors (for all but the highest-power term) change too, so the new estimates, despite lower collinearity, are not necessarily more precise.

Figures 5.2–5.6 show relations between Y and one X variable. Additional X variables complicate analysis, but the basic ideas of these graphs still hold. Consider the multiple regression equation

$$\log \hat{Y}_i = b_0 + b_1 \log X_{i1} + b_2(-(X_{i2}^{-1})) + b_3 X_{i3}^2$$

At any given levels of X_2 *and* X_3, the relation between Y and X_1 reduces to a simple log-log model like Equation [5.4]:

$$\log \hat{Y}_i = c + b_1 \log X_{i1}$$

where c is a constant equal to $b_0 + b_2(-(X_{i2}^{-1})) + b_3 X_{i3}^2$. The relation between Y and X_2 likewise reduces to

$$\log \hat{Y}_i = d + b_2(-(X_{i2}^{-1}))$$

(where $d = b_0 + b_1 \log X_{i1} + b_3 X_{i3}^2$), and that between Y and X_3 reduces to

$$\log \hat{Y}_i = f + b_3 X_{i3}^2$$

($f = b_0 + b_1 \log X_{i1} + b_2(-(X_{i2}^{-1}))$), for any given levels of the other two X variables. Conditional plots (described later in this chapter) use such reductions graphically.

Choosing Transformations

Theory or scatterplots may dictate the initial choice of transformations. If not, a reasonable starting point is to seek transformations that symmetrize each variable's univariate distribution. Considering variables in the water-use regression, we have:

Y (postshortage water use): made symmetrical and approximately normal by the $Y^* = Y^{.3}$ transformation (Figure 1.17).
X_1 (income): $X_1^* = X_1^{.3}$ proves again to be the best transformation (Figure 2.14).
X_2 (preshortage water use): $X_2^* = X_2^{.3}$ works best.
X_3 (education): the distribution is asymmetrical but in a complicated way (multiple peaks and gaps but no outliers) not readily improved by transformations.
X_4 (retirement): transformations do not change the shape of a dummy variable's distribution.
X_5 (number of people): positively skewed, more so than income or water use. $X_5^* = \log_e X_5$ pulls in the upper tail.

We earlier used change in the number of people living in the household (X_6) as our sixth X variable. Let:

X_5 = number of people resident in summer 1981

X_0 = number of people resident in summer 1980

Then

$$X_6 = X_5 - X_0$$

But both X_5 and X_0 are positively skewed, and X_6 consequently is heavy tailed. A better-behaved measure is change in the logarithm of the number of people:

$$X_7^* = \log_e X_5 - \log_e X_0$$
$$= \log_e\left(\frac{X_5}{X_0}\right)$$

Note that this is not the same as $\log_e X_6 = \log_e(X_5 - X_0)$. Our new variable is based not on $X_5 - X_0$, absolute change in people, but on X_5/X_0, *proportional* change in people.

These transformations produce a new set of variables, employed in the regression analysis of Table 5.2. The model is intrinsically linear:

$$\hat{Y}_i^* = 1.856 + .516X_{i1}^* + .626X_{i2}^* - .036X_{i3} + .101X_{i4} + .715X_{i5}^* + .916X_{i7}^* \quad [5.13a]$$

Table 5.2 Curvilinear regression–water-use regression with transformed variables

Source	SS	df	MS	
Model	1310.1171	6	218.35285	Number of obs = 496
Residual	509.636645	489	1.04220173	$F(6, 489)$ = 209.51
				Prob > F = 0.0000
Total	1819.75374	495	3.67627019	R-square = 0.7199
				Adj R-square = 0.7165
				Root MSE = 1.0209

Variable	Coefficient	Std. Error	t	Prob > $\|t\|$	Mean
wtr81_3					9.776982
inc_3	.51572	.1297219	3.976	0.000	2.474998
wtr80_3	.6255023	.0290827	21.508	0.000	10.29697
educat	−.0361339	.0160111	−2.257	0.024	14.00403
retire	.1013897	.1189905	0.852	0.395	.2943548
logpeop	.7146849	.1104854	6.469	0.000	.9750739
clogpeop	.9156944	.2627408	3.485	0.001	−.0187373
_cons	1.856262	.3849294	4.822	0.000	1

Regression coefficients, R^2, hypothesis tests, and so forth have their usual meanings *but refer to the transformed variables.* For example, we see that the logarithm of the number of people (X_5^*) has a significant effect on the .3 power of water use (Y^*). Each one-unit increase in log(number of people) leads to a .715-unit increase in predicted .3 power of water use, other things being equal. In [5.13a] the effect of log(proportional change in people), X_7^*, is significant, but retirement, X_4, is not. The six X variables together explain about 72% of the variance of the .3 power of water use.[2]

These interpretations sound awkward. In terms of the original variables, relations are not linear:

$$\hat{Y}_i^{.3} = 1.856 + .516X_{i1}^{.3} + .626X_{i2}^{.3} - .036X_{i3} + .101X_{i4} + .715\log_e X_{i5} + .916\log_e X_{i7} \qquad [5.13b]$$

Later sections describe a graphical approach to understanding such equations. First we'll return to some problems noticed in Chapter 4, to see whether transformations have improved them.

Evaluating Consequences of Transformation

Criticism uncovered several distributional problems with the raw-data water-use regression (Chapter 4):

1. mild curvilinearity
2. heteroscedasticity
3. nonnormal residuals
4. influential cases

To see how transformations affect these problems, we can subject the transformed-variables regression ([5.13]) to similar critical scrutiny.

An e-versus-$\hat{Y}$ plot for the raw-data regression, at top in Figure 5.7, reveals heavy-tailed errors, heteroscedasticity, and influential cases. (Point sizes in Figure 5.7 are proportional to the influence statistic Cook's D.[3]) In contrast, the e-versus-$\hat{Y}$ plot for the transformed-variables regression (at bottom in Figure 5.7) shows lighter tails and no sign of heteroscedasticity or unduly influential cases. Unlike the raw-data residuals, which are in units of Y (cubic feet of water), the transformed-data residuals are in the cryptic units of Y^* (the .3 power of cubic feet of water).

Figure 5.8 takes a closer look at residuals from [5.13]. Their distribution is nearly symmetrical, as seen in the symmetry plot at lower left. (The mean roughly equals the median: $\bar{e} = 0$ and Md $= .03$.) It has slightly heavier-than-normal tails, however ($s_e = 1.02$, IQR/1.35 $= .85$). A few high and low outliers depart from the diagonal line in the quantile-normal plot at lower left in Figure 5.8. Otherwise, the residuals follow an approximately normal distribution, indicated by quantiles along the diagonal (normal-distribution) line.

Before transformation, the normal-errors and homoscedasticity assumptions were implausible (Figures 4.3, 4.4, and 4.9). Judging from Figures 5.7 and 5.8, however, we could reasonably assume normal errors and homoscedasticity with regard to the transformed-variables regression.

Two influential households account for about 40% of the magnitude of the coefficient on income in the raw-data regression (Figure 4.11). Another household

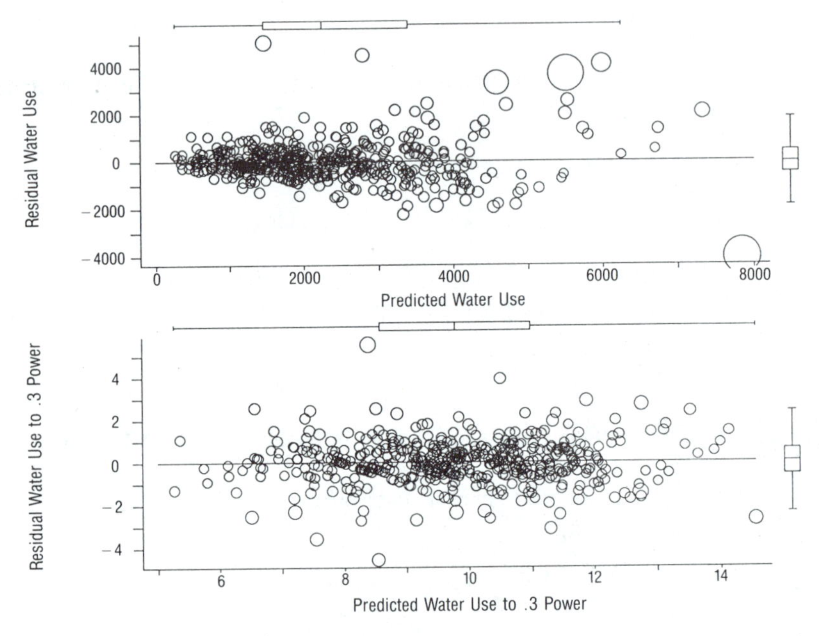

Figure 5.7 e-versus-$\hat{Y}$ plots with points proportional to scaled Cook's D, for raw-data (top) and transformed-variables (bottom) regressions.

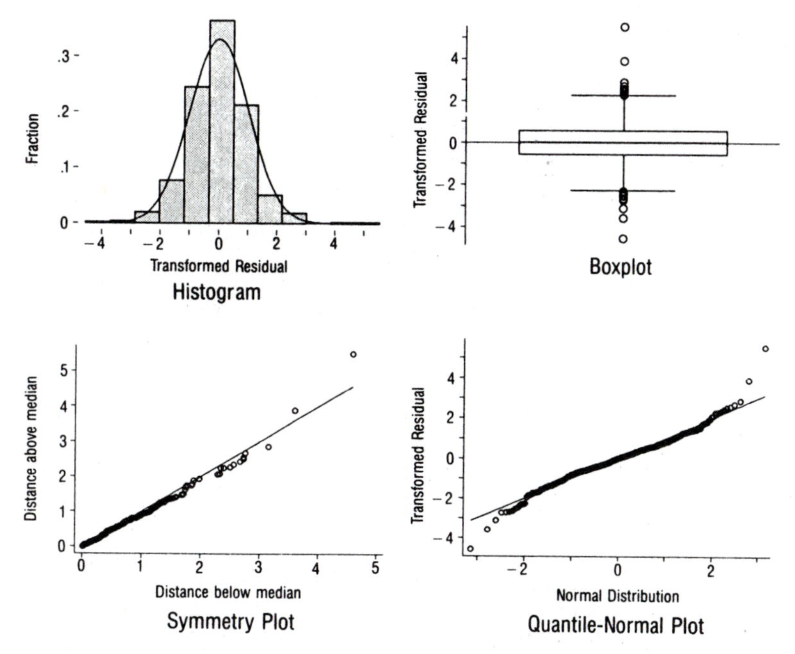

Figure 5.8 Distribution of residuals from transformed-variables regression (Table 5.2).

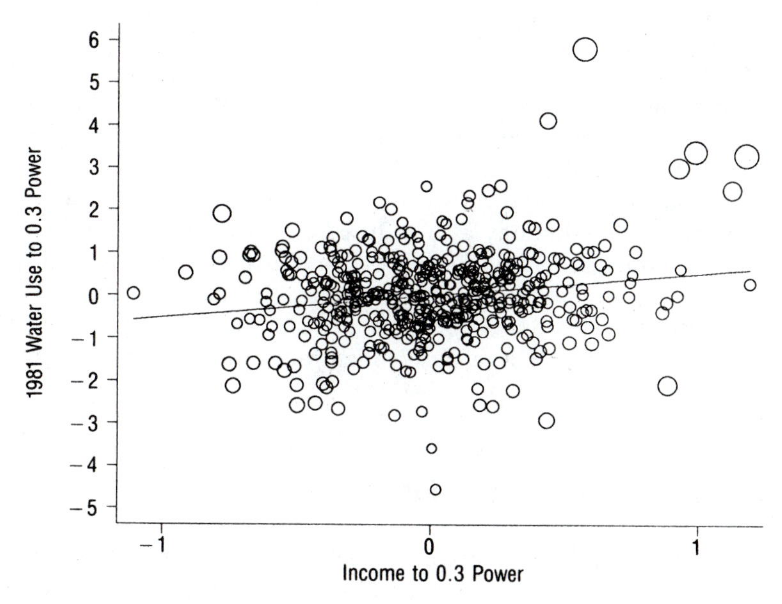

Figure 5.9 Proportional leverage plot for transformed-variables regression: 1981 water use versus income.

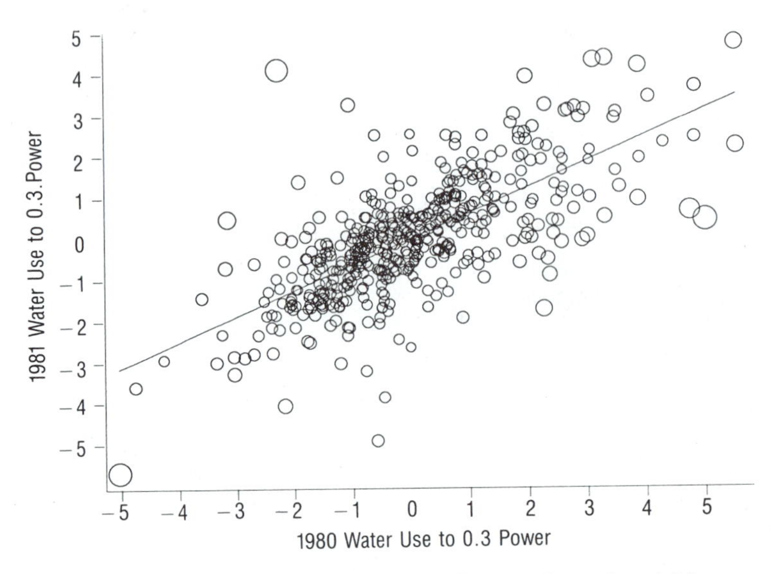

Figure 5.10 Proportional leverage plot for transformed-variables regression: 1981 water use versus 1980 water use.

held down the coefficient on preshortage water use (Figure 4.12). Influence problems are less acute for the transformed-variables regression, however. By pulling in outliers, the transformations create distributions without high-leverage or influential cases. No DFBETAS for [5.13] exceed $\pm.5$, and the largest Cook's D is .06. In contrast, the raw-data regression produced three DFBETAS beyond ± 1 and Cook's D values as high as .31. Proportional leverage plots confirm the absence of exceptionally influential cases (compare Figure 5.9 with 4.11 and Figure 5.10 with 4.12).

These analyses show that transformations originally chosen to symmetrize univariate distributions also reduced heteroscedasticity, nonnormality, and influence problems in multiple regression. Furthermore, they address the problem of curvilinearity, as seen in the following section. We cannot always find transformations that solve so many problems at once, but this simple approach works surprisingly often. A drawback is that transformed-variables regressions like [5.13] are harder to understand.

Conditional Effect Plots

A *conditional effect plot* graphs predicted values of Y against one X variable, with the other X variables held constant. This process visualizes the meaning of curvilinear equations like [5.13].

The regression equation is

$$\hat{Y}_i^* = 1.856 + .516X_{i1}^* + .626X_{i2}^* - .036X_{i3} + .101X_{i4} + .715X_{i5}^* + .916X_{i7}^*$$

with Y^*, X_1^*, and so on indicating transformed variables. Means of the six X variables are:

$$\bar{X}_1^* = 2.475$$
$$\bar{X}_2^* = 10.297$$
$$\bar{X}_3 = 14.004$$
$$\bar{X}_4 = .294$$
$$\bar{X}_5^* = .975$$
$$\bar{X}_7^* = -.019$$

The relationship between Y^* and X_1^*, when the other X variables equal their means, is

$$\begin{aligned}\hat{Y}_i^* &= 1.856 + .516X_{i1}^* + .626(10.297) - .036(14.004) \\ &\quad + .101(.294) + .715(.975) + .916(-.019) \\ &= 8.507 + .516X_{i1}^* \end{aligned} \qquad [5.14a]$$

or, in terms of the original variables,

$$\hat{Y}_i^{.3} = 8.507 + .516X_{i1}^{.3} \qquad [5.14b]$$

A different set of X_2^*–X_7^* *values would change only the intercept* in [5.14].

For any level of X_1 (income), Equation [5.14] implies a corresponding predicted value $\hat{Y}^*$. To express Y^* in the original units of Y, we apply an *inverse transformation* (Chapter 2). Since the original transformation raised Y to the .3 power ($Y^* = Y^{.3}$), the appropriate inverse transformation raises $\hat{Y}^*$ to the 1/.3 power:

$$\hat{Y} = \hat{Y}^{*\frac{1}{.3}}$$

For example, suppose a household earned \$20,000 ($X_1 = 20$, so $X_1^* = 20^{.3} = 2.456$) and was exactly average in all other respects. From [5.14a] we predict its transformed water use to be

$$\begin{aligned}\hat{Y}^* &= 8.507 + .516(2.456) \\ &= 9.774\end{aligned}$$

If the predicted *.3 power of water use* is 9.774, then predicted *water use* is $9.774^{1/.3} = 1996.4$ cubic feet. Similar calculations obtain $\hat{Y}$ values for other values of X_1. Connecting a set of such $(X_1, \hat{Y})$ points produces a conditional effect plot, Figure 5.11.

Although the relation between Y^* and X_1^* is linear ([5.14a]), the relation between Y and X_1 ([5.14b]) is curved. Other things being equal, the higher the income, the

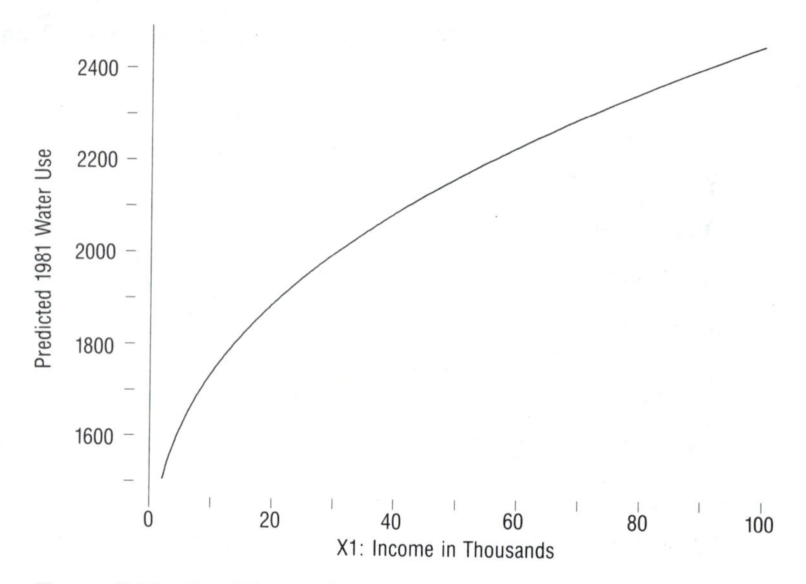

Figure 5.11 Conditional effect plot showing curvilinear relation between 1981 water use and income, with other X variables at means.

higher the postshortage use—and hence the *lower* the water conservation. Postshortage water use goes up most steeply as income increases from low to moderate. The impact diminishes as income further increases from moderate to high.

This finding makes substantive sense. A household with $30,000 income probably has a green lawn and all the usual middle-class appliances. Poorer households with $10,000 income may lack such luxuries and so use less water. On the other hand, households with $100,000 and $80,000 incomes may not differ as greatly. Thus a $20,000 difference matters more among lower-income households than among higher-income households.

Figure 5.12 gives three views of the Y–X_1 relationship. The middle curve is the same as in Figure 5.11. At top and bottom are curves obtained when other X variables equal their extremes. The top curve's equation is

$$\begin{aligned}\hat{Y}_i^* &= 1.856 + .516X_{i1}^* + .626(17.027) - .036(6)\\ &\quad + .101(1) + .715(2.3026) + .916(0)\\ &= 14.046 + .516X_{i1}^*\end{aligned}$$

The bottom curve is

$$\begin{aligned}\hat{Y}_i^* &= 1.856 + .516X_{i1}^* + .626(4.9013) - .036(20)\\ &\quad + .101(0) + .715(0) + .916(0)\\ &= 4.204 + .516X_{i1}^*\end{aligned}$$

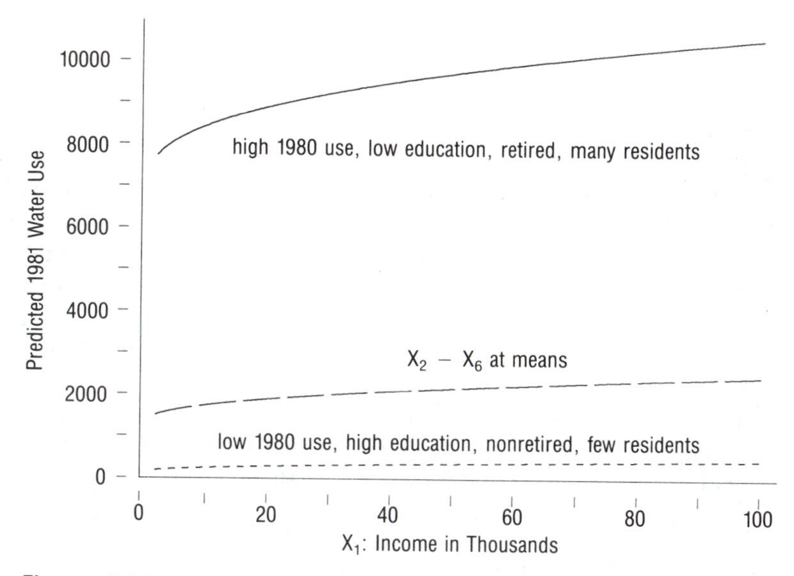

Figure 5.12 Conditional effect plot with three levels of other X variables.

Income has a much greater effect on water use for households with high preshortage water use, low education, retired heads, and many residents (top curve). Under the opposite conditions, the water-use/income relationship is nearly flat (bottom curve). Equations for the three curves in Figure 5.12 have identical slopes and differ only in their intercepts, but the intercepts affect the steepness of the curves.

Interactions are X-variable effects that depend on the levels of other X variables (Chapter 3). Transforming Y makes all effects interactions, which complicates interpretation. Conditional effect plots with curves for several levels of the other X variables (like Figure 5.12) help clarify what an interaction model implies.

Comparing Effects

The middle curve in Figure 5.12 is the same one shown in Figure 5.11, but they look different because the Y scales have changed. Individual graphs are most readable if their scales extend no farther than needed (Figure 5.11). On the other hand, comparisons are easier between graphs with the same Y scale.

Figure 5.13 shows conditional plots for the six predictors of [5.13]. For each plot, five of the six X variables are set equal to their respective means, so the equations collapse to approximately:

$$\hat{Y}_i^{.3} = 8.507 + .516X_{i1}^{.3}$$

$$\hat{Y}_i^{.3} = 3.338 + .626X_{i2}^{.3}$$

$$\hat{Y}_i^{.3} = 10.288 - .036X_{i3}$$

$$\hat{Y}_i^{.3} = 9.755 + .101X_{i4}$$

$$\hat{Y}_i^{.3} = 9.087 + .715\log_e X_{i5}$$

$$\hat{Y}_i^{.3} = 9.802 + .916\log_e X_{i7}$$

All six plots in Figure 5.13 have the same Y scale. X scales span each variable's range.

Outliers control range, which depends only on the high and low extremes. We may get a more stable picture by looking at how $\hat{Y}$ changes toward the middle of X's range, rather than from one extreme to the other. Vertical lines in Figure 5.13 mark the 10th and 90th percentiles of each X, enclosing the middle 80%. We can visually judge the strength of each X variable's effect by comparing the movement of curves over this middle range of X values.

The top-center plot shows that, as preshortage water use increases from its 10th to 90th percentile, with other variables at their means, predicted postshortage water use increases by about 3800 cubic feet. In contrast, it rises only about 1000 cubic feet across the full range of income, and 400 cubic feet from income's 10th to 90th percentiles. Education and retirement have relatively weak effects. The number of people in most households did not change, so 10th and 90th percentile of proportionate change (lower right) are the same—1.

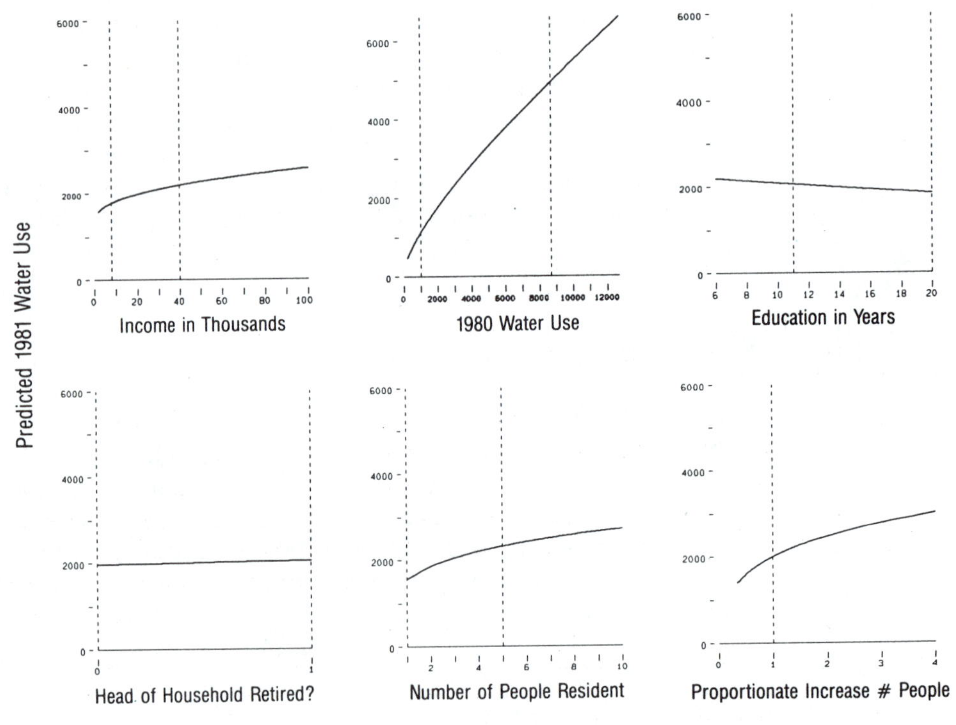

Figure 5.13 Conditional effect plots for X variables of Equation [5.13], each with other X variables at means.

Conditional effect plots with the same Y scales provide one way to answer the question "Which effects are stronger?" They can serve this purpose for linear, curvilinear, or nonlinear models.

Nonlinear Models

The preceding sections have focused on *curvilinear* regression, a relatively simple extension of linear models and statistical techniques. *Nonlinear* regression departs more substantially from linear methods and requires its own models and techniques. Two basic arguments for nonlinear modeling are:

empirical—the data suggest a specific nonlinear model;
mechanistic—a nonlinear equation expresses our understanding of the mechanism causing changes in Y.

The distinction between empirical and mechanistic modeling sometimes gets blurred, as when mechanistic explanations are proposed after a model has been empirically chosen.

We have many nonlinear models to choose from. This section describes just a few of the possibilities. Further examples and discussion appear in a good introduction by Rawlings (1988) or in the more comprehensive presentation by Seber and Wild (1989). Other texts include Bates and Watts (1988), Draper and Smith (1981), and Gallant (1987).

Nonlinear *exponential* models have the form

$$Y_i = \alpha e^{\beta X_i} + \varepsilon_i \qquad [5.15]$$

where α and β are parameters, ε is an error term, and $e = 2.71828\ldots$, the base number for natural logarithms. Often the X variable is time. α equals Y-intercept or initial level, and β controls rate of change in steepness. If $\beta > 0$, [5.15] describes exponential growth that starts slowly, then accelerates. Figure 5.14 depicts an exponential growth curve going up to right:

$$\hat{Y}_i = 4e^{.02X_i} \qquad [5.16]$$

If $\beta < 0$, we get exponential decay, like the down-to-right curve in Figure 5.14:

$$\hat{Y}_i = 25e^{-.03X_i} \qquad [5.17]$$

As noted earlier, curves similar to Figure 5.14 (see right graph in Figure 5.2) result from an intrinsically linear (semilog) exponential model assuming multiplicative errors.

Their explosive, unlimited increase makes exponential growth models unrealistic over the long run, although many physical, natural, and social phenomena exhibit short-term exponential growth. Better long-term models generally include

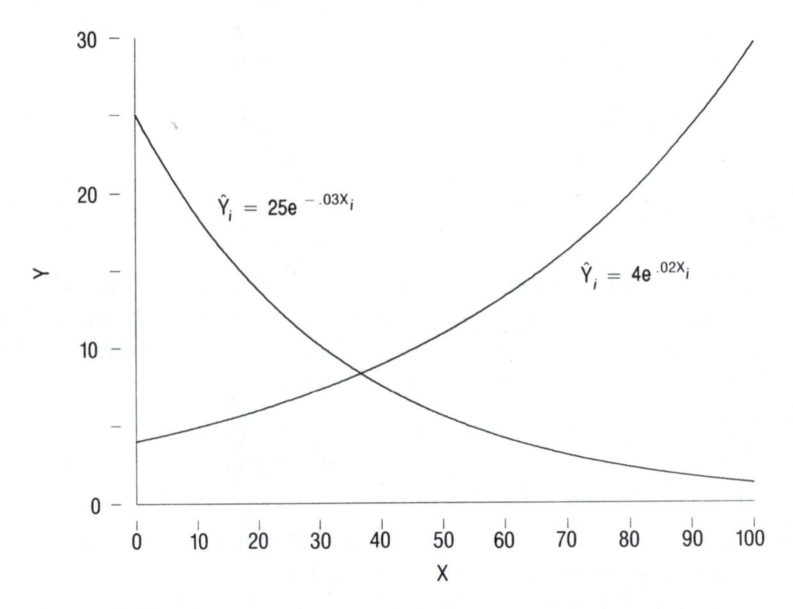

Figure 5.14 Exponential growth and decay curves.

some limitation on growth. For example, *negative exponential* curves rise rapidly at first, then level off as they approach an asymptotic limit α. Rate of change in steepness is controlled by β:

$$Y_i = \alpha(1 - e^{-\beta X_i}) + \varepsilon_i \qquad [5.18]$$

Figure 5.15 shows two negative exponential curves approaching upper limits $\alpha = 10$:

$$\hat{Y}_i = 10(1 - e^{-.02X_i})$$

$$\hat{Y}_i = 10(1 - e^{-.07X_i})$$

The $\beta = -.07$ curve is notably steeper than the $\beta = -.02$. Exercise 17 at the end of this chapter employs Equation [5.18] to model a predator/prey system.[4]

Two-term exponential curves rise steeply from zero, then fall again to approach zero gradually. Two parameters, θ_1 and θ_2, control details of shape:

$$Y_i = \frac{\theta_1}{\theta_1 - \theta_2}(e^{-\theta_2 X_i} - e^{-\theta_1 X_i}) + \varepsilon_i \qquad [5.19]$$

Figure 5.16 shows two two-term exponential curves; the higher one has $\theta_1 = .05$ and $\theta_2 = .04$:

$$\hat{Y}_i = \frac{.05}{.05 - .04}(e^{-.04X_i} - e^{-.05X_i})$$

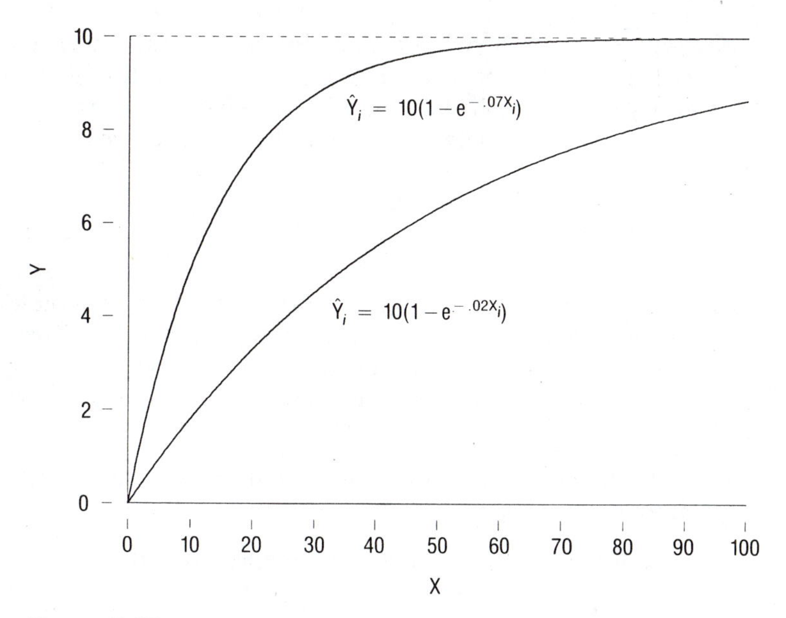

Figure 5.15 Negative exponential curves.

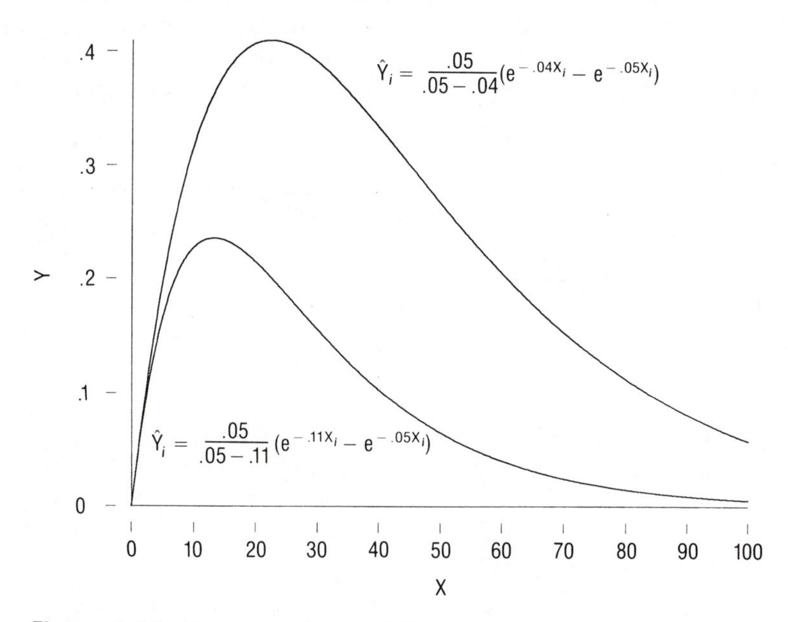

Figure 5.16 Two-term exponential curves.

Growth rates for many processes increase at first, then decrease as limits approach. Such behavior, combining a start like the up-to-right curve in Figure 5.14 with an asymptotic finish like Figure 5.15, produces an *S*-shaped or *sigmoid* growth curve. The best-known sigmoid model is the *logistic* curve:

$$Y_i = \frac{\alpha}{1 + \gamma e^{-\beta X_i}} + \varepsilon_i \qquad [5.20]$$

When $X = 0$, $\hat{Y} = \alpha/(1 + \gamma)$. Logistic curves approach asymptotes of α as X approaches $+\infty$, and they approach 0 as X approaches $-\infty$. They provide models for phenomena as diverse as the growth of yeast in a bottle and the spread of rumors. Figure 5.17 shows three curves, illustrating effects of the three logistic parameters:

α is the asymptotic upper limit;
γ determines horizontal position or "takeoff point"; and
β controls steepness.

Gompertz curves describe asymmetric *S*-shaped growth:

$$Y_i = \alpha e^{-\gamma e^{-\beta X_i}} + \varepsilon_i \qquad [5.21]$$

As with logistic curves, the α parameter in [5.21] sets the asymptotic upper limit, γ determines horizontal position, and β controls steepness. Gompertz curves cross the Y-axis at height $\alpha e^{-\gamma}$. Figure 5.18 shows three examples. Unlike perfectly symmetrical logistic curves (Figure 5.17), Gompertz curves (Figure 5.18) accelerate and decelerate at different rates.

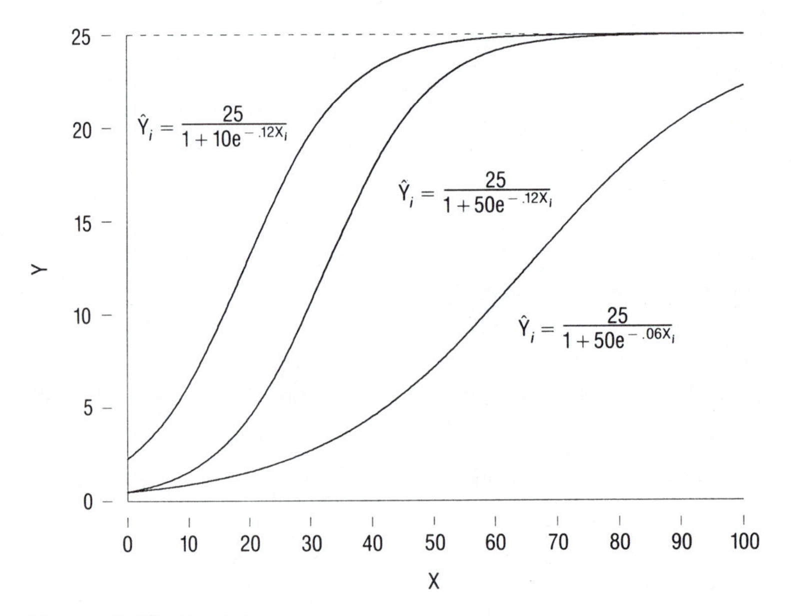

Figure 5.17 Logistic curves.

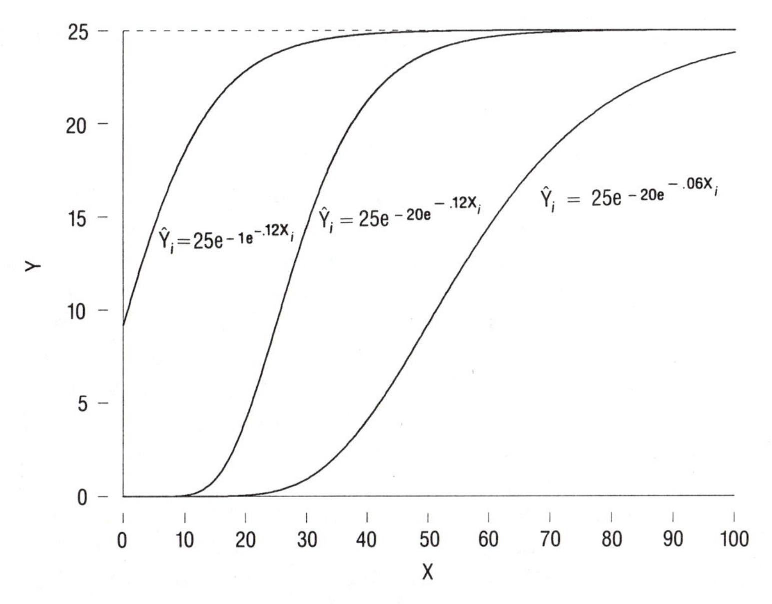

Figure 5.18 Gompertz curves.

Estimating Nonlinear Models

The fit of nonlinear models is commonly judged by the same least squares criterion used with linear regression: we seek parameter estimates that minimize the residual sum of squares, RSS $= \Sigma(Y_i - \hat{Y}_i)^2$. Unlike linear regression, however, the equations for nonlinear models generally cannot be solved to yield unique parameter estimates that guarantee minimum RSS. Instead, we resort to computer programs that estimate parameters *iteratively*, through a series of gradually improved guesses.

There exist several ways to do this, but none of them foolproof. They begin with a set of initial parameter estimates, often user supplied, then try to find better estimates that reduce RSS. One common method, called *modified Gauss-Newton*, uses linear approximations to estimate how RSS changes with small shifts from the present set of parameter estimates.[5] Problems with any method can arise when the RSS surface is complex, having more than one local minimum. Then the iterations can lead to a "solution" that is far from optimal, if not physically impossible.

Good initial estimates, close to the correct parameter values, improve the prospects for a successful nonlinear fit. Sometimes theory or previous experience with similar data suggests appropriate initial estimates. At other times, analytical work is needed. For example, initial estimates for a nonlinear exponential model (Equation [5.15]) could be obtained from a curvilinear semilog regression (Equation [5.3]). Scatterplots supply useful information, such as initial estimates of α (upper limit) for negative exponential, logistic, or Gompertz curves. If the Y-intercept for a logistic or Gompertz curve appears measurably different from zero, we can use it together with our estimated α to obtain an initial estimate of γ:

logistic: $\gamma = \alpha/(Y\text{-intercept}) - 1$;
Gompertz: $\gamma = \log_e \alpha - \log_e(Y\text{-intercept})$.

Some software packages make it easy to seek reasonable initial estimates graphically. On the same graph we plot the scatter of the data and a connected series of points defined by applying our nonlinear model (with arbitrary initial parameter estimates) to a realistic range of X values. By varying the parameter estimates (as in Figures 5.17 and 5.18) and redrawing the graph, we learn what changes shift the model closer to the data. This graphical approach also provides an intuitive demonstration of what the nonlinear model parameters mean.

Table 5.3 lists British census data on the percentage of women with at least one child. By age 20, 17% of the women born in 1945 had at least one child; by age 25, 60% of the 1945 cohort had at least one child; and so forth. (The 1987 source for these data of course had no information on women who were born in 1945 and were more than 40 years old.) Within each cohort, graphs of percentage versus age suggest asymmetric, Gompertz-type S-curves (Equation [5.21]).

Table 5.4 shows modified Gauss-Newton iterations fitting a Gompertz model to the 1945 cohort, based on initial values taken from the final estimates for the 1940 cohort: $\alpha = 89$, $\gamma = 942$, and $\beta = .31$. (I obtained initial values for fitting 1940 from a 1930 fit, and 1930 from 1920; 1920 initial values were found graphically as described above.) Each iteration refines the three parameter estimates and reduces the residual sum of squares. Iterations stop when change no longer brings much improvement; we then say that the process has *converged*. The specific stopping rule can be adjusted by the user. Final estimates for the 1945 cohort yield the model

$$\hat{Y}_i = 90.4e^{-468.1e^{-.28X_i}} \qquad [5.22]$$

Figure 5.19 shows a near-perfect fit between model and data ($R^2 \approx 1$).

Table 5.3 Percentage of women with at least one child, by women's age and year of birth (England and Wales)[1]

Women's Age	Women's Year of Birth 1920	1930	1940	1945	1950	1955	1960
15	0%	0%	0%	0%	0%	0%	0%
20	7	9	13	17	19	18	13
25	39	48	59	60	53	45	39
30	67	75	82	82	75	68	—
35	76	83	87	88	83	—	—
40	78	86	89	90	—	—	—
45	—	86	89	—	—	—	—

[1] Data from Office of Population Censuses and Surveys, 1987.

Table 5.4 Iterative least squares (modified Gauss-Newton method) fitting of Gompertz curve to 1945 cohort data from Table 5.3

Model: *cohort45* = alpha*exp(−gamma*exp(−beta**age*))

Initial value for alpha: 89
Initial value for gamma: 942
Initial value for beta: .31

Iteration n.	alpha	gamma	beta
1	90.387712	235.94488	.2797197
2	91.108602	154.92555	.23153695
3	89.950566	314.85895	.27864886
4	90.611953	359.54244	.26978088
5	90.402229	453.66425	.28162348
6	90.427092	466.77297	.2815985
7	90.425316	468.06244	.28170342
8	90.425334	468.05911	.28170291
9	90.425334	468.05914	.28170291
10	90.425334	468.05914	.28170291

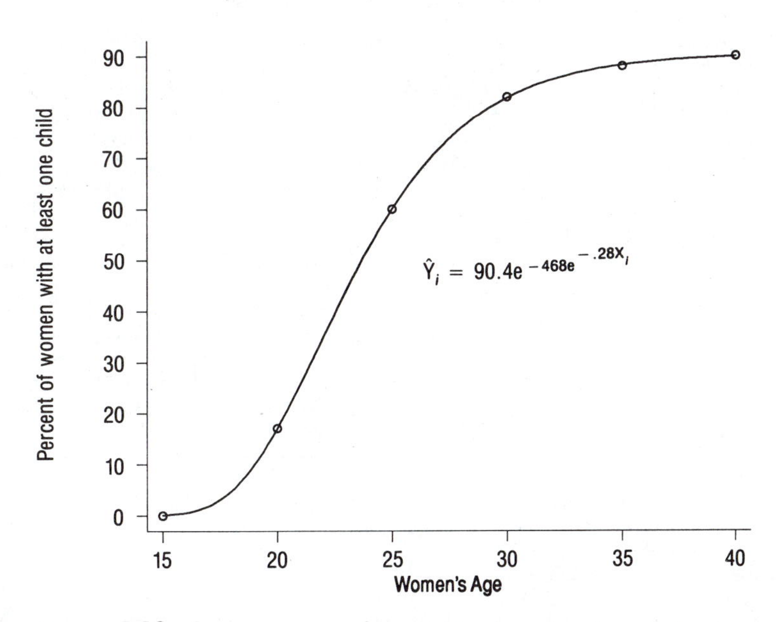

Figure 5.19 Gompertz curve fit to 1945 cohort data from Table 5.3.

Table 5.5 contains typical nonlinear-regression output, including the following sums of squares:

$$\text{Model SS} = \Sigma \hat{Y}_i^2 \quad [5.23]$$

$$\text{Residual SS} = \Sigma (Y_i - \hat{Y}_i)^2 \quad [5.24]$$

$$\text{Total SS} = \Sigma Y_i^2 \quad [5.25]$$

$$= \text{Model SS} + \text{Residual SS}$$

$$\text{Corrected Total SS} = \Sigma (Y_i - \bar{Y})^2 \quad [5.26]$$

Equations [5.24] and [5.26] correspond to OLS residual and total sums of squares, respectively (Equations [2.7] and [1.2]). For nonlinear regression, we define R^2 as

$$R^2 = 1 - \left(\frac{\text{Residual SS}}{\text{Corrected Total SS}} \right) \quad [5.27]$$

Table 5.5 Results from nonlinear regression fitting Gompertz curve to 1945 cohort data (Tables 5.3 and 5.4)

n of iterations: 10

Model: *cohort45* = 90.425334*exp(−468.05914*exp(−.28170291*age*))

Residual Statistics:
Residual Average = −.01438449 Std. Dev. = .15308897
Skewness = −.16997136 Kurtosis = 1.9255805

Variation	d.f.	SS	MS
Model	3	160379.95	53459.983
Residual	3	.11842265	.03947422
Total	6	160380.07	26730.011
Corr Total	5	7528.8333	1505.7667

$R^2 = .9999$

			Asymptotic 95% c.i.	
Parameter	**Estimate**	**Asymptotic Std. Error**	**lower**	**upper**
alpha	90.425334	.16068386	89.91404	90.93663
gamma	468.05914	22.592213	396.1707	539.9476
beta	.28170291	.00222295	.2746295	.2887763

Asymptotic Correlations among Parameters

	alpha	gamma	beta
alpha	1.0000		
gamma	−0.5872	1.0000	
beta	−0.6343	0.9927	1.0000

Equation [5.27] could, in principle, yield negative values if the model fits worse than the mean does.

Table 5.5 also gives skewness and kurtosis of the residuals, which help assess their normality (a normal distribution has skewness = 0 and kurtosis = 3, by the definitions used here). Nonlinear regression, like OLS, generally assumes normal i.i.d. errors.

Standard errors, 95% confidence intervals, and correlations among parameter estimates in Table 5.5 derive from asymptotic (large-sample) theory. With so small a sample ($n = 6$), they provide only rough approximations. Approximate confidence intervals use the t-distribution:

$$\text{parameter estimate} \pm t_{n-K} \times \text{standard error}$$

where K equals the number of parameters estimated. Appropriate null hypotheses depend on the specific parameter's role in a nonlinear model, so confidence intervals provide more useful information than the standard regression t-tests. For example, if $\beta = 1$ in a Gompertz model:

$$\hat{Y}_i = \alpha e^{-\gamma e^{-\beta X_i}}$$

then a simpler model with only two parameters would fit equally well. The 95% confidence interval for β in Table 5.5 does not include 1, so we can reject H_0: $\beta = 1$ at the 5% significance level.

Very high correlations among parameters may suggest that a model is unnecessarily complex. Although one high estimate appears in Table 5.5 ($r_{\gamma\beta} = .9927$), the confidence intervals, near-perfect fit, and overall success of the Gompertz model (demonstrated further in Table 5.6) all argue against the effort of exploring simpler alternatives here.

Interpretation

We should fit a formal nonlinear model, rather than just drawing an exploratory graph like Figure 5.1, if we are interested in any of the following:

1. evaluating how well a theoretically derived (*mechanistic*) nonlinear model fits the data;
2. predicting Y within or beyond the X-range of the data; or
3. substantively or comparatively interpreting model parameters.

For purely empirical reasons, we choose Gompertz curves to model Table 5.3's fertility data. No mechanistic theory was proposed to explain why such data *should* follow a Gompertz growth curve. This example better illustrates the second and third purposes: prediction and parameter interpretation/comparison.

Table 5.6 lists parameter estimates for Gompertz curves fit to six cohorts from Table 5.3. These results summarize the way in which percentage of women with at

Table 5.6 Gompertz parameter estimates for fertility data (Table 5.3)

Women Born in:	Parameter Estimates α	γ	β
1920	79.8	461.2	.26
1930	86.5	538.0	.27
1940	89.1	942.0	.31
1945	90.4	468.1	.28
1950	87.5	144.9	.23
1955	88.9	60.3	.18

All $R^2 > .99$

least one child increases with age, for English and Welsh women born in 1920, 1930, 1940, 1945, 1950, or 1955. The curves fit almost perfectly (all $R^2 > .99$). Figure 5.20 graphs curves for the 1945, 1950, and 1955 cohorts over the age range of the data.

Parameter estimates in Table 5.6 tell a story of social change. Upper limits (α) predict the maximum percent of women with at least one child. This is lowest for women born in 1920 ($\alpha = 79.8$), who were in their twenties during World War II. It is highest for the generation born in 1945, whose peak childbearing years occurred during the better times of the 1960s and 1970s ($\alpha = 90.4$). For post-1945 cohorts, α declines again, although not drastically. Perhaps the decline reflects changing values

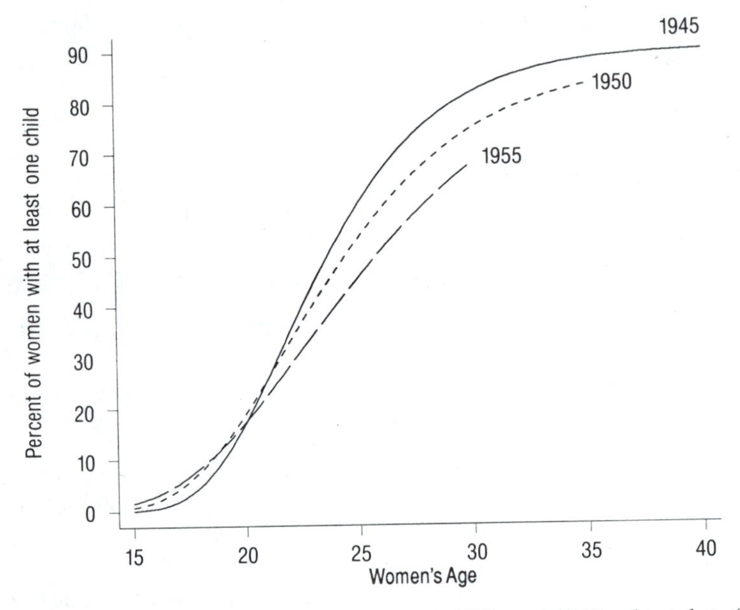

Figure 5.20 Gompertz curves for 1945, 1950, and 1955 cohort data (see Table 5.6).

regarding careers and families. As Figure 5.20 shows, we lack data on the actual upper limits of 1950 and 1955 cohorts. The predicted upper limits (1950 $\alpha = 87.5$; 1955 $\alpha = 88.9$) seem reasonable, however. Confidence in such out-of-sample predictions is strengthened if similar models perform well when the data *are* complete, as is clearly the case for the 1920–1940 cohorts.

The Gompertz β parameter affects the curve's steepness. Table 5.6 shows that the percentage of women with at least one child grew most steeply among the 1940 cohort ($\beta = .31$). In the 1950 and 1955 cohorts we see β substantially declining (1950 $\beta = .23$; 1955 $\beta = .18$), a change also pictured in Figure 5.20. Women born in postwar years did not rush so quickly into motherhood. For example, 60% of women born in 1945 had at least one child by the time they were 25 years old; only 39% of women born in 1955 had a child by this age. The β estimates in Table 5.6, declining since 1940, reflect a general trend toward postponed childbearing. As changing values, career opportunities, and economic pressures made early motherhood less attractive, women began having their first child later in life.

Conclusion

Data analysis typically begins with straight-line models because they are simplest, not because we believe reality is inherently linear. Theory or data may suggest otherwise, which would call for one or more of the curve-fitting methods discussed in this chapter.

Exploratory regression smooths the data, showing how the center of the Y distribution changes with the level of X. Unlike with parametric regression methods, we do not specify or estimate parameters for any particular model and make few assumptions about underlying distributions. Cross-medians resist the influence of outliers. Exploratory regression graphs can serve for data description or as precursors to more formal modeling.

Transformed-variables regression is implicitly curvilinear. We can choose transformations for the curves they allow, or the curves may be a byproduct of transformations chosen for other reasons. Transformations address several problems:

1. curvilinearity
2. influential cases
3. nonnormal errors
4. heteroscedasticity

With luck, we can simultaneously reduce more than one problem. The more severe the problems are, the less trustworthy the raw-data regression and the greater the need for transformations.

Nonlinear regression iteratively estimates parameters in models that cannot be transformed to linearity. The process requires more effort than OLS. First, the analyst must choose an appropriate model from the many that are available. Second, the estimation procedure often proves sensitive to initial values. It then falls to the analyst to specify appropriate initial values, not too far from the final values he or she is trying to estimate. Different computational procedures may also lead to different results. Finally, interpretation of nonlinear parameters may require careful thinking. Graphs showing the model with a variety of parameter values help both in guessing initial values and in interpreting final ones.

Exercises

Figure 5.1 graphed data on chromium pollution in New Hampshire's Great Bay. The pollution originated from a leather tannery along the Cocheco River. Further data on this region follow, from a study by Capuzzo and Anderson (1973). The authors proposed measuring the chromium accumulation as a new way to estimate the rate of natural sedimentation.

Cocheco River chromium contamination

Station	Distance from Discharge Point (km)	Highest Chromium Concentration (ppm)
1	2.6	221
2	5.6	77
3	4.4	72
4	5.1	56
5	7.2	44
6	9.9	56
7	10.8	47

Exercises 1–2 use the chromium data.

1. Regress chromium (Y) on distance (X), and graph the line on a scatterplot. Describe the results.
2. To fit the chromium data, we need a curve that drops steeply at first, then more slowly. We can achieve this through regressions of the form $\hat{Y} = b_0 + b_1X^*$, where X^* is a power transformation of X with $q < 1$. For example, we might regress Y on $\log X$ (see down-to-right curve at left in Figure 5.2) or on $-(X^{-1})$. Try out and graphically compare the success of $q = 0$ (log), $q = -2$, and $q = -3$ transformations in linearizing the Y/X^* relationship and in fitting a Y-versus-X (not X^*) scatterplot. Calculate the squared correlation between observed and predicted Y values for each curve. Does this confirm your visual impressions of which curve fits best?[6]

The next table lists deforestation rates for 50 nations: the percentage of their forests destroyed annually over 1981–1985 (from Council on Environmental Quality, 1985). Also listed are population growth rates (births minus deaths per 1000 people) and per capita gross national product (GNP), 1985. Exercises 3–4 use these data.

Population growth rate, GNP, and deforestation in 50 nations

Nation	X_1 Population Growth Rate	X_2 Per Capita GNP (dollars)	Y Annual Deforestation Rate (percent)
1 Angola	25.1	940	1.0
2 Benin	28.5	310	2.6
3 Bolivia	28.1	570	.2
4 Brazil	22.2	2240	.4
5 Burma	25.2	190	.3
6 Cambodia	25.9	120	.3
7 Cameroon	25.4	890	.4
8 C.A.R.	22.9	310	.1
9 Colombia	23.3	1460	1.7
10 Congo	25.9	1180	1.0
11 Costa Rica	26.3	1430	3.9
12 Cuba	10.4	710	.1
13 Ecuador	31.7	1350	2.3
14 Ethiopia	27.7	140	.2
15 Gabon	16.5	4000	.1
16 Ghana	32.4	360	.9
17 Guatemala	29.1	1130	2.0
18 Guinea	23.3	310	1.7
19 Guyana	22.6	670	0
20 Honduras	33.8	660	2.4
21 India	19.9	260	.2
22 Indonesia	17.7	580	.5
23 Ivory Coast	28.0	950	5.9
24 Kenya	41.1	390	.7
25 Laos	25.1	80	1.2
26 Liberia	31.5	490	2.2
27 Madagascar	27.9	320	1.2
28 Malaysia	22.9	1890	1.2
29 Mexico	26.8	2270	1.2
30 Mozambique	27.6	150	.8
31 Nepal	23.3	170	3.9
32 Nicaragua	34.5	920	2.7
33 Nigeria	33.3	860	4.0
34 Pakistan	27.4	380	.2
35 Panama	22.6	2120	.9
36 Papua N.G.	26.8	820	.1
37 Paraguay	28.8	1610	4.6
38 Peru	26.0	1310	.4
39 Philippines	25.4	820	.7
40 Somalia	25.2	290	.2
41 Sri Lanka	20.3	320	2.1
42 Sudan	28.5	440	.2
43 Tanzania	35.1	280	.4
44 Thailand	21.0	790	2.4
45 Togo	28.5	340	.7
46 Uganda	35.2	230	1.1
47 Venezuela	29.6	4140	.4
48 Vietnam	21.1	190	.6
49 Zaire	29.4	190	.2
50 Zambia	33.0	640	1.2

3. To improve symmetry and linearity, take natural logs of per capita GNP

$$X_2^* = \log_e X_2$$

and negative reciprocal roots of deforestation rate

$$Y^* = -(Y^{-.5})$$

Regress Y^* on X_1 and X_2^*. Construct a conditional effect plot showing the curvilinear relationship between deforestation and growth rate, with log GNP held at its median:

   a. Use your regression equation to find $\hat{Y}^*$ for each value of X_1, with X_2^* set at its median.
   b. Apply the appropriate inverse transformation, $\hat{Y} = (-\hat{Y}^*)^{-2}$, to obtain natural-units predicted values.
   c. Graph these $\hat{Y}$ values against X_1, and connect them with line segments or a smooth curve.

4. Construct a conditional effect plot showing the curvilinear relationship between deforestation and GNP, with growth rate held at its median:
   a. Use your regression equation to find $\hat{Y}^*$ for each value of X_2^*, with X_1 set at its median.
   b. Apply the appropriate inverse transformation.
   c. Graph these $\hat{Y}$ values against X_2 (not X_2^*), and connect them with line segments or curves.

Solar-energy scientists have observed a linear relation:

$$\hat{Y} = c + dX$$

where Y is the diffuse fraction of solar radiation and X is the ratio of global to extraterrestrial radiation.[7] For some time they believed that this relation is the same at any location; the values of c and d do not systematically change with latitude, for instance. Predictions of monthly average diffuse radiation, used for evaluating a location's solar power potential, were based upon this assumption of constant c and d. An article by Alfonso Soler (1990) questions this assumption. He reports the following estimates of c and d, from solar radiation experiments at 27 European locations. These data form the basis for Exercises 5–10.

Estimates of c and d at 27 European locations

	Location	Latitude	c	d
1	Bergen	60.4	1.224	−1.706
2	Lerwick	60.1	1.078	−1.140
3	Hamburg	53.6	1.043	−1.037
4	Cambridge	52.2	.937	−.841
5	Aberporth	52.1	1.064	−1.140
6	Valentia	51.9	.958	−.851

	Location	Latitude	c	d
7	London	51.5	.990	−1.103
8	Kew	51.5	.980	−1.026
9	Bracknell	51.3	.995	−.990
10	Trappes	48.8	1.004	−1.050
11	Locarno Monti	46.2	.886	−.897
12	Carpentras	44.1	.934	−.985
13	Uccle	50.8	.971	−.926
14	Palermo	38.1	1.049	−1.325
15	Macerata	43.3	.992	−1.034
16	Genova	44.4	.727	−.777
17	Madrid	40.5	.859	−.903
18	Oporto	41.1	1.035	−1.090
19	Coimbra	40.2	1.144	−1.400
20	Lisboa	38.7	1.013	−1.217
21	Wien	48.2	1.059	−1.193
22	Salzburg	47.8	.854	−.862
23	Bergen	60.4	1.040	−1.210
24	Aas	59.7	.970	−1.030
25	Athens	38.0	1.260	−1.410
26	Rodos	36.4	1.310	−1.530
27	Kythos	37.4	1.230	−1.390

5. Regress d on latitude. Does latitude explain much of the variance of d? Is its effect significant? Construct a residual versus predicted Y plot, and examine it for potential problems.
6. If we performed only a linear regression, we might conclude that d is unrelated to latitute. The e-versus-$\hat{Y}$ plot suggests otherwise, however. Regress d on latitude and latitude squared. Compare R^2 and F test for this second-order polynomial regression with those for your linear regression of Exercise 5. Construct a scatterplot, and draw the regression curve.
7. Use the regression equation from Exercise 6 to calculate the latitude at which predicted d is highest (apply Equation [5.11]). What is $\hat{d}_m$ (maximum $\hat{d}$)? Mark this point on your scatterplot.
8. The relation between latitude and latitude squared is obviously not linear. Transformed variables nonetheless often correlate strongly with their original values. In polynomial regressions, which include original and transformed versions of the same variable, this may cause multicollinearity problems. Graph latitude squared versus latitude, and determine their correlation. Compare standard errors for the coefficient on latitude in linear (Exercise 5) and polynomial (Exercise 6) regressions, and comment on the extent and consequences of collinearity here.
9. Center latitude by subtracting the mean from each case ($X_i^* = X_i - \bar{X}$).
   a. Graph X^{*2} versus X^*, and determine their correlation. Did centering reduce collinearity?
   b. Regress d on X^* and X^{*2}. Compare your results with the regression of d on X and X^2 (Exercise 6). How did centering affect coefficients? Standard errors? t-tests? Predicted Y? R^2?

10. Analyze the relationship between c and latitude, by the same steps followed in Exercises 5–8.
11. Bache, Serum, Youngs, and Lisk (1972; reprinted in Bates and Watts, 1988) report the following data on concentrations of PCBs (a cancer-causing pollutant) in trout from Lake Cayuga, New York. Fit the curvilinear model:

$$\log_e \hat{Y}_i = b_0 + b_1 X_i^{\frac{1}{3}}$$

where X is age and Y is PCB concentration. Graph and summarize your results.

PCB Concentrations in Lake Cayuga Trout

Trout	Age (years)	PCB Concentration (ppm)
1	1	.6
2	1	1.6
3	1	.5
4	1	1.2
5	2	2.0
6	2	1.3
7	2	2.5
8	3	2.2
9	3	2.4
10	3	1.2
11	4	3.5
12	4	4.1
13	4	5.1
14	5	5.7
15	6	3.4
16	6	9.7
17	6	8.6
18	7	4.0
19	7	5.5
20	7	10.5
21	8	17.5
22	8	13.4
23	8	4.5
24	9	30.4
25	11	12.4
26	12	13.4
27	12	26.2
28	12	7.4

12. Risk and Sammarco (1991) found that skeletal density of the coral *Porites lobata* increases with distance from the Australian shore, due to differences between inshore and offshore environments. They summarized this relation using a second-order polynomial:

$$\hat{Y}_i = b_0 + b_1 X_i + b_2 X_i^2$$

Is the coefficient on X^2 significantly different from zero? How well does this curve fit? Graph and discuss your results.

Density of Great Barrier Reef Coral Heads (*Porites lobata*)

Sample	Reef	Distance to Shore (km)	Coral Head Density (g/cm^3)
1	Middle Reef	3.5	1.337
2	Middle Reef	3.5	1.216
3	Middle Reef	3.5	1.309
4	Alma Bay	14.3	1.053
5	Alma Bay	14.3	1.082
6	Alma Bay	14.3	1.079
7	Orpheus Is.	15.4	1.236
8	Orpheus Is.	15.4	1.190
9	Orpheus Is.	15.4	1.299
10	Pandora Reef	15.9	1.246
11	Pandora Reef	15.9	1.298
12	Pandora Reef	15.9	1.301
13	Great Palm Is.	27.8	1.375
14	Great Palm Is.	27.8	1.384
15	Great Palm Is.	27.8	1.307
16	Morinda Shoals	29.6	1.436
17	Morinda Shoals	29.6	1.493
18	Morinda Shoals	29.6	1.469
19	Little Broadhurst	49.5	1.387
20	Little Broadhurst	49.5	1.437
21	Little Broadhurst	49.5	1.433
22	Bowden Reef	67.9	1.406
23	Bowden Reef	67.9	1.402
24	Bowden Reef	67.9	1.428
25	Grub Reef	74.5	1.437
26	Grub Reef	74.5	1.589
27	Grub Reef	74.5	1.461

13. Following are estimates of the hippopotamus population (1970 to 1983) along sections of Zambia's Luangwa River, from Tembo (1987). Regress natural logarithms of hippo population on year; graph and describe the resulting curve.

Year	Hippopotamus Population Estimate
1970	2815
1972	2919
1975	2342
1976	4501
1977	5147
1978	4765
1979	5151
1981	4884
1982	6293
1983	6544

14. If the hippopotamus trend seen in Exercise 13 continued, what population would we predict for the year 2000? The year 2100? What is wrong with these predictions?
15. Familiarize yourself with available nonlinear regression software by fitting Gompertz (or other) curves to the fertility data of Table 5.3. Try out a variety of initial values (including defaults, if the software supplies them), and comment on the results. Graph fitted curves in the manner of Figure 5.19.
16. DeBlois and Leggett (1991) set up experimental aquaria to study dynamics of a simple predator/prey system. They found that rates of predation by amphipods (a small crustacean) were nonlinearly related to the density of their prey, capelin eggs (capelin are small marine fish). For theoretical and empirical reasons the researchers chose a negative exponential model for the relation between predation rate (capelin eggs eaten per amphipod) and prey abundance (initial egg density).

    Estimate parameters for a negative exponential model (Equation [5.18]) fit to these data. Graph the results.

Invertebrate predators (Amphipods) and fish eggs (Capelin)[1]

Experiment	X: Initial Egg Density (eggs/cm^3)	Y: Capelin Eggs Eaten/ Amphipod
1	0	0.0
2	2	1.0
3	4	2.0
4	4	2.8
5	7	4.2
6	7	4.0
7	8	5.0
8	12	5.0
9	12	6.2
10	12	6.6
11	12	4.5
12	22	6.0
13	23	10.0
14	23	7.0
15	23	8.5
16	23	4.8
17	32	4.5
18	36	9.0
19	37	12.5
20	37	10.5
21	37	9.0
22	70	8.0
23	70	10.0
24	70	10.5
25	71	14.0
26	72	11.0

[1] Data estimated from Figure 4 of DeBlois and Leggett (1991).

Notes

1. Tukey (1977) proposed a straight-line version of band regression, based on cross-medians from three vertical bands. This approach, popularized by Velleman and Hoaglin (1981) (also see Emerson and Hoaglin in Hoaglin et al., 1983), has been called *Tukey regression* or the *resistant line*.

   Some authors have proposed similar mean-based band regression methods as a way of coping with measurement error (reviewed in Theil, 1971: 610–612). Tukey's approach seems better suited for protection against occasional large errors.

2. The usual R^2 or R_a^2 statistics *do not provide a basis for comparing regressions with differently transformed Y variables*, like those in Tables 3.2 and 5.2. We are explaining two different things: variance in water use (Table 3.2) and variance in the .3 power of water use (Table 5.2).

   To compare fit of linear and curvilinear OLS models, find the squared correlation between Y and $\hat{Y}$. For linear models, or if Y was not transformed, the squared correlation between Y and $\hat{Y}$ equals R^2. If Y was transformed, first apply an inverse transformation to $\hat{Y}^*$ to obtain $\hat{Y}$ values; correlate these with Y and square the result. Like R^2, $r_{Y\hat{Y}}^2$ measures the proportion of the variance of Y explained by the regression model.

   Nonlinear regression calls for a different approach (Equation [5.27]).

3. For such graphs it helps to rescale Cook's D:

$$D_i^* = (99/4)D_i(D_i + 1)^2 + 1 \qquad \text{if } D_i \leq 1$$

$$D_i^* = 100 \qquad \text{if } D_i > 1$$

   D^* theoretically ranges from 1 (if $D_i = 0$) to 100 (if $D_i \geq 1$). Weighting a scatterplot by D^* therefore could result in highly influential cases appearing with up to 100 times the area of noninfluential cases. Note 12 in Chapter 4 describes similarly motivated rescaling for DFBETAS.

   Figure 5.7 uses this rescaling. Since the largest D_i in the raw-data regression (top) is .31, the largest plotted point has about 14 times the area of the smallest.

4. A three-parameter equation known to chemists as the *monomolecular growth model* and to agricultural researchers as *Mitscherlich's law* also produces curves with general shapes resembling those in Figure 5.15:

$$Y_i = \alpha\{1 - e^{-\beta(X_i + \delta)}\} + \varepsilon_i$$

5. Rawlings (1988) gives a relatively nontechnical description of this and other estimation methods. See Seber and Wild (1989) for a more extended treatment.

6. See Note 2. The squared correlation between Y and $\hat{Y}$ provides a general summary curvilinear (OLS) fit.

   Since the $q = -3$ curve fits best in this example, we arrive at a model in which chromium contamination drops off with the inverse cube of distance from its source. Such models are common and are often theoretically expected in pollution-dispersal research.

7. Among solar scientists the customary notation for this equation is actually

$$\bar{H}_d/\bar{H} = c + d\bar{H}/\bar{H}_0$$

where $\bar{H}_d$, $\bar{H}$, and $\bar{H}_0$ are the monthly average daily diffuse, global, and extraterrestrial radiation.

6 Robust Regression

Under ideal conditions, OLS performs better than other regression methods. Under certain nonideal conditions, however, OLS breaks down. Careful diagnostic analysis (Chapter 4) helps to work around OLS weaknesses but demands some effort and sophistication. *Robust regression* is designed to perform well under a broader range of conditions than OLS. Researchers who skip diagnostic efforts would, in principle, often be safer using robust methods instead of OLS.

Although robust methods should particularly benefit unsophisticated researchers, until recently these methods were mainly the province of experts. Calculations for robust regression are more complicated than those for OLS, and few standard statistical packages provided them. Theoretical expositions are also less accessible. Robustness literature exhibits the controversies and continual updating typical of "cutting edge" research. Such challenges have slowed general acceptance, despite the obvious value of robustness.

OLS is the best unbiased estimator given normally distributed (Gaussian) errors. But what estimator is best when errors are not Gaussian? Unfortunately there is no simple answer. Nonnormality takes countless forms, and one estimator cannot be optimal for all of these possibilities. Robustness research traditionally narrowed the question by focusing on one common and damaging kind of nonnormality: heavy-tailed error distributions. More recently, research attention has widened to encompass leverage (X-outlier) problems too.

A Two-Variable Example

Figure 6.1 shows a scatterplot of death rates versus air pollution in 60 U.S. metropolitan areas.[1] It graphs age-adjusted mortality rates (Y) against the natural logarithm of hydrocarbon pollution potential (X_1).[2] OLS finds a weak positive relationship:

$$\hat{Y}_i = 918.4 + 7.97 \log_e X_{i1} \qquad [6.1]$$

$$\text{SE}_b: \quad 6.87$$

$$t: \quad 1.16$$

Equation [6.1] explains only about 2% of the variance in mortality rates ($R^2 = .02$). From this analysis we might conclude that hydrocarbon pollution has little impact on mortality.

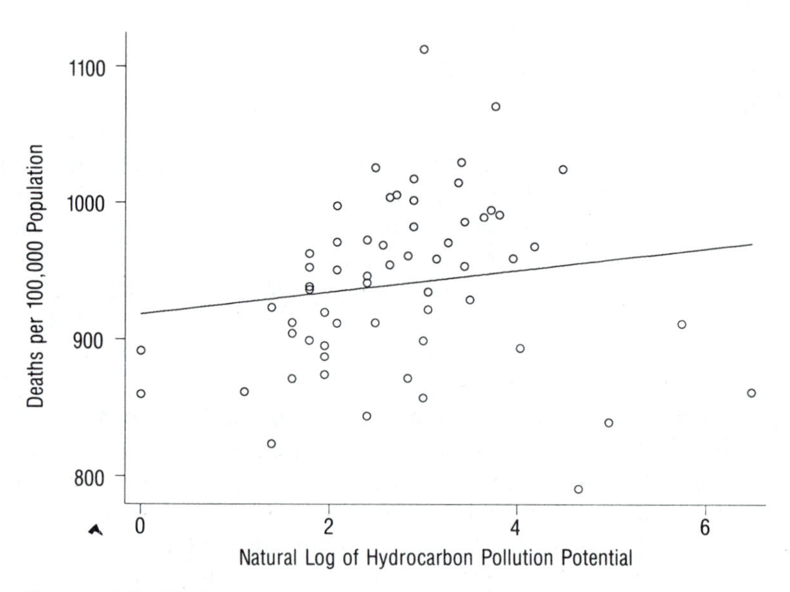

Figure 6.1 OLS regression of mortality rate on log hydrocarbon pollution.

Four outliers, cities with high pollution but low mortality rates, appear at the lower right in Figure 6.1: Los Angeles, San Diego, San Francisco, and San Jose. They pull the OLS line down, making it less steep than it would be based on the remaining 56 cases. Since all four are Californian, we should suspect that their low death rates reflect omitted variable(s), rather than just random errors around the true air pollution/mortality relationship.

We might try to deal with this problem by identifying and including the omitted variables or (if that cannot be done) by reestimating the equation with the four outliers deleted. Deleting these four cities will obviously strengthen the positive relationship. Do we *want* to claim a stronger relationship between air pollution and mortality? When decisions about outlier deletion are made subjectively, opportunities increase for researchers' biases to affect their conclusions.

Besides being subjective and potentially controversial, outlier deletion requires extra work—diagnostics, reestimation, and then further diagnostics to check that new outliers are not created by deleting the old (as often happens). If we have many variables, many models, or not much experience, the effort of careful analysis may seem too great. We may be tempted just to assume problems away and apply OLS blindly.

Robust regression provides an alternative that is somewhat safer than OLS when applied blindly. Outliers automatically get lower weights, which lessens their influence—we need not individually notice and do something about them. A robust child might play out in the rain without catching a cold. Similarly, a robust statistical method might be applied to outlier-filled data yet still produce reasonable results. This does not mean that robust methods *should* be used blindly or that the child should play in the rain; we would still be better off using careful judgment.

Figure 6.2 shows the same scatter as Figure 6.1, with a robust regression line

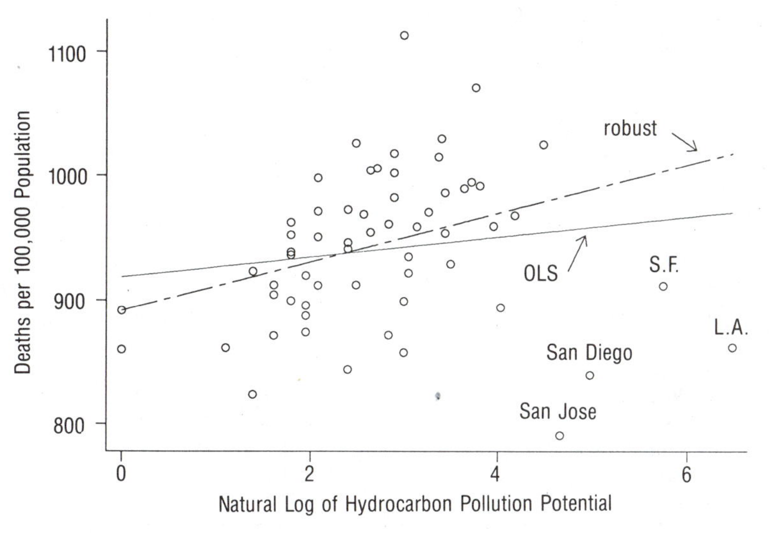

Figure 6.2 OLS and robust regressions of mortality rate on log hydrocarbon pollution.

added:

$$\hat{Y}_i = 891.7 + 19.46 \log_e X_{i1} \qquad [6.2]$$

SE_b: 6.55

t: 2.97

The robust line is steeper, with a slope significantly different from zero ($P = .002$). The four Californian cities do not pull it down as they did the OLS line in Figure 6.1.

Manual outlier deletion is all-or-nothing. In effect, we assign each case a weight of 1 (if included in the analysis) or 0 (if deleted). In contrast, robust regression permits weights between 1 and 0, using *weighted least squares* (WLS). Outlying cases get *downweighted* (given weights less than one) gradually: the farther out a case lies, the lower its weight.

Table 6.1 lists the data for all 60 cities, with predicted values, residuals, and robust weights. The larger the absolute value of the residual, the lower the weight. Mortality in San Jose ($Y_1 = 790.7$) is much lower than predicted ($\hat{Y}_1 = 982.3$), resulting in a large residual ($e_1 = -191.6$). Such outliers ordinarily exert strong

Table 6.1 Hydrocarbon Pollution Potential, Mortality Rate, and Results from Robust Regression of Mortality on $\log_e$ (Hydrocarbon Pollution)

Metropolitan area	Pollution X_i	Mortality Y_i	Predicted $\hat{Y}_i$	Residual e_i	Weight w_i
1 San Jose	105	790.7	982.3	−191.6	.08
2 New Orleans	20	1113.0	950.0	163.0	.23
3 Los Angeles	648	861.8	1017.7	−155.9	.27
4 San Diego	144	839.7	988.4	−148.7	.32
5 Baltimore	43	1071.0	964.9	106.1	.60
6 Wichita	4	823.8	918.7	−94.9	.68
7 Lancaster	11	844.1	938.4	−94.3	.68
8 Minneapolis	20	857.6	950.0	−92.4	.69
9 San Francisco	311	911.7	1003.4	−91.7	.70
10 Richmond	12	1026.0	940.1	85.9	.73
11 Portland	56	894.0	970.1	−76.1	.79
12 Denver	17	871.8	946.9	−75.1	.79
13 Birmingham	30	1030.0	957.9	72.1	.81
14 Chattanooga	18	1018.0	945.0	70.0	.82
15 Albany	8	997.9	932.2	65.7	.84
16 Memphis	15	1006.0	944.4	61.6	.86
17 Wilmington	14	1004.0	943.1	60.9	.86
18 Philadelphia	29	1015.0	957.3	57.7	.87
19 Rochester	7	874.3	929.6	−55.3	.88
20 Buffalo	18	1002.0	948.0	54.0	.89
21 Grand Rapids	5	871.3	923.1	−51.8	.90
22 Miami	3	861.4	913.1	−51.7	.90
23 Seattle	20	899.3	950.0	−50.7	.90

Table 6.1 (*Continued*)

Metropolitan area	Pollution X_i	Mortality Y_i	Predicted $\hat{Y}_i$	Residual e_i	Weight w_i
24 Chicago	88	1025.0	978.9	46.1	.92
25 Hartford	7	887.5	929.6	−42.1	.93
26 Greensboro	8	971.1	932.2	38.9	.94
27 Allentown	6	962.4	926.6	35.8	.95
28 Atlanta	18	982.3	948.0	34.3	.95
29 Toledo	11	972.5	938.4	34.1	.95
30 Worcester	7	895.7	929.6	−33.9	.96
31 Dallas	1	860.1	891.8	−31.7	.96
32 New York	41	994.6	964.0	30.6	.96
33 Milwaukee	33	929.2	959.8	−30.6	.96
34 Akron	21	921.9	951.0	−29.1	.97
35 Canton	12	912.3	940.1	−27.8	.97
36 Cleveland	31	986.0	958.6	27.4	.97
37 Bridgeport	6	899.5	926.6	−27.1	.97
38 Indianapolis	13	968.7	941.7	27.0	.97
39 Louisville	38	989.3	962.5	26.8	.97
40 Houston	6	952.5	926.6	25.9	.97
41 Pittsburgh	45	991.3	965.8	25.5	.97
42 York	8	911.8	932.2	−20.4	.98
43 Springfield	5	904.2	923.1	−18.9	.99
44 Syracuse	8	950.7	932.2	18.5	.99
45 Boston	21	934.7	951.0	−16.3	.99
46 Cincinnati	26	970.5	955.1	15.4	.99
47 Nashville	17	961.0	946.9	14.1	.99
48 Providence	6	938.5	926.6	11.9	.99
49 Youngstown	14	954.4	943.1	11.3	.99
50 Utica	5	912.2	923.1	−10.7	1.00
51 Kansas City	7	919.7	929.6	−9.9	1.00
52 Dayton	6	936.2	926.6	9.6	1.00
53 Detroit	52	959.2	968.6	−9.4	1.00
54 Reading	11	946.2	938.4	7.8	1.00
55 Columbus	23	958.8	952.8	6.0	1.00
56 Washington	65	967.8	973.0	−5.2	1.00
57 St. Louis	31	953.6	958.6	−5.0	1.00
58 New Haven	4	923.2	918.7	4.5	1.00
59 Flint	11	941.2	938.4	2.8	1.00
60 Fort Worth	1	891.7	891.8	−.1	1.00

influence on a regression, but here San Jose receives a near-zero weight ($w_1 = .08$) and has little effect.

Figure 6.3 repeats the scatterplot of Figure 6.2, but with each case's position marked by its weight. Cases near the regression line get weights near 1, and cases farther away (like the four at lower right) get systematically lower weights.

Computer programs perform WLS automatically, based on a user-specified weight variable. Robust regression derives weights from equations called *weight functions.* Robustness, efficiency, and other properties depend on the specific weight

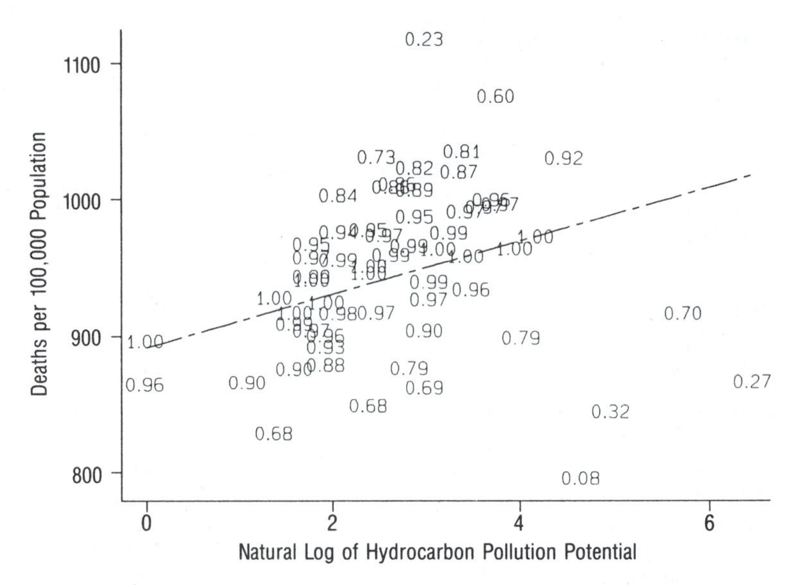

Figure 6.3 Robust weights and regression line; weights decrease with vertical distance from line (residual).

function chosen. The standard errors and hypothesis tests printed by most WLS programs are incorrect for robust regression, however. Robust standard errors and tests require further calculation.

Weighted least squares (WLS) regression employs an $n \times n$ weighting matrix $\mathbf{W}$, with case weights w_i on the major diagonal and zeros elsewhere. Instead of the OLS $\mathbf{Y} = \mathbf{X}\boldsymbol{\beta} + \boldsymbol{\varepsilon}$, the WLS model is

$$\mathbf{W}^{\frac{1}{2}}\mathbf{Y} = \mathbf{W}^{\frac{1}{2}}\mathbf{X}\boldsymbol{\beta} + \mathbf{W}^{\frac{1}{2}}\boldsymbol{\varepsilon} \qquad [6.3]$$

or, equivalently,

$$\mathbf{Y}^* = \mathbf{X}^*\boldsymbol{\beta} + \boldsymbol{\varepsilon}^*$$

where $\mathbf{Y}^* = \mathbf{W}^{\frac{1}{2}}\mathbf{Y}$, $\mathbf{X}^* = \mathbf{W}^{\frac{1}{2}}\mathbf{X}$, and $\boldsymbol{\varepsilon}^* = \mathbf{W}^{\frac{1}{2}}\boldsymbol{\varepsilon}$. The $K \times 1$ vector of estimated coefficients, $\mathbf{B}$, is

$$\mathbf{B} = (\mathbf{X}^{*\prime}\mathbf{X}^*)^{-1}\mathbf{X}^{*\prime}\mathbf{Y}^* \qquad [6.4]$$

Standard errors for coefficient estimates depend upon how the weights were chosen.

WLS can also be performed "by hand," using transformed variables and OLS. From case weights w_i:

$$\text{generate } Y_i^* = Y_i\sqrt{w_i}$$

$$\text{generate } X_{ik}^* = X_{ik}\sqrt{w_i} \qquad \text{for } k = 1 \text{ to } K - 1$$

$$\text{generate } X_{i0}^* = \sqrt{w_i}$$

Regress Y^* on X_0^* and all X_k^*, specifying *no constant.* The resulting coefficient estimates equal those from WLS. The coefficient on X_0^* equals the WLS Y-intercept.

Goals of Robust Estimation

Outliers affect more than just OLS slopes. They often have even greater, though less obvious, effects on standard errors, hypothesis tests, R^2, and other statistics. The prevalence and broad consequences of outliers make them a major problem for OLS and related methods. Robust estimation originated with the need for analytical methods that would be less affected by outliers.

An estimate is *resistant* if its value is not much affected by small changes in sample data. "Small changes" refer either to large changes in a small fraction of the cases (such as introducing a few gross measurement errors) or to small changes in a large fraction of the cases (such as many minor rounding-off errors). The concept of *robustness* is theoretically distinct. A robust estimator performs well even when there are small violations of assumptions about the underlying population—for example, an error distribution is not really Gaussian. Most resistant estimators are also distributionally robust.[3]

Ordinary least squares is not resistant or robust. A single case can have an arbitrarily large impact on sample estimates. Although OLS is an efficient estimator given normally distributed errors, it loses efficiency when error distributions have heavier-than-normal tails.[4] Unfortunately, heavy-tailed distributions are common, judging by the frequency of outliers in many kinds of data. Thus, practicing researchers often face situations in which OLS performs poorly; they need better alternatives.

Peter Huber (1981) writes that a good robust procedure should be:

1. consistent and reasonably efficient when the assumed model is true;
2. only slightly impaired by small departures from the model; and
3. not drastically affected by somewhat larger departures from the model.

Since "departures from the model" could be anything, we want procedures that perform well with a variety of underlying distributions.

Early robustness research focused on measures of center (mean, median, and so on) for symmetrical but heavy-tailed distributions. Skewed distributions, which do not have clear-cut centers, were necessarily ignored. This led to the mistaken perception that robust methods in general require symmetrical distributions.[5] A good robust regression procedure should be able to handle both skewed and symmetrical error distributions.

Three broad families of robust methods have received the most study:

1. *M-estimators* are maximum-likelihood estimators, a family that includes the arithmetic mean and OLS regression as special cases. They minimize the sum of a function of the residuals. *M*-estimation is often approximated through weighted least squares (WLS), an approach sometimes called *W-estimation* (illustrated earlier in Figure 6.3).
2. *R-estimators* are based on ranks. For example, one approach is to minimize a sum that employs weighting by ranks of the residuals. Although theoretically attractive, *R*-estimators are harder to work with and do not perform quite as well as the best *M*-estimators.
3. *L-estimators* are linear functions of sample order statistics (quantiles). Trimmed means are the best-known example.[6]

Each family includes many subtypes. The next section describes the specific *M*-estimators used for Figure 6.3.

M-Estimation and Iteratively Reweighted Least Squares

Robust regression by weighted least squares (*W-estimation*) typically follows these steps:

1. Obtain *starting values*, the preliminary regression parameter estimates, perhaps by OLS. Find the residuals.
2. Use these residuals to calculate a set of case weights. Cases with large residuals receive lower weights.
3. Apply WLS to obtain a second set of regression parameter estimates, and calculate the new residuals.
4. Repeat step 2: calculate a new set of case weights based on the new residuals, and apply WLS yet again.

Continue in this manner until there is negligible change from one iteration to the next. The process is called iteratively reweighted least squares (IRLS), to distinguish it from one-step WLS. In theory, *W*-estimates obtained through IRLS are equivalent to corresponding *M*-estimates.

Scaled residuals (u_i) are residuals (e_i) divided by a scale estimate (s):

$$u_i = \frac{e_i}{s} \qquad [6.5]$$

OLS applications commonly employ the residual standard deviation ([2.11]) for a scale estimate. Equation [2.11] is not resistant, however; one outlier can control it. A simple resistant alternative uses the median absolute deviation, or MAD:

$$\text{MAD} = \text{median}|e_i - \text{median}(e_i)| \quad [6.6]$$

The resulting scale estimate is

$$s = \frac{\text{MAD}}{.6745} \quad [6.7]$$

Applied to normally distributed errors, both [2.11] and [6.7] estimate the population standard deviation (σ_ε). Resistance gives [6.7] the advantage when we are working with nonnormal errors.

M-estimators for β_k in the regression model

$$\mathbf{Y} = \mathbf{X}\boldsymbol{\beta} + \boldsymbol{\varepsilon}$$

seek estimates b_k that minimize an *objective function* of the residuals:

$$\Sigma_i \rho\{u_i\} = \Sigma_i \rho\{(Y_i - \Sigma_k b_k X_{ik})/s\} = \min! \quad [6.8]$$

Equivalently, we seek b_k values that solve the set of K simultaneous equations:

$$\Sigma_i \psi\{u_i\} = \Sigma_i \psi\{(Y_i - \Sigma_k b_k X_{ik})/s\} X_{ik} = 0 \quad [6.9]$$

where ψ is the first derivative of ρ (disregarding any multiplicative constant). Robustness and other properties of the estimator depend on the specific ρ or ψ function chosen.

Robust weights derive from the ψ function:

$$w_i = \psi\{u_i\}/u_i \quad [6.10]$$

Alternatively, we could view $\psi\{u_i\}$ as a *weighted residual*: the ith-case weight times the scaled residual:

$$\psi\{u_i\} = w_i u_i \quad [6.11]$$

Ordinary least squares is a simple *M*-estimator, assigning all cases the same weight (1):

$$w_i = 1 \qquad \text{for all } i \quad [6.12]$$

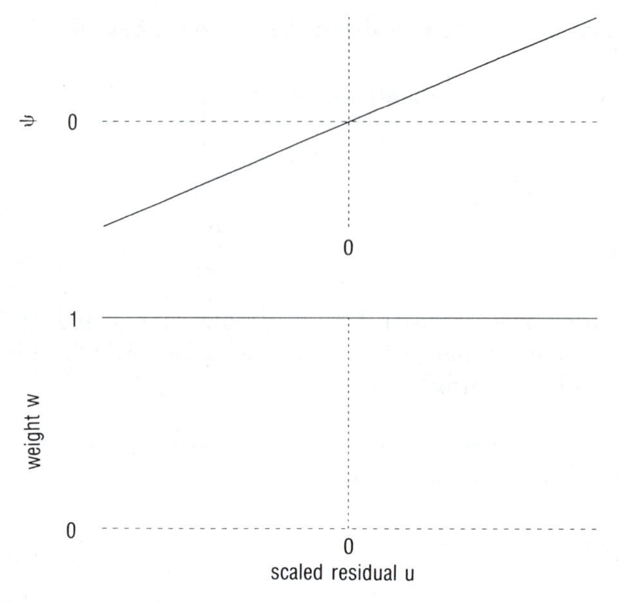

Figure 6.4 OLS ψ and weight functions.

Figure 6.4 shows ψ and weight functions for OLS. The horizontal axes are scaled residuals.

OLS is defined by:

$$\rho\{u_i\} = u_i^2 \qquad [6.13]$$

or

$$\psi\{u_i\} = u_i \qquad [6.14]$$

The *objective function* for OLS is the sum of squared residuals. OLS estimates solve

$$\Sigma_i \rho\{u_i\} = \Sigma_i u_i^2 = \text{min!}$$

or

$$\Sigma_i \psi\{u_i\} X_{ik} = \Sigma_i u_i X_{ik} = 0$$

One robust M-estimator, *Huber estimation*, gradually downweights cases with scaled residuals larger than some *tuning constant*, c:

$$w_i = 1 \qquad \text{if } |u_i| \leq c$$

$$w_i = \frac{c}{|u_i|} \qquad \text{if } |u_i| > c \qquad [6.15]$$

If the case-i residual lies no more than c standard deviations from the median, it receives a weight of 1. (Standard deviations should be estimated robustly, as by [6.7].) Scaled residuals larger than c result in progressively lower weights (Huber, 1981).

Huber estimation is defined by the functions

$$\rho\{u_i\} = u_i^2 \qquad \text{if } |u_i| \le c$$
$$\rho\{u_i\} = c(2|u_i| - c) \qquad \text{if } |u_i| > c \qquad [6.16]$$

or

$$\psi\{u_i\} = u_i \qquad \text{if } |u_i| \le c$$
$$\psi\{u_i\} = c\,\text{sgn}(u_i) \qquad \text{if } |u_i| > c \qquad [6.17]$$

The larger the tuning constant c, the more closely Huber estimation resembles OLS. For maximum protection against nonnormal errors, we want a small c; for maximum normal-distribution efficiency, we want a large c. Huber estimation with $c = 1.345$ is 95% as efficient as OLS when applied to Gaussian errors, yet it gives very good protection against outliers. Then we are downweighting cases if

$$|u_i| > 1.345$$

or—more easily remembered—if

$$|e_i| > 2\text{MAD}$$

(since $s = \text{MAD}/.6745$, $1.345s \approx 2\text{MAD}$). Figure 6.5 shows Huber ψ and weight functions for $c = 1.345$.

John Tukey's *biweight* estimator (also called a *bisquare*) provides even greater protection against heavy-tailed error distributions. Weights are:

$$w_i = [1 - (u_i/c)^2]^2 \qquad \text{if } |u_i| \le c$$
$$w_i = 0 \qquad \text{if } |u_i| > c \qquad [6.18]$$

As absolute scaled residuals approach c, case weights get gradually smaller. Absolute residuals of c or greater result in weights of zero; such ill-fit cases are effectively dropped from the analysis (Mosteller and Tukey, 1977).

Biweight estimation is defined by the functions

$$\rho\{u_i\} = (c^2/3)(1 - [1 - (u_i/c)^2]^3) \qquad \text{if } |u_i| \le c$$
$$\rho\{u_i\} = c^2/3 \qquad \text{if } |u_i| > c \qquad [6.19]$$

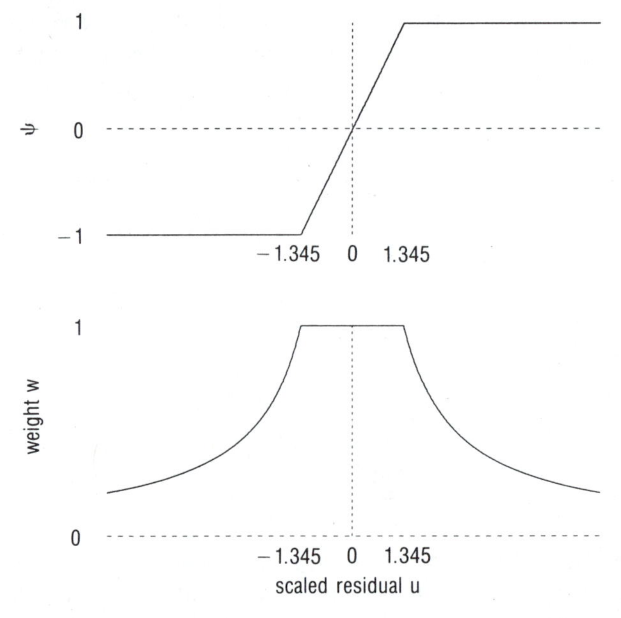

Figure 6.5 Huber ψ and weight functions, $c = 1.345$.

or

$$\psi\{u_i\} = u_i[1 - (u_i/c)^2]^2 \qquad \text{if } |u_i| \le c$$
$$\psi\{u_i\} = 0 \qquad \text{if } |u_i| > c \qquad [6.20]$$

Biweight estimation achieves 95% Gaussian efficiency when the tuning constant $c = 4.685$. Then we are downweighting cases if $0 < |u_i| < 4.685$ and dropping them entirely if

$$|u_i| \ge 4.685$$

or, equivalently, if

$$|e_i| \ge 7\text{MAD}$$

Figure 6.6 shows biweight ψ and weight functions with $c = 4.685$.

Unlike Huber ψ functions, biweight ψ functions return to zero; estimators with this property are called *redescending*. Here are two other contrasts between Figures 6.5 and 6.6:

1. Huber estimation does not downweight cases with small or moderate residuals. In this central region it behaves just like OLS. Downweighting begins suddenly at $|u_i| = c$. Biweight estimation downweights every case with a nonzero residual, in a smooth progression.
2. Biweights go to zero at $|u_i| = c$; Huber weights never reach zero, so no observations are completely dropped.

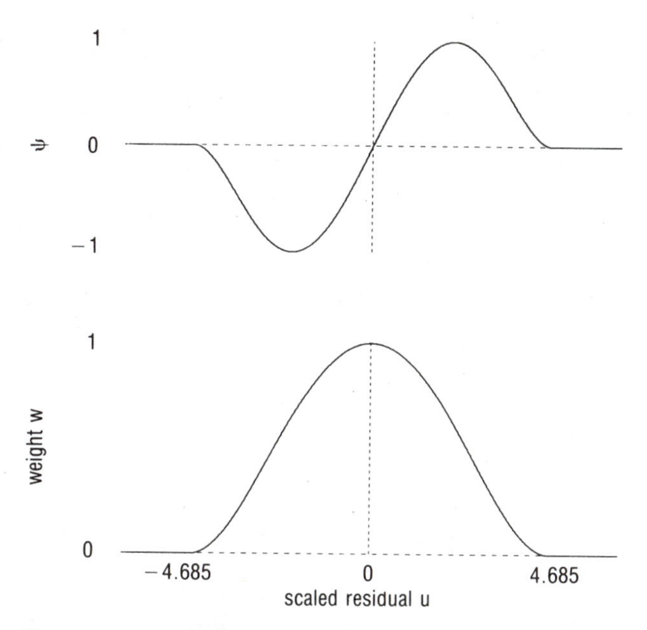

Figure 6.6 Biweight ψ and weight functions, $c = 4.685$.

Because the biweight more radically downweights or drops extreme outliers, it tends to be more efficient than Huber estimation when applied to very heavy-tailed error distributions. Furthermore, the smooth biweight function makes it less sensitive to small errors.

Unfortunately, biweight ρ functions (and those of other redescending estimators) may have more than one minimum. In practice, this means that biweight estimation via IRLS is sensitive to the initial parameter estimates (starting values), from which the first set of residuals are calculated. "Bad" starting values could cause the iteration procedure to become trapped in a local minimum, resulting in a poor fit and misleading coefficient estimates. Huber estimation does not share this problem. The Huber ρ function is convex, with only one minimum. IRLS estimation should converge on this minimum regardless of the starting values chosen.

The next section describes an iterative procedure that combines the advantages of OLS, Huber, and biweight estimation.

Calculation by IRLS

The robust regression of Figure 6.3 involved the following steps (subscripts denote iteration numbers; for simplicity I omit case subscripts):

Iteration 0. OLS calculates starting values, initial residuals (e_0), and a MAD-based scale estimate (s_0).

Iteration 1. A Huber function ([6.15]) with $c = 1.345$ defines case weights (w_1):

$$w_1 = 1 \qquad \text{if } |e_0|/s_0 \leq 1.345$$
$$w_1 = 1.345/(|e_0|/s_0) \qquad \text{if } |e_0|/s_0 > 1.345$$

Weighted least squares then calculates the next set of residuals (e_1) and a new MAD-based scale estimate (s_1).

Iteration 2. Find new Huber weights for each case:

$$w_2 = 1 \qquad \text{if } |e_1|/s_1 \leq 1.345$$
$$w_2 = 1.345/(|e_1|/s_1) \qquad \text{if } |e_1|/s_1 > 1.345$$

and again perform WLS, obtaining e_2 and s_2.

Iteration continues until the maximum change in weights from one iteration to the next is small (less than .05). For the air-pollution data, this happens on iteration 4. Then we shift:

Iteration 5. Case weights are calculated by applying a biweight function ([6.18]) with $c = 4.685$:

$$w_5 = \{1 - [(e_4/s_4)/4.685]^2\}^2 \qquad \text{if } |e_4|/s_4 \leq 4.685$$
$$w_5 = 0 \qquad \text{if } |e_4|/s_4 > 4.685$$

WLS then obtains new residuals (e_5) and scale estimate (s_5).

Iteration 6. Calculate new biweights:

$$w_6 = \{1 - [(e_5/s_5)/4.685]^2\}^2 \qquad \text{if } |e_5|/s_5 \leq 4.685$$
$$w_6 = 0 \qquad \text{if } |e_5|/s_5 > 4.685$$

and use in a new WLS regression, obtaining e_6 and s_6.

When the maximum change in weights from one iteration to the next becomes negligible (less than .01), iteration stops. This occurs on iteration 11. WLS using the w_{11} weights (Table 6.1) provides the final set of robust parameter estimates (Equation [6.2]).

Figure 6.7 graphically depicts the changes in weights from one iteration to the next. In the mth-iteration plot, the vertical axis shows weights w_m, and the horizontal axis shows weights from the previous iteration, w_{m-1}. Changes are greatest on the first iteration and then rapidly decrease until we shift to the biweight (iteration 5); then, again the changes rapidly decrease, becoming almost invisible by iteration 9.

Figure 6.8 graphs final robust weights (biweights) against scaled residuals. (Weights and raw residuals e_i appeared in Table 6.1. Scaled residuals are $u_i = e_i/s = e_i/(\text{MAD}/.6745) = e_i/48.09$.) No $|u_i| \geq c = 4.675$, so weights do not quite fall to zero. Weights in Figure 6.8 follow the general biweight function shown in Figure 6.6.

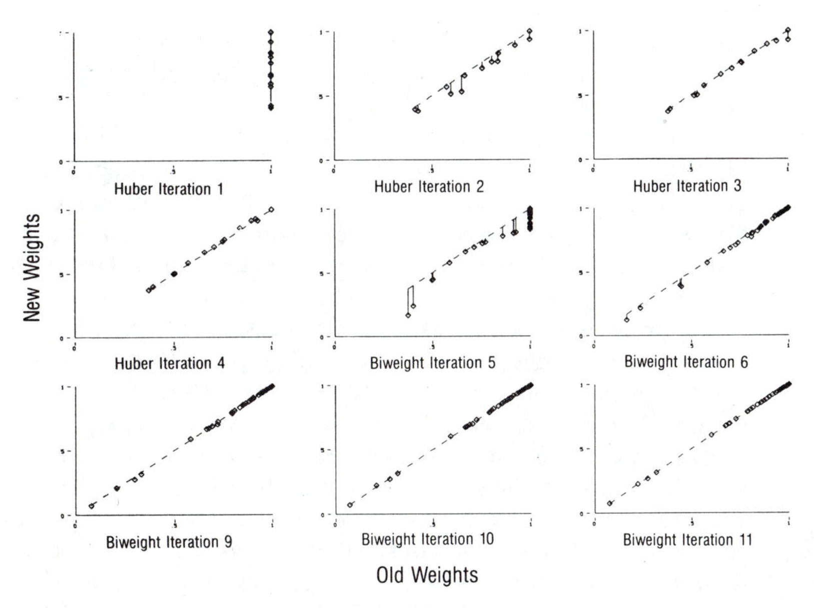

Figure 6.7 Changes in weights during IRLS estimation

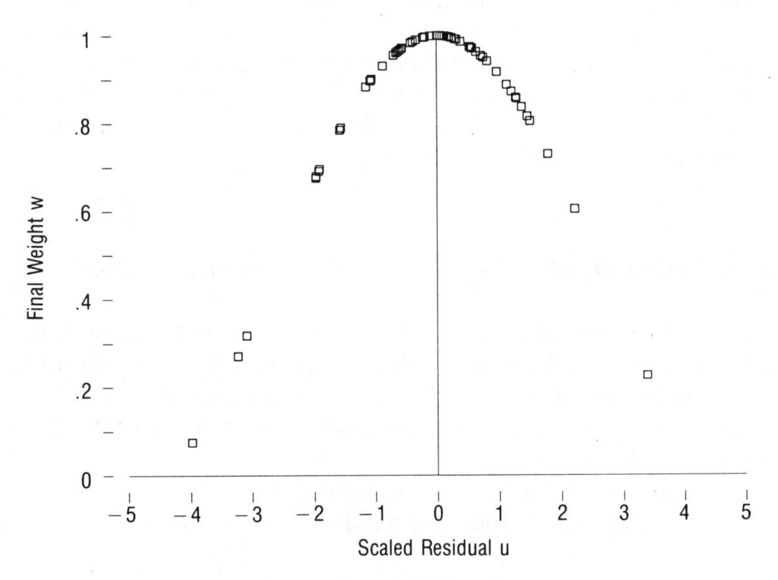

Figure 6.8 Weights versus scaled residuals from regression of mortality on log pollution.

Most cases, clustered at the top of Figure 6.8, received weights near 1. Four outliers received weights below .5: San Jose, Los Angeles, and San Diego, at lower left, and New Orleans, at lower right.

To summarize, we started with OLS, used Huber weighting to obtain more robust estimates, and then finished using a biweight. This approach (proposed by Guoying Li, in Hoaglin, Mosteller, and Tukey, 1985) takes advantage of the simplicity of OLS starting values, the convergence of Huber estimation, and the smoothness and severe-outlier resistance of biweight estimation. The Huber iterations provide robust starting values for the biweight.

Any program with WLS capability can calculate coefficient estimates and robust weights in this manner. Some of the WLS output may be misleading, however. With iterative reweighting, the usual WLS standard errors and hypothesis tests are wrong. (The next section describes how to correct them.) Furthermore, the WLS sums of squares, residual standard deviation (root mean squared error), and R^2 do not compare with their OLS counterparts. We cannot use R^2, for example, to support statements like "the robust equation fits the data better/worse than the OLS equation." The WLS R^2 pertains to the data *after weighting*, which is a different data set from the one OLS sees. A WLS R^2 tells us how well *the robust equation fits the weighted data*, whereas the OLS R^2 tells us how well *the OLS equation fits the unweighted data.*

We might summarize how well the robust equation fits the *un*weighted data by finding the squared correlation between Y and robust $\hat{Y}$ values. However, this number will always be lower than the OLS R^2, leading to the conclusion that robust equations always "fit worse" (as judged by R^2) than OLS. OLS by definition obtains a higher R^2 than any other estimation strategy applied to the same model and data. But, like OLS itself, R^2 is not resistant and therefore is not always a good measure of fit.

Standard Errors and Tests for *M*-Estimates

Most WLS programs estimate standard errors on the assumption that weights are fixed and assigned a priori. This assumption is false for robust regression, in which weights are a random variable derived from sample residuals. Consequently, robust standard errors cannot be obtained directly through WLS; we need a more specialized method.

The method described below, adapted from Street, Carroll, and Ruppert (1988), allows calculation of robust regression standard errors with an OLS program.

If a robust regression model

$$Y_i = \mathbf{X}_i\boldsymbol{\beta} + \varepsilon_i$$

is correct, with $\mathbf{X}_i$ denoting the ith X (row) vector, coefficient (column) vector $\boldsymbol{\beta}$ estimated by $\mathbf{B}$, and errors ε_i i.i.d. with scale σ estimated by s, then the final estimate $\mathbf{B}$

is asymptotically normally distributed with mean $\boldsymbol{\beta}$ and covariance

$$\Lambda = \sigma^2 E[\psi^2\{\varepsilon\}]((E[\psi^{\cdot}\{\varepsilon\}])^2 \textstyle\sum_i \mathbf{X}_i'\mathbf{X}_i)^{-1} \qquad [6.21]$$

where $\psi^{\cdot}$ is the derivative of ψ. For biweight estimation:

$$\psi^{\cdot}\{u_i\} = [1 - (u_i^2/c^2)][1 - (5u_i^2/c^2)] \qquad |u_i| \leq c$$

$$\psi^{\cdot}\{u_i\} = 0 \qquad |u_i| > c$$

For Huber estimation:

$$\psi^{\cdot}\{u_i\} = 1 \qquad |u_i| \leq c$$

$$\psi^{\cdot}\{u_i\} = 0 \qquad |u_i| > c$$

1. To estimate Λ, we find

$$a = \frac{\Sigma_i \psi^{\cdot}\{u_i\}}{n} \qquad [6.22]$$

   and

$$\lambda = 1 + \frac{(K/n)(1 - a)}{a} \qquad [6.23]$$

   where u_i are scaled residuals, c is the robust regression's tuning constant, and K is the number of parameters in the regression. Also, n represents sample size before weighting.
2. Next define a set of *pseudovalues* of Y:

$$\tilde{Y}_i = \hat{Y}_i + (\lambda s/a)\psi\{u_i\} \qquad [6.24]$$

   where $\hat{Y}_i$ are the Y values predicted by the robust regression equation, s is the scale estimate, and $\psi\{u_i\}$ is defined as in [6.20].
3. Employ OLS to regress $\tilde{\mathbf{Y}}$ on $\mathbf{X}$. The covariances, standard errors, and tests from this regression are asymptotically correct, with degrees of freedom based on unweighted sample size. Coefficient estimates equal those obtained from the robust regression.

This procedure obtains $\text{SE}_{b_1} = 6.55$ as the estimated standard error of the coefficient on $\log_e X_1$ ($b_1 = 19.46$) in Equation [6.2]. We test $H_0\!: \beta_1 = 0$ in the usual manner:

$$t = \frac{b_1}{\text{SE}_{b_1}} = \frac{19.46}{6.55} = 2.97$$

with df $= n - K = 58$, yielding $P = .002$. Confidence intervals likewise follow the familiar pattern, $b_1 \pm t(\text{SE}_{b_1})$. For example, a 95% interval requires $t = 2.0$:

$$19.46 \pm 2(6.55)$$

$$6.36 \leq \beta_1 \leq 32.56$$

If our model is correct, we could expect intervals constructed in this manner to contain the population parameter β_1 about 95% of the time.

Robust estimation encourages the question "What if our model is not correct?" (Really this should be asked of any analysis, but such doubts motivate the use of robust estimation in the first place.) If the model *is* correct, then both OLS and robust methods provide consistent estimates of the population parameter β; OLS offers greater efficiency with Gaussian errors, and robust estimation offers greater efficiency with heavy-tailed error distributions. But, if the model is not correct, β as defined by the model does not really exist. Then what are we estimating?[7] The best answer is somewhat circular: β is the limiting value of our estimator b:

$$\beta = \lim_{n \to \infty} E[b]$$

In other words, we define the estimated parameter in terms of the estimator we are using, rather than in terms of an assumed model.

Using Robust Estimation

A robust regression analysis provides the following:

1. Robust coefficient and standard error estimates. Discrepancies between OLS and robust results reveal the effects of outliers and warn that OLS may be untrustworthy.
2. Robust predicted values and residuals. Robust predictions may better fit the majority of the data, since they have less tendency to track a few influential points. For the same reason robust residuals identify which cases are really unusual.
3. Robust weights. As diagnostic statistics, weights highlight outliers.

Robust and OLS regression thus complement each other.

OLS is simpler, and it is preferable if both methods produce the same results. Then we can view the robust findings as confirmation of OLS's validity. The larger the discrepancy, the more we should lean toward the robust results. As a rough guideline, we might check whether any of the OLS coefficients are more than one (robust) standard error from the corresponding robust coefficient. The air-pollution analysis fails this test. The OLS coefficient (from [6.1], $b_1 = 7.97$) and the robust coefficient (from [6.2], $b_1 = 19.46$) differ by $(19.46 - 7.97)/6.55 = 1.75$ robust standard errors, due to the influence of outliers upon OLS.

Robust estimation does not eliminate the need for careful exploratory and diagnostic work. Although robust regression protects against some problems, it remains vulnerable to others. For example, if the true relationship is curvilinear, fitting a robust straight line makes no more sense than fitting a nonrobust straight line. Less obviously, M-estimation is not immune to the problem of leverage.

A high-leverage case is potentially influential due to an unusual combination of X values (see Chapter 4). If it also has an unusual Y value, a leverage case may exert so much influence that it controls OLS estimates and has a relatively small residual. Procedures that use OLS starting values will not see this case as an outlier and will not downweight it. Other cases that are actually less exceptional may be downweighted instead, compounding the distortion.

The leverage of case i, h_i, equals the ith diagonal element of the hat matrix $\mathbf{H}$. For weighted least squares:

$$\mathbf{H} = \mathbf{X}^*(\mathbf{X}^{*\prime}\mathbf{X}^*)^{-1}\mathbf{X}^{*\prime} \tag{6.25}$$

where $\mathbf{X}^*$ is the matrix of X values premultiplied by the square root of the weight matrix $\mathbf{W}$:

$$\mathbf{X}^* = \mathbf{W}^{\frac{1}{2}}\mathbf{X}$$

Except for this multiplication, [6.25] resembles an OLS hat matrix ([4.10]).

M-estimators may fail to detect or downweight even severe outliers with high h_i (risky when $\max(h_i) > .2$).

Figures 6.1 and 6.2 regressed mortality rates on logarithms of hydrocarbon pollution. Logarithms were employed because preliminary analysis found the pollution distribution to be severely skewed. Figure 6.9 illustrates what happens if we bypass preliminary exploratory/transformation steps and just regress mortality on hydrocarbon pollution.

Both OLS and robust regression produce lines with little relation to most of the data. Instead, they track two leverage points (San Francisco, $h = .17$, and Los Angeles, $h = .76$). Such tracking leads to small residuals, so San Francisco and Los Angeles never get downweighted. Robust estimation offers little improvement over OLS in this example.

Deleting San Francisco and Los Angeles dramatically alters the robust fit (Figure 6.10). Although OLS still sees a weak negative relationship, robust regression now finds a substantial positive one—consistent with earlier transformed-variable results. With San Francisco and Los Angeles removed, San Jose and San Diego become outliers and leverage points. However, they lie far enough from the OLS line that the robust procedure spots them as outliers and downweights them slightly on its first iteration. With this downweighting, the robust line steepens, making the two

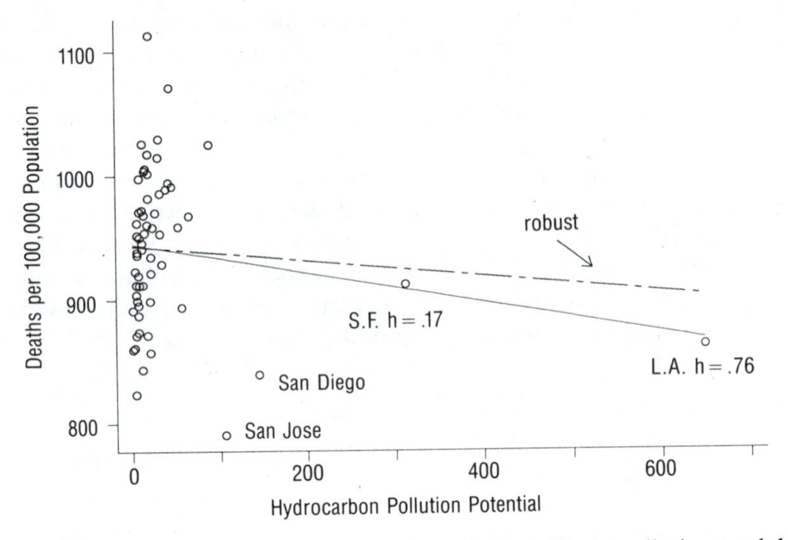

Figure 6.9 OLS and robust regressions of mortality on pollution track high-leverage cases.

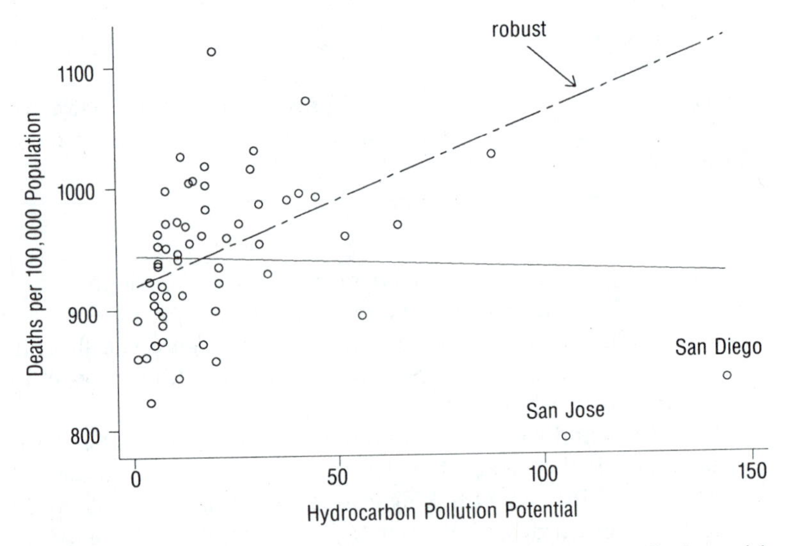

Figure 6.10 OLS and robust regression of mortality on pollution, with two highest-leverage cases dropped.

cities' residuals larger, which leads to further downweighting, and so on. Robust estimation now has a chance to succeed, and it produces a line that better reflects the general up-to-right trend of the data.

Figures 6.9 and 6.10 illustrate three general points:

1. *M*-estimation protects against unusual Y values but not necessarily against unusual X values (leverage).
2. Diagnostic work (like examining graphs and case statistics) is still needed.
3. Exploratory analysis and symmetrizing transformations reduce the likelihood that problems will arise.

Points 2 and 3 apply to other problems besides leverage. For example, both Huber and biweight estimation require an estimate of the errors' standard deviation or scale, s. If errors do not all have the same scale (heteroscedasticity), then this approach loses its legitimacy. Transformations may help by reducing heteroscedasticity.

This chapter began with the claim that robust methods are, by design, safer than classical methods. As a relative statement this is true, but Figure 6.9 illustrates the hazards of applying even robust methods blindly. No method is safe when we use it without thinking or looking closely at the data.

A Robust Multiple Regression

The obvious outliers in Figures 6.2, 6.9, and 6.10 have something in common: California. Some omitted variable(s), related to their California location, presumably accounts for these cities' low mortality despite high pollution. We can assess possible explanations by including other variables in a robust multiple regression.

Table 6.2 lists 13 demographic and environmental variables that might predict urban mortality rates. Transformations improve linearity or symmetry in several instances. For example, the graph at the upper right in Figure 6.11 shows how logarithms reduce the skew of hydrocarbon pollution.

Table 6.2 Possible Predictors of Urban Mortality Rates

X_1: relative hydrocarbon pollution potential (natural log)
X_2: mean yearly precipitation in inches
X_3: mean January temperature in degrees Fahrenheit
X_4: median years of education for population > 25 years old
X_5: percent of population that is non-White (square root)
X_6: population per household
X_7: percent of population $\geq$ 65 years old
X_8: percent of housing units that are sound with all facilities
X_9: population per square mile of urbanized area (natural log)
X_{10}: mean July temperature in degrees Fahrenheit
X_{11}: percent employment in white-collar occupations
X_{12}: percent of families with incomes below poverty (negative reciprocal root)
X_{13}: mean annual relative humidity as a percentage

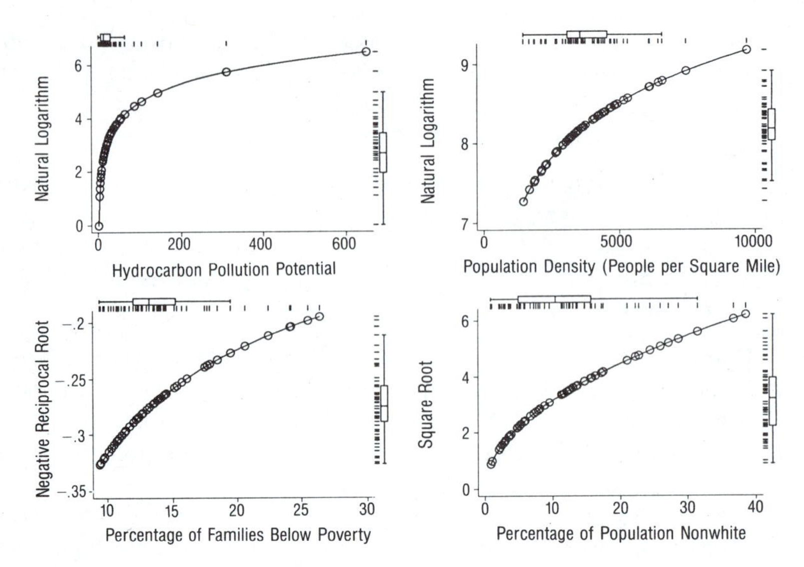

Figure 6.11 Transforming possible predictors of mortality rates.

To identify the most useful predictors, in the absence of strong theoretical guidance, we resort to backward elimination. First we robustly regress mortality on all 13 X variables. After dropping the variable with the lowest t-statistic, we perform a second robust regression with the remaining 12 variables. Elimination continues in this manner until, for all remaining variables, $|t| \geq 2$. The log of hydrocarbon pollution (X_1) survives this elimination, as do four other predictors: annual precipitation (X_2), mean January temperature (X_3), median education (X_4), and square root of percent non-White (X_5). Final robust coefficient estimates, standard errors, and t-tests are:

$$\begin{array}{lcccccc} \hat{Y}_i = & 1001.8 & +\,17.77X_{i1} & +\,2.32X_{i2} & -\,2.11X_{i3} & -\,19.1X_{i4} & +\,26.2X_{i5} \\ \text{SE}_b: & 82.5 & 4.63 & .64 & .50 & 6.2 & 4.4 \\ t: & 12.1 & 3.84 & 3.63 & -4.20 & -3.1 & 6.0 \\ P<: & .001 & .001 & .001 & .001 & .005 & .001 \end{array} \qquad [6.26]$$

Because we used statistical criteria to select X variables, the t-tests yield misleadingly low P-values (see Chapter 3). Even if actual probabilities are 50 times higher than shown in [6.26], however, all coefficients (except on median education) remain distinguishable from zero at $\alpha = .05$.

Equation [6.26] implies that death rates are highest in cold, rainy, and polluted cities with poorly educated, largely non-White populations. Adjusting for $X_1 - X_5$, effects of the remaining eight variables in Table 6.2 are indistinguishable from zero.

The four California cities identified earlier as bivariate outliers (Los Angeles, San

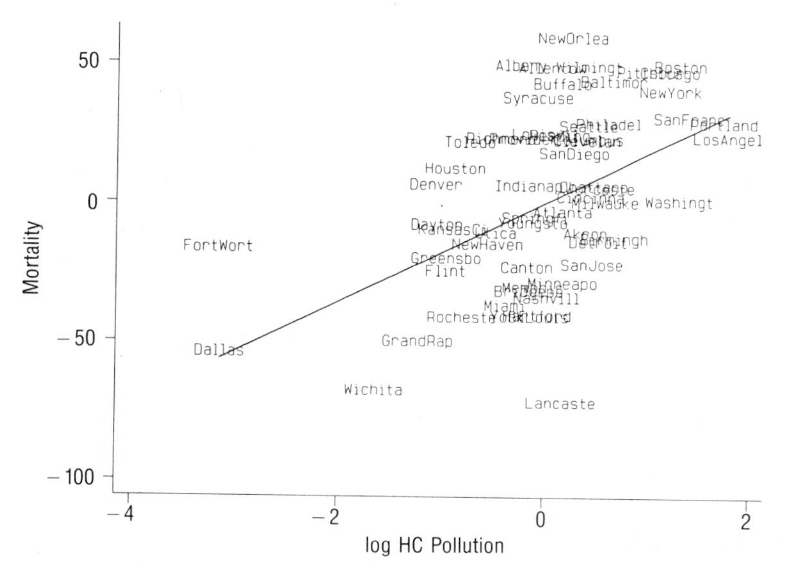

Figure 6.12 Leverage plot of mortality versus log pollution, adjusting for four other predictors.

Diego, San Francisco, and San Jose) are warmer in January, and much drier, than most U.S. metropolitan areas. These characteristics lower their predicted death rates, counterbalancing the effects of high pollution. Figure 6.12 shows a leverage plot of mortality versus log pollution, adjusting for January temperature, precipitation, education, and percent non-White. In this plot the California cities are not outliers. After adjustment for X_2 through X_5, their positions conform to the overall pattern of mortality increasing with pollution.

The IRLS estimation strategy described earlier applies equally well to bivariate and multiple regression, and it obtained Equation [6.26]. Figure 6.13 shows final-iteration weights. In contrast to the bivariate fit, which required 11 iterations and weights as low as .08 (Figures 6.7–6.8), the multivariate fit requires only four iterations and relatively mild reweighting. Equation [6.26] consequently does not differ much from its nonrobust OLS counterpart:

$$\begin{array}{lcccccc} \hat{Y}_i = & 986.3 & + 17.47X_{i1} & + 2.35X_{i2} & - 2.13X_{i3} & - 18.0X_{i4} & + 27.3X_{i5} \\ \text{SE}_b: & 82.7 & 4.64 & .64 & .50 & 6.2 & 4.4 \\ t: & 11.9 & 3.77 & 3.68 & -4.23 & -2.9 & 6.2 \end{array} \qquad [6.27]$$

In these multiple regressions, unlike the bivariate regressions seen earlier, the robust results support the OLS results. For simplicity, we might choose to report only the OLS results when writing up our research; a footnote saying that similar results were obtained by robust methods could increase readers' confidence in our conclusions.

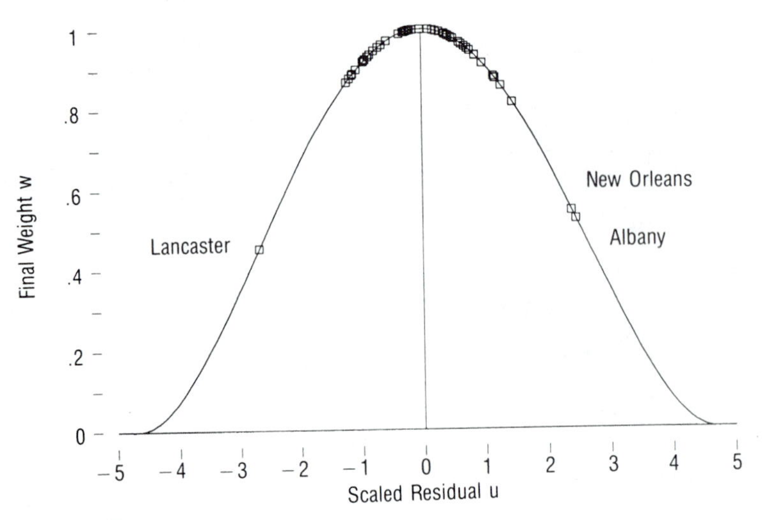

Figure 6.13 Weights versus scaled residuals from robust regression of mortality on five predictors.

This chapter reports four estimates of the coefficient on log hydrocarbon pollution:

	OLS	Robust	
One X variable	7.97	19.46	(from [6.1] and [6.2])
Five X variables	17.47	17.77	(from [6.27] and [6.26])

Four outliers kept the bivariate OLS estimate ($b_1 = 7.97$) low. After adjusting for climate and education, both OLS and robust regression produce similar, higher estimates, $b_1 = 17.47$ or $b_1 = 17.77$. Notice that the robust bivariate regression produced an estimate much closer to the multivariate results ($b_1 = 19.46$) because it downweighted outliers whose large residuals resulted from omitted variables (weather).

To complete the analysis, we should check for leverage. Table 6.3 lists the five cities with $h_i > .2$ (not high, but "risky"). Leverage reflects an unusual combination of X values. For example, Miami has the most rain and the warmest Januaries of

Table 6.3 The five highest-leverage cases in robust regression of mortality on X_1–X_5

Metropolitan Area	h_i (leverage)
Fort Worth	.24
Dallas	.25
San Diego	.26
Los Angeles	.28
Miami	.36

any city in the data. Los Angeles and San Diego are the least rainy and most polluted cities; Dallas and Fort Worth are the least polluted (in these 1963 data).

To see what effect these five leverage points have, we can redo the robust regression without them:

$$\hat{Y}_i = 954.2 + 22.3X_{i1} + 3X_{i2} - 1.94X_{i3} - 17.9X_{i4} + 23.3X_{i5}$$

SE_b:	89.3	6.6	0.8	.72	6.5	4.9
t:	10.7	3.4	3.8	−2.70	−2.8	4.8

[6.28]

Throwing out nearly 10% of the data is a drastic step, done here just for comparison with [6.26]. All coefficients shift a bit, but none become nonsignificant or change sign. The coefficient on log hydrocarbon pollution (X_1) actually increases about 25%. Standard errors for the $n = 55$ analysis ([6.28]) exceed those for the $n = 60$ analysis ([6.26]), but not by much.[8] Deleting the highest-leverage cases thus brings minor changes but does not alter substantive conclusions regarding mortality and air pollution.

Bounded-Influence Regression

The classic M-estimators described so far are robust against large errors in Y, unless these errors occur at extreme (high-leverage) X values. Unfortunately, leverage can undermine robust M-estimation just as it does OLS. *Bounded-influence* methods are designed to resist leverage. These methods tend to be less efficient than the best M-estimators but afford a higher degree of safety.

Bounded-influence regression represents the state of the art, with new developments being published even as this text is written.[9] Several different approaches have been proposed. This section describes a quick-and-dirty bounded-influence technique that can be accomplished via IRLS with a standard statistical package.[10] Methods promising still greater robustness exist but require more elaborate programs.[11]

Huber and biweight estimation downweight cases with unusual Y values (large residuals). To downweight unusual X values, we could also assign weights on the basis of leverage. For example, let c^H be the 90th sample percentile of h_i (leverage). That is, only the highest-leverage 10% of the sample have $h_i > c^H$. We perform an initial OLS regression and define leverage-based weights w_i^H:

$$w_i^H = 1 \qquad \text{if } h_i \le c^H$$
$$w_i^H = (c^H/h_i)^2 \qquad \text{if } h_i > c^H \qquad [6.29]$$

We proceed with IRLS estimation as usual, but, instead of applying just the Huber or biweight weights, w_i, at each step we weight by

$$w_i w_i^H \qquad [6.30]$$

Note that w_i changes with each iteration, whereas w_i^H remains constant.

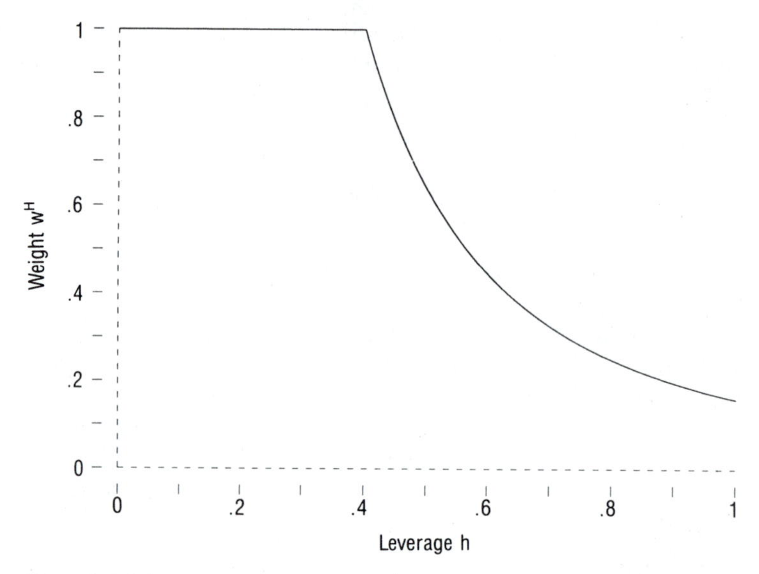

Figure 6.14 Bounded-influence weights versus leverage, $c^H = .4$.

For 90% of the cases, $w_i^H = 1$ and [6.30] is no different from [6.15] or [6.18]. For higher-leverage cases, w_i^H declines steeply. Figure 6.14 shows how [6.29] behaves with $c^H = .4$. Actual case weights depend on the combination of leverage-based weight ([6.29]) and residual-based weight ([6.15]) or [6.18]).

Standard errors and hypothesis tests for bounded-influence regression are more complicated than for the M-estimators described earlier. We will not consider them here but will view bounded-influence estimation only as a descriptive tool.

Table 6.4 contains data on PCB levels (hazardous industrial chemicals), measured in 1984 and 1985, in sediments of 37 U.S. estuaries (from the Council on Environmental Quality, 1988). If PCB levels are fairly stable, we might expect the regression of 1985 concentration (Y) on 1984 concentration (X) to yield a slope not far below 1. We actually obtain $\hat{Y} = 85 + .04X$ (OLS) or $\hat{Y} = 8.3 + .04X$ (robust M-estimate), because of the leverage case at right in Figure 6.15 (Boston Harbor, $h = .997$).[12] Both lines track this case and miss the steeper pattern of the other 36 points. Judged by residuals, or vertical distance from the lines, Boston appears less unusual than other cases—fooling the M-estimator into downweighting the wrong cases.

Faced with an extraordinary X value, M-estimation (which downweights extraordinary residuals) works no better than OLS. Bounded-influence regression, on the other hand, downweights Boston Harbor (to 0), finding a line that better fits most of the data (Figure 6.16):

$$\hat{Y} = 0.24 + .94X$$

This example is extreme; only a careless analyst would blunder ahead without noticing the problem. Someone who did notice might first try logarithms, which

Table 6.4 PCB contamination in sediments of 37 U.S. estuaries

Estuary	1984 PCB level (ppb)	1985 PCB level (ppb)
1 Casco Bay, ME	95.28	77.55
2 Merrimack River, MA	52.97	29.23
3 Salem Harbor, MA	533.58	403.10
4 Boston Harbor, MA	17,104.86	736.00
5 Buzzards' Bay, MA	308.46	192.15
6 Narragansett Bay, RI	159.96	220.60
7 East Long Island Sound, NY	10.00	8.62
8 West Long Island Sound, NY	234.43	174.31
9 Raritan Bay, NJ	443.89	529.28
10 Delaware Bay, DE	2.50	130.67
11 Lower Chesapeake Bay, VA	51.00	39.74
12 Pamilico Sound, NC	0	0
13 Charleston Harbor, SC	9.10	8.43
14 Sapelo Sound, GA	0	0
15 St. Johns River, FL	140.00	120.04
16 Tampa Bay, FL	0	0
17 Apalachicola Bay, FL	12.00	11.93
18 Mobile Bay, AL	0	0
19 Round Island, MS	0	0
20 Mississippi River Delta, LA	34.00	30.14
21 Barataria Bay, LA	0	0
22 San Antonio Bay, TX	0	0
23 Corpus Christi Bay, TX	0	0
24 San Diego Harbor, CA	422.10	531.67
25 San Diego Bay, CA	6.74	9.30
26 Dana Point, CA	7.06	5.74
27 Seal Beach, CA	46.71	46.47
28 San Pedro Canyon, CA	159.56	176.90
29 Santa Monica Bay, CA	14.00	13.69
30 Bodega Bay, CA	4.18	4.89
31 Coos Bay, OR	3.19	6.60
32 Columbia River Mouth, OR	8.77	6.73
33 Nisqually Beach, WA	4.23	4.28
34 Commencement Bay, WA	20.60	20.50
35 Elliott Bay, WA	329.97	414.50
36 Lutak Inlet, AK	5.50	5.80
37 Nahku Bay, AK	6.60	5.08

make the data more manageable.[13] Figure 6.17 shows the log/log relationship and uncovers another outlier: Delaware Bay. *M*-estimation and bounded-influence methods now produce similar lines, but outliers keep the OLS line less steep. They also make OLS estimation less efficient: the standard error of the OLS slope in Figure 6.17 is $SE_b = .054$, almost four times the standard error of the corresponding *M*-estimate ($SE_b = .015$). (Recall that outliers affect standard errors even more than they do slopes.) The section on bootstrap methods in Appendix 2 takes another look at this analysis.

OLS assigns all cases equal weight, which allows wild sample-to-sample fluctuations in the presence of outliers. *M*-estimation downweights cases with

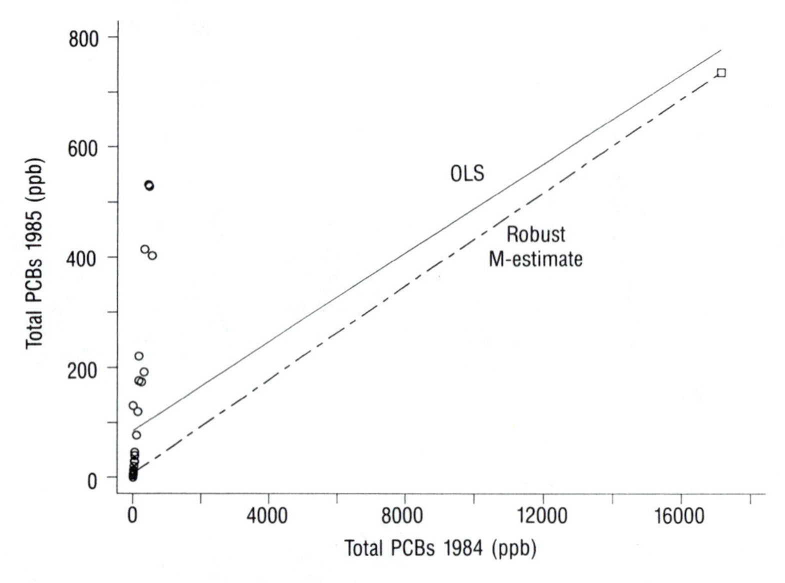

Figure 6.15 OLS and robust regression of 1985 on 1984 PCB levels; both lines track a high-leverage case (Boston Harbor).

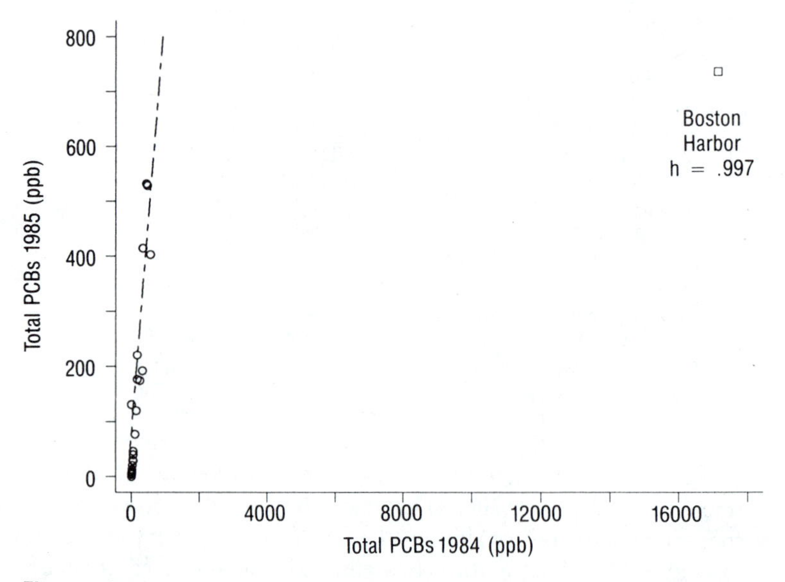

Figure 6.16 Bounded-influence regression resists pull of X-outlier (leverage case).

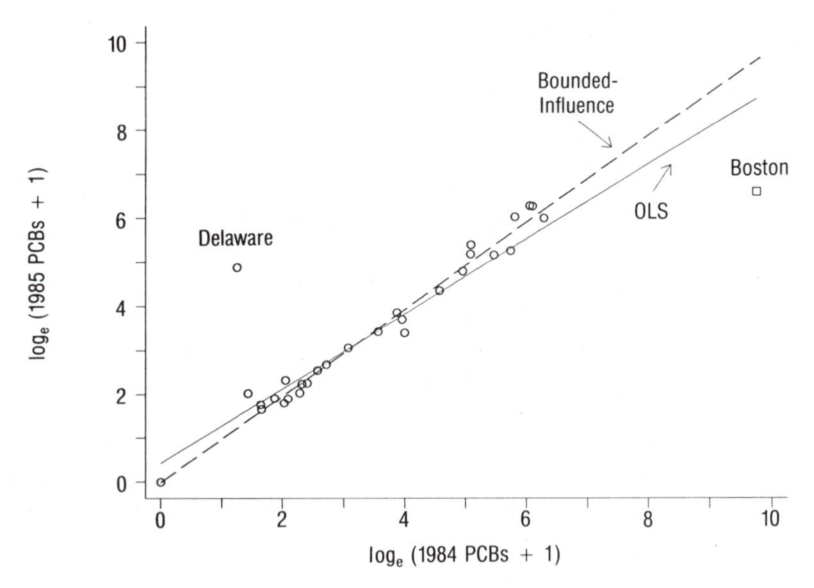

Figure 6.17 OLS and bounded-influence regressions with both variables as logarithms.

exceptional e values, making it more efficient when outliers result from large errors in Y. Bounded-influence regression downweights cases with exceptional e or X values and so is better when either Y or X contains large errors. In this respect, bounded influence offers greater safety against bad data. On the other hand, if the X values are *not* bad, bounded-influence regression may disregard valuable information and perform less efficiently than M-estimation.

A simple modification reduces M-estimators' susceptibility to leverage: on the first iteration, calculate Cook's D and drop cases with $D_i > 1$. Although this process lacks the smoothness and versatility of bounded-influence regression, it does offer improved robustness with little extra effort. Such screening would have eliminated Boston Harbor ($D_i = 6117$) from the regressions of Figure 6.15, making M-estimation produce much better results.

Conclusion

Applied to outlier-filled data, robust methods have better statistical properties than OLS. Their theoretical advantages include efficiency (less sample-to-sample variation) and more accurate confidence intervals and tests, since they resist extreme values and do not assume normality.

Robust regression can be employed routinely to critique OLS. If robust and OLS results agree, that's good—OLS passes one diagnostic check, enhancing confidence in our conclusions. If robust and OLS results disagree, on the other hand, we have a problem. Robust results are generally more convincing, since they should reflect the bulk of the data, not just a few outliers. We can examine the outliers (cases given low weights) for clues about the source of the problem and what to do about it. Thus, as a critical tool, robust regression contributes three things:

1. It alerts us that a problem exists.
2. It provides a model that better fits most of the data.
3. It more reliably identifies outliers.

If we are in a hurry, unable to carefully replicate and study each analysis, robust regression is a safer choice than OLS. Because they are safer, robust methods might be recommended to inexperienced analysts. The degree of protection robust estimation affords is limited, though, and should not be exaggerated.

Robust M-estimation protects mainly against unusual Y values. Bounded-influence methods add protection against unusual X values as well. (Less systematically, deleting cases with high D_i or h_i also adds such protection.) Against other statistical problems, including curvilinearity, heteroscedasticity, and autocorrelation, these methods may be no safer than OLS. Alternative robust methods have been devised specifically for some of these problems, but they too have weaknesses. No single method can be universally robust, so careful diagnostic work remains valuable.

Exercises

When oil prices rose during the 1970s, wood stoves came back into fashion for heating in some parts of the country. Although it often is cheaper than other heat sources, wood burning pollutes both outdoor and indoor air. The next table, from experiments by Traynor, Apte, Carruthers, Dillworth, Grimsrud, and Gundel (1987), includes measures of the peak indoor carbon monoxide (CO) levels during 11 tests of wood-burning stoves. Robust methods are particularly appropriate here due to two unusual tests (9 and 10): the stove overheated, possibly due to overfilling with wood, and experimenters reduced airflow by using a damper that caused the house to fill with smoke. Such incidents are common with nonairtight stoves, especially with inexperienced operators. Robust estimation allows us to look for further patterns in the data without being blinded by the extreme Y values of these two tests. Exercises 1–4 use these data.

Indoor carbon monoxide pollution from wood-burning stoves

Test	Stove Type	Burning Time (hours)	Amount of Wood Burned (kg)	Peak House CO (ppm)
1	Airtight	14.8	37.3	2.8
2	Airtight	8.8	38.4	1.2
3	Airtight	13.0	21.2	1.6
4	Airtight	13.7	27.2	2.0
5	Airtight	18.5	40.6	1.2
6	Airtight	18.0	43.2	1.4
7	Airtight	16.1	24.2	3.8
8	Nonairtight	8.7	24.4	7.7
9	Nonairtight	10.4	32.4	35.0
10	Nonairtight	5.4	23.2	43.0
11	Nonairtight	9.5	38.6	3.5

1. Using OLS, regress peak CO on burning time. Is the relation significant? Does it make sense?
2. Using robust M-estimation (and screening for $D_i > 1$), regress peak CO on burning time. Is this relation significant? Construct a scatterplot with OLS and robust lines. Refer to this plot and the robust weights to explain why OLS and robust results differ.
3. Regress peak CO on amount of wood burned, using both OLS and M-estimation. Graph and compare the results.
4. With appropriate dummy X variables, robust regression becomes robust analysis of variance or a two-sample t-test (see Chapter 3). Regress (using first OLS, then M-estimation) peak CO on stove type, as a dummy variable coded $0 =$ airtight, $1 =$ nonairtight. Interpret the regression equations; then graph and compare them. Why is the robust mean for nonairtight stoves so low?

The next table contains data from Nigerian field trials of the drug Immobilon, used to temporarily immobilize wild animals for research (from Okaeme, Agbelusi, Mshelbwala, Wari, Ngulge, and Haliru, 1988). Researchers shot kob (a short-horned antelope) with darts containing Immobilon and then observed how much distance the animal covered before dropping. Subsequently the animals were revived with another drug.

Immobilon trials on nine western kob

Animal	Dart dosage relative to weight (ml/kg)	Distance covered before dropping (m)
1	.020	850
2	.011	950
3	.022	750
4	.023	3000
5	.033	100
6	.020	350
7	.026	450
8	.021	1300
9	.017	1150

Exercises 5–6 refer to these data.

5. Perform OLS and robust regression of distance covered (Y) on relative dose (X). Graph the results.
6. With kob #4, the tranquilizer dart failed to administer the intended dosage; the animal actually received only about half the relative dose shown in the data. Hence the unusual Y value for this case reflects an experimental error. How do your OLS and robust analyses differ in their treatment of kob #4?

The following data on 20 New York river basins were originally collected by Haith (1976) to explore relationships between nonpoint-source water pollution (nitrogen concentration) and various types of land use. Haith's data were subsequently reprinted by Allen and Cady (1982) and reanalyzed by Simpson, Ruppert, and Carroll (1989). Simpson et al. discuss this "particularly vexing dataset" and present results based on several robust methods. Variables are:

X_1—the percentage of land in active agriculture
X_2—the percentage of land forested, brushland, or plantation
X_3—the percentage of land urban (including residential, commercial, and industrial)
Y—the nitrogen concentration in river water

Land use and nitrogen content in 20 river basins

Basin	% Land in Agriculture	% Land Forests	% Land Urban	Nitrogen (mg/1)
1 Olean	26	63	1.49	1.10
2 Cassadaga	29	57	.79	1.01
3 Oatka	54	26	2.38	1.90
4 Neversink	2	84	3.88	1.00
5 Hackensack	3	27	32.51	1.99
6 Wappinger	19	61	3.96	1.42
7 Fishkill	16	60	6.71	2.04
8 Honeoye	40	43	1.54	1.65
9 Susquehanna	28	62	1.25	1.01
10 Chenango	26	60	1.13	1.21
11 Tioughnioga	26	53	1.08	1.33
12 West Canada	15	75	.86	.75
13 East Canada	6	84	.62	.73
14 Saranac	3	81	1.15	.80
15 Ausable	2	89	1.05	.76
16 Black	6	82	.65	.87
17 Schoharie	22	70	1.12	.80
18 Raquette	4	75	.58	.87
19 Oswegatchie	21	56	.63	.66
20 Cohocton	40	49	1.23	1.25

Exercises 7–10 use these data.

7. Using OLS, regress nitrogen concentration (Y) on percentage of land in agriculture (X_1), percentage of land forested (X_2), and percentage of land urban (X_3). Calculate leverage and Cook's D, and construct an e-versus-$\hat{Y}$ plot. On this plot, identify any cases with $h_i > .5$ or $D_i > 1$.
8. Delete cases with $D_i > 1$, and perform a robust regression. Compare robust and OLS results.
9. Construct a weight-versus-residual plot. Which cases were most downweighted, and why?
10. The three types of land use do not quite add up to 100%, because there are other possible uses. Nonetheless, we may wonder whether multicollinearity is a problem with these data. Standard robust techniques are not robust against multicollinearity. Use the diagnostic methods of Chapter 4 (correlations between coefficients; tolerance) to explore for possible multicollinearity here. What are its likely effects?

11. Exercises 12–14 of Chapter 3 used data on beryllium-exposed mine workers. OLS regression of LT ratio on age, employment, and exposure detects no significant relations. Repeat this regression with M-estimation, and again test for significance. Why the change?
12. Perform a bounded-influence regression of mortality on hydrocarbon pollution, using data from Table 6.1. Graph the resulting line, and compare your results with Figures 6.9 and 6.10.

Notes

1. Data from McDonald and Ayers (1978) and McDonald and Schwing (1973); reprinted in Goodall (1983).
2. The pollution potential is determined as the product of the tons of pollutant emitted per day per square kilometer and a dispersion factor that accounts for rising height, wind speed, number of episode days, and dimensions of each metropolitan area. Figures are for the year 1963.
3. As Peter Huber notes, "'distributionally robust' and 'outlier resistant,' although conceptually distinct, are practically synonymous notions" (1981:4).
4. "Efficient" means that an estimator has lower variance ([A1.26]) than other estimators for a given model. Monte Carlo experiments in Appendix 2 compare the efficiencies of OLS and robust regression.
5. See Carroll and Welsh (1988).
6. A 25% trimmed mean, for example, is the sample mean recalculated after the highest 25% and the lowest 25% of the cases are deleted. Trimming attempts to provide simple protection against outliers, but it has drawbacks. If we trim too much, or trim both tails of a distribution with only one heavy tail, we lose efficiency. If we don't trim enough, we lose protection against outliers. These problems arise because a case must either be fully included or completely excluded; there is no smooth transition of in-between weights, as there is with the robust M-estimators described in this chapter.
7. As Rey (1983:9) notes, "what is estimated when the model assumption is erroneous may be obscure. This is the conceptual difficulty which lies at the root of all robustness principles."
8. Leverage points—especially "good" outliers (see Chapter 4)—can *decrease* standard errors. Leverage points that are "bad" outliers may either increase or decrease standard errors. Other things being equal, deleting cases (reducing sample size) increases standard errors.
9. For example, see Simpson, Ruppert, and Carroll (1989) and the literature they cite.
10. Suggested by David Ruppert, personal communication, 5/30/90.
11. For example, we might gain robustness by replacing h_i in [6.29] with a robust measure of leverage. Outliers affect h_i, as they do other OLS statistics. *Masking* can occur: a case exerts so much influence that the model fits it closely, and diagnostic statistics like h_i fail to detect its unusualness. Several authors have

proposed robust alternatives to OLS diagnostic statistics that could be utilized in robust regression (see Rousseeuw and Van Zomeren, 1990; Ruppert and Simpson, 1990).

12. Boston Harbor's pollution problems became an issue during the 1988 presidential campaign, when George Bush used them to attack the environmental record of his rival, Massachusetts Governor Michael Dukakis.
13. Since the raw data contain zeros, take $\log(X + 1)$ and $\log(Y + 1)$.

7 Logit Regression

The linear regression models of Chapters 2–6 assume measurement Y variables. Categorical Y variables require a different approach. Several alternatives exist; *logit regression* (also called *logistic regression*) is perhaps the most widely known. Like linear regression, logit regression provides a flexible, general-purpose modeling strategy with straightforward interpretation. In practical use the two methods have many similarities, although their underlying mathematical bases are different.

Many research questions, especially in social and medical science, involve trying to predict the probability that something will happen—for example, the probability that people will vote for Smith or contract AIDS. Such questions involve two-category (*dichotomous*) Y variables: vote/not vote or AIDS/no AIDS. Techniques for modeling dichotomies extend to Y variables with three or more categories (*polytomous* or *polychotomous*), but with some increase in complexity. This chapter looks only at dichotomous examples. [For polytomous examples, consult Agresti (1990) or Hosmer and Lemeshow (1989).] After presenting basic ideas of logit regression, with an emphasis on interpretation, the chapter concludes with a discussion of diagnostic statistics and graphs.

Limitations of Linear Regression

OLS is sometimes used for "quick and dirty" analysis of models with dichotomous Y variables. We can regress a dummy $\{0, 1\}$ Y variable on X variables in the usual manner; then we interpret coefficients and predicted values with reference to the predicted probability that $Y = 1$. Although attractively simple, this approach runs into several problems.

Figure 7.1 illustrates dichotomous linear regression, using survey data from a small Vermont town where toxic wastes had contaminated the grounds of two public schools. Some townspeople thought the schools should be closed until proven safe. Others opposed this costly step and said that the schools should stay open. The Y variable in Figure 7.1 is opinion about the schools (1 = close, 0 = stay open); the X variable is the length of time the respondent had lived in town. The regression line

$$\hat{Y}_i = .594 - .008X_{i1} \tag{7.1}$$

indicates a weak negative relationship ($R^2 = .08$) between Y and X_1: long-term residents were less likely to favor closing the schools. Since all data points fall along two horizontal lines, the scatterplot itself is hard to read; parallel boxplots (Figure 7.2) show the situation more clearly.

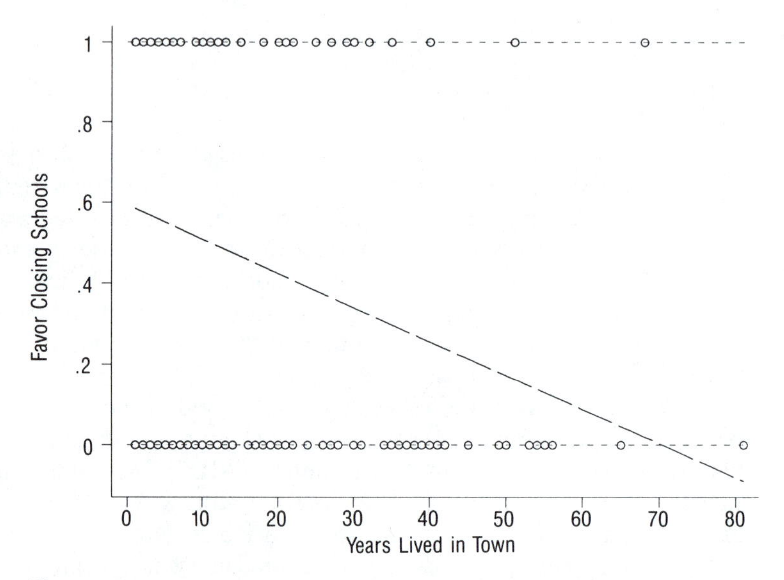

Figure 7.1 Linear regression of a dichotomous Y variable (0 = open schools, 1 = close schools) on a measurement X variable (years lived in town).

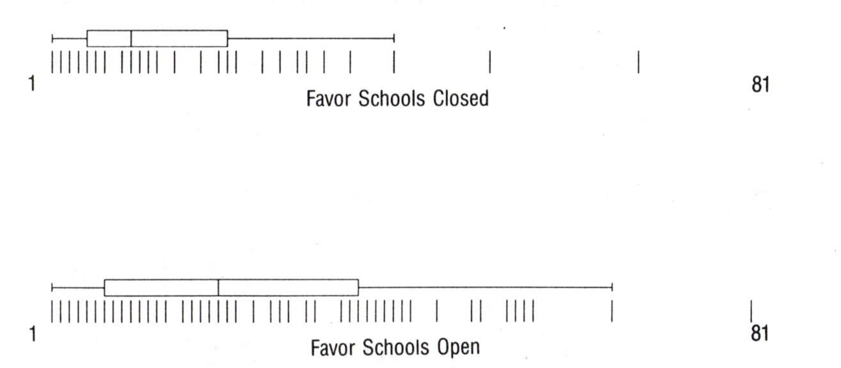

Figure 7.2 Boxplots and oneway scatterplots of years lived in town, for respondents favoring closed and open schools.

For any value of X_1 there exist only two possible residuals:

If $Y_i = 1$, then $e_i = 1 - (.594 - .008X_{i1})$

If $Y_i = 0$, then $e_i = -(.594 - .008X_{i1})$

By similar reasoning, with any dichotomous Y variable, errors (ε_i) have only two possible values for each X_i. If a linear model $E[Y_i] = \beta_0 + \beta_1 X_{i1}$ is correct, the variance of these errors depends upon $E[Y_i]$ and therefore upon X_i, making a constant-variance (homoscedasticity) assumption untenable.[1] OLS coefficient estimates are then unbiased but not efficient. Furthermore, standard error estimates are biased, invalidating hypothesis tests and confidence intervals.

A weighted least squares (WLS) technique may overcome the problems caused by heteroscedasticity. Linear models also have a more fundamental limitation, however.

Predicted values from [7.1] might be interpreted as probabilities. For a person who has lived 10 years in this town:

$$\hat{Y} = .594 - .008(10)$$
$$= .514$$

We estimate that the probability equals .514 that a 10-year resident, selected at random, will favor closing the schools. This probability falls to .034 for a 70-year resident. For each additional year lived in town, the predicted probability decreases by $b_1 = -.008$. These straightforward interpretations break down at the extreme, however. For 80-year residents the predicted "probability" equals $.594 - .008(80) = -.046$—but by definition, probabilities *cannot* be less than 0 (or greater than 1).

Any linear model with nonzero slope eventually predicts values greater than 1 or less than 0. Impossible predictions derived from reasonable X values tell us that the model is unrealistic, which implies that we are estimating parameters that do not exist. To more realistically model probabilities, we need a function that approaches, but never exceeds, the $\{0, 1\}$ boundaries.

The Logit Regression Model

$P(Y = 1)$ denotes the probability that a $\{0, 1\}$ Y variable equals 1. The probability that Y does not equal 1 is

$$P(Y \neq 1) = P(Y = 0) = 1 - P(Y = 1)$$

The *odds* favoring $Y = 1$ are

$$\mathcal{O}(Y = 1) = \frac{P(Y = 1)}{1 - P(Y = 1)}$$

Odds range from 0 (when $P(Y = 1) = 0$) to ∞ (when $P(Y = 1) = 1$).

Suppose $Y = 1$ indicates that it rains today and $Y = 0$ indicates that it does not. If the probability of rain today is $P(Y = 1) = .2$, then the probability of no rain is $1 - P(Y = 1) = .8$. The odds of rain today are

$$\begin{aligned}\mathcal{O}(Y = 1) &= \frac{P(Y = 1)}{1 - P(Y = 1)} \\ &= \frac{.2}{.8} \\ &= .25\end{aligned}$$

These odds could be stated as .25 to 1, or 1 to 4. Thus a .2 probability amounts to 1-to-4 odds. Alternatively, we could specify the odds *against* rain today:

$$\begin{aligned}\mathcal{O}(Y \neq 1) &= \frac{1}{\mathcal{O}(Y = 1)} \\ &= \frac{1}{.25} \\ &= 4\end{aligned}$$

or 4 to 1. The following discussion shortens $P(Y = 1)$ to P and $\mathcal{O}(Y = 1)$ to $\mathcal{O}$.

By taking the natural logarithm of the odds, we obtain a *logit*:

$$\begin{aligned}L &= \log_e \mathcal{O} \\ &= \log_e\{(P)/(1 - P)\}\end{aligned} \qquad [7.2]$$

Logits range from $-\infty$ (when $P = 0$) to ∞ (when $P = 1$).

Logit regression refers to models with a logit as left-hand-side variable:

$$L_i = \beta_0 + \beta_1 X_{i1} + \beta_2 X_{i2} + \cdots + \beta_{K-1} X_{i,K-1} \qquad [7.3]$$

If the logit (L) is a linear function of X variables, then probability (P) is a nonlinear, S-shaped function like that in Figure 7.3. Predicted probabilities approach, but never reach or exceed, the boundaries of 0 and 1. Thus logit regression provides a more realistic model for probabilities than does linear regression.[2]

Given a set of X values and estimated coefficients, we can estimate logits ($\hat{L}$) much as we do $\hat{Y}$ in a linear regression. Reversing the logit transformation yields predicted probabilities that $Y = 1(\hat{P})$:

$$\hat{P} = \frac{1}{1 + e^{-\hat{L}}} \qquad [7.4]$$

This reexpression is useful for graphing.

Unless X strongly affects Y, graphing $\hat{P}$ over the data's X range will not show a complete S-curve. Instead, we will see a partial curve like that in Figure 7.4, which graphs

$$\hat{L}_i = .460 - .041 X_{i1} \qquad [7.5]$$

where $\hat{L}$ is the predicted logit of favoring school closings because of toxic-waste contamination and X_1 is years lived in town. (A negative logit coefficient implies that the curve goes down to right, as in Figure 7.4; Figure 7.3 depicts a positive relation.) For respondents who just moved to town ($X_1 = 0$), the predicted logit is $L = .460 - .041(0) = .460$. Their predicted probability of favoring school closing

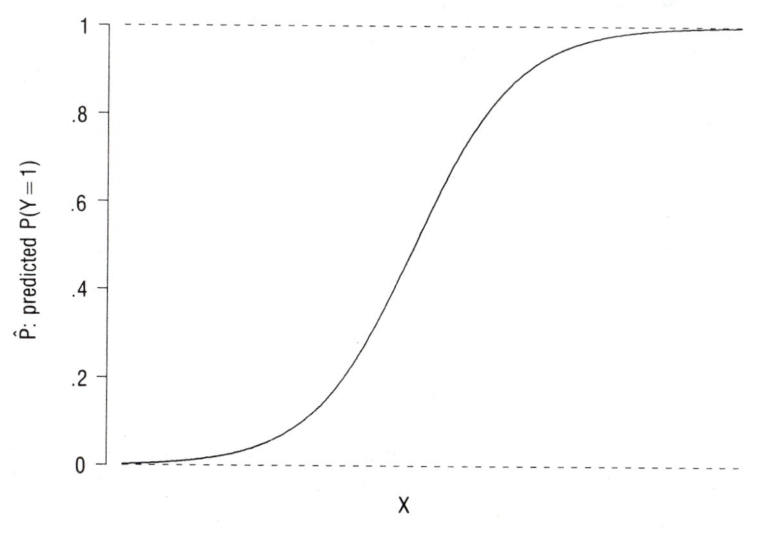

Figure 7.3 Predicted probability as a logit function of X.

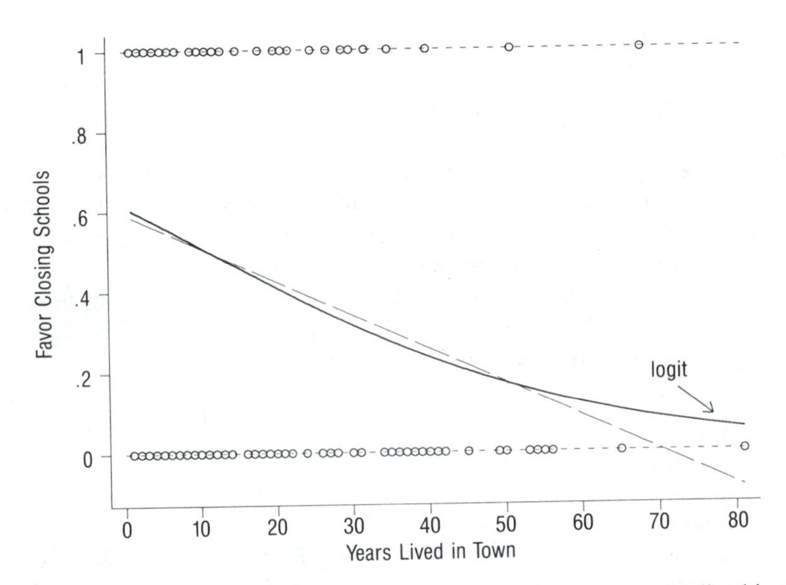

Figure 7.4 Logit regression of school-closing opinion on years lived in town, also showing linear regression line.

is therefore $\hat{P} = 1/[1 + e^{-(.460)}] = .613$. For 80-year residents we predict $\hat{L} = .460 - .041(80) = -2.82$, or $\hat{P} = 1/[1 + e^{-(-2.82)}] = .056$. Thus, over the X_1 range from 0 to 80 years, the probability of favoring closed schools declines from .613 to .056.

For comparison, an OLS line (Figure 7.1) appears again in Figure 7.4. The line and curve begin close together but separate as they near probability 0 (impossible). The line continues downward, predicting probabilities even lower than "impossible," whereas the curve's descent slows to approach, but never reach, this boundary. In general, logit and linear predictions become increasingly different as probabilities near 0 or 1.

Linear models are easy to interpret:

$$E[Y_i] = \beta_0 + \beta_1 X_{i1} + \beta_2 X_{i2}$$

implies that $E[Y] = \beta_0$ when X_1 and X_2 equal 0. $E[Y]$ rises by β_1 units with each one-unit increase in X_1, if X_2 does not change. Although logit models are linear with respect to the logits, they are nonlinear with respect to odds or probabilities. Parameters in a logit model

$$L_i = \beta_0 + \beta_1 X_{i1} + \beta_2 X_{i2}$$

could be interpreted as follows:

β_0: When $X_1 = X_2 = 0$, the odds favoring $Y = 1$ are $\mathcal{O} = e^{\beta_0}$ (where $e = 2.71828\ldots$ is the base number of natural logarithms). The probability that $Y = 1$ is $P = 1/(1 + e^{-\beta_0})$.

β_1: Each one-unit increase in X_1 multiplies the odds favoring $Y = 1$ by e^{β_1}, if X_2 stays the same. Another way to say this is that the odds favoring $Y = 1$ change by $100(e^{\beta_1} - 1)$ percent with each one-unit increase in X_1. The probability that $Y = 1$ changes by an amount that depends not only on β_1 but also on all other terms in the equation: β_0, X_1, β_2, and X_2.

The equation $\hat{L}_i = .460 - .041X_{i1}$ indicates that every one-year increase in X_1 multiplies predicted odds of wanting closed schools by $e^{-.041} = .96$. In other words, the odds change (decrease) by about $100(e^{-.041} - 1) = -4\%$ per year. To express this in terms of probabilities, select specific X values or refer to a graph. As Figure 7.4 shows, the probability of wanting closed schools decreases most steeply for about 0–40 years residency. Additional years have a diminishing effect.[3]

Logit models have theoretical and statistical advantages over linear models for dichotomous Y variables. The next section describes their estimation.

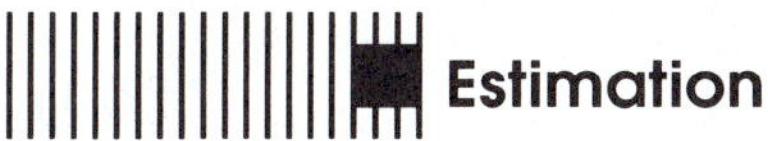

Estimation

Logit models are commonly estimated by maximum likelihood rather than by least squares. A *likelihood function* expresses the probability of obtaining the observed sample as a function of model parameters. Maximum likelihood methods ask: What parameter values make our sample most likely?

Let X_i stand for the ith combination of X values. Based on a logit model, the conditional probability that $Y_i = 1$ is

$$P_i = \frac{1}{(1 + e^{-L_i})} \qquad [7.6]$$

where

$$L_i = \beta_0 + \sum_{k=1}^{K-1} \beta_k X_{ik} \qquad [7.7]$$

The contribution of the ith case to the likelihood function equals P_i if $Y_i = 1$, and it equals $1 - P_i$ if $Y_i = 0$. We could write this contribution as

$$P_i^{Y_i}(1 - P_i)^{1 - Y_i}$$

Assuming that the cases are independent (no autocorrelation), the likelihood function itself is the product of these individual contributions:

$$\mathscr{L} = \Pi\{P_i^{Y_i}(1 - P_i)^{1 - Y_i}\} \qquad [7.8]$$

Π is a *multiplication operator*, analogous to the summation operator Σ.

We seek estimates of the β parameters that yield the highest possible values for the likelihood function, Equation [7.8]. Equivalently, we maximize the logarithm of [7.8], called the *log likelihood*:

$$\log_e \mathscr{L} = \Sigma\{Y_i \log_e P_i + (1 - Y_i)\log_e(1 - P_i)\} \qquad [7.9]$$

Logarithms convert multiplication into addition, making the log likelihood easier to work with.

To find maximum likelihood estimates, take first derivatives of the log likelihood with respect to each of the estimated parameters, and then set these derivatives equal to zero. This results in simultaneous equations:

$$\Sigma(Y_i - P_i) = 0 \qquad [7.10]$$

and

$$\Sigma(Y_i - P_i)X_{ik} = 0 \qquad \text{for } k = 1, 2, 3, \ldots, K - 1 \qquad [7.11]$$

These equations are nonlinear in the parameters and cannot be solved directly (unlike the normal equations for OLS). Instead, we resort to an iterative procedure, in which the computer finds successively better approximations for β_k values that satisfy [7.10]–[7.11].

Table 7.1 shows computer output for the school-closing example described earlier (Equation [7.5]). Logit regression output resembles OLS output, but note the list of log likelihoods ([7.9]) at upper left. At iteration 0, no parameters except the intercept have been estimated. On successive iterations, parameter estimates improve, causing the log likelihood to increase. The process stops when the relative change in each coefficient drops below .0001.[4]

Table 7.1 Logit regression of school-closing opinion on years lived in town.

Iteration 0: Log Likelihood = −104.60578
Iteration 1: Log Likelihood = −97.80942
Iteration 2: Log Likelihood = −97.634236
Iteration 3: Log Likelihood = −97.633571

Logit Estimates — Number of obs = 153; chi2(1) = 13.94; Prob > chi2 = 0.0002

Log Likelihood = −97.633571

Variable	Coefficient	Std. Error	t	Prob > $\|t\|$	Mean
close					.4313725
lived	−.0409876	.01214	−3.376	0.001	19.26797
_cons	.4599786	.2625656	1.752	0.082	1

Although the estimation strategies differ, logit regression requires some of the same assumptions as OLS:

1. The model is specified correctly. For logit regression this means that true conditional probabilities are a logistic function (or, logits are a linear function) of the X variables. No important variables are omitted, and no extraneous variables are included. X variables are measured without error.
2. The cases are independent.
3. None of the X variables are linear functions of the others. Perfect multicollinearity makes estimation impossible; strong multicollinearity makes estimates imprecise.

Influential cases also present problems for logit regression, as they do for OLS.

If these conditions are met, maximum likelihood estimates of logit parameters should, theoretically, have the desirable properties of unbiasedness, efficiency, and normality—in large enough samples. There exist several rules of thumb for what constitutes a large-enough sample (for example, $n - K$ should exceed 100). The issue of sample size is tricky, however, because the statistical properties of logit estimates depend not just on the overall sample size but also on the number of cases with a given combination of X and Y values. Skewed Y distributions are particularly troublesome. A sample of 200 cases, but only 5 cases with $Y = 1$, provides little information about the partial effects of several X variables.

Hypothesis Tests and Confidence Intervals

Hypothesis tests for logit regression often employ a *nested-models* strategy similar to F-tests in OLS. Let $\log_e \mathscr{L}_K$ represent the log likelihood ([7.9]) of a logit model with K parameters ($K - 1$ X variables). To test whether this model significantly improves upon a simpler model with H fewer predictors ($0 < H < K$), compare $\log_e \mathscr{L}_K$ with $\log_e \mathscr{L}_{K-H}$ (the log likelihood of the simpler model). The null hypothesis that the H omitted X variables have no effect,

$$H_0: \beta_1 = \beta_2 = \cdots = \beta_H = 0$$

is tested by

$$\chi^2_H = -2(\log_e \mathscr{L}_{K-H} - \log_e \mathscr{L}_K) \qquad [7.12]$$

which follows a theoretical χ^2 (chi-square) distribution with H degrees of freedom. If this test rejects H_0, the more complex model fits significantly better.

To test the null hypothesis that all X variables' coefficients are zero (similar to an overall F-test in OLS), apply [7.12] with the iteration-0 log likelihood as $\log_e \mathscr{L}_{K-H}$ and the final-iteration log likelihood as $\log_e \mathscr{L}_K$. For example, Table 7.2 displays the logit regression of school-closing opinion on four predictors:

lived (X_1): years respondent had lived in the town;
educ (X_2): respondent's education, in years;
contam (X_3): whether respondent believed his/her own property or water had been affected by the chemical contamination; and
hsc (X_4): whether respondent attended meetings of the Health and Safety Committee, a citizen's group that organized in response to the contamination crisis.

The log likelihood at iteration 0, with no parameters estimated except the intercept, is $\log_e \mathscr{L} = -104.60578$. After the final iteration the log likelihood is $\log_e \mathscr{L} = -74.690816$. Following [7.12]:

$$\chi^2 = -2[-104.60578 - (-74.690816)]$$
$$= 59.83$$

Table 7.2 prints this statistic at center right. The probability of a greater χ^2, with four degrees of freedom (the final model includes four more parameters than the initial intercept-only model), is $P < .00005$. We reject the null hypothesis that coefficients on all four variables are zero.

Equation [7.12] can test the significance of any individual coefficient or any set of coefficients. Table 7.3 shows a model like Table 7.2 but with three further X

Table 7.2 Logit regression of school-closing opinion on years lived in town, education, contamination, and HSC meetings

Iteration 0: Log Likelihood = −104.60578
Iteration 1: Log Likelihood = −76.104878
Iteration 2: Log Likelihood = −74.725772
Iteration 3: Log Likelihood = −74.690849
Iteration 4: Log Likelihood = −74.690816

Logit Estimates — Number of obs = 153

chi2(4) = 59.83

Log Likelihood = −74.690816 — Prob > chi2 = 0.0000

Variable	Coefficient	Std. Error	t	Prob > $\|t\|$	Mean
close					.4313725
lived	−.0464826	.0149263	−3.114	0.002	19.26797
educ	−.1659221	.0899317	−1.845	0.067	12.95425
contam	1.208137	.465396	2.596	0.010	.2810458
hsc	2.17289	.4641194	4.682	0.000	.3071895
_cons	1.731439	1.301999	1.330	0.186	1

Table 7.3 Logit regression of school-closing opinion on seven background variables

Iteration 0: Log Likelihood = −104.60578
Iteration 1: Log Likelihood = −73.307756
Iteration 2: Log Likelihood = −70.718684
Iteration 3: Log Likelihood = −70.526461
Iteration 4: Log Likelihood = −70.52469
Iteration 5: Log Likelihood = −70.524689

Logit Estimates

Log Likelihood = −70.524689

Number of obs = 153
chi2(7) = 68.16
Prob > chi2 = 0.0000

Variable	Coefficient	Std. Error	t	Prob > $\|t\|$	Mean
close					.4313725
lived	−.0466422	.0169751	−2.748	0.007	19.26797
educ	−.2060233	.093197	−2.211	0.029	12.95425
contam	1.282082	.4813682	2.663	0.009	.2810458
hsc	2.418002	.5096638	4.744	0.000	.3071895
female	−.0515618	.5571215	−0.093	0.926	.6078431
kids	−.6706227	.5656146	−1.186	0.238	.5882353
nodad	−2.225988	.9991178	−2.228	0.027	.1699346
_cons	2.893725	1.602985	1.805	0.073	1

variables:

female (X_5): dummy variable coded 1 for female, 0 for male;
kids (X_6): dummy variable coded 1 if respondent has children under 19 living in town, 0 otherwise; and
nodad (X_7): interaction of term, coded 1 if respondent is male *and* has no children in town, 0 otherwise.

The χ^2 statistic at upper right ($\chi^2 = 68.16$, $P < .00005$) indicates that we can again reject the hypothesis that all coefficients are zero. But the immediate question is whether the coefficients added to the model since Table 7.2 are zero. That is, do the three new predictors bring a significant improvement in fit?

To test the hypothesis that the coefficients on all three added variables are zero, compare the final log likelihood from Table 7.3 with that for the simpler model of Table 7.2. Applying [7.12]:

$$\chi^2 = -2[-74.690816 - (-70.524689)]$$
$$= 8.33$$

With three degrees of freedom (because there are three more parameter estimates in Table 7.3 than in Table 7.2), we obtain $P < .05$ and so reject the hypothesis that all three are zero.

Table 7.4 Reduced model with male/nonparent interaction term

Iteration 0: Log Likelihood = – 104.60578
Iteration 1: Log Likelihood = –73.813367
Iteration 2: Log Likelihood = –71.47445
Iteration 3: Log Likelihood = –71.327206
Iteration 4: Log Likelihood = –71.326227
Iteration 5: Log Likelihood = –71.326227

Logit Estimates — Number of obs = 153; chi2(5) = 66.56; Prob > chi2 = 0.0000

Log Likelihood = –71.326227

Variable	Coefficient	Std. Error	*t*	Prob > |*t*|	Mean
close					.4313725
lived	–.0396488	.0154812	–2.561	0.011	19.26797
educ	–.1966667	.0926128	–2.124	0.035	12.95425
contam	1.298551	.4766294	2.724	0.007	.2810458
hsc	2.27855	.4903703	4.647	0.000	.3071895
nodad	–1.730948	.7252746	–2.387	0.018	.1699346
_cons	2.182273	1.330141	1.641	0.103	1

Might one or two of these coefficients be zero? Experimentation reveals that the interaction term *nodad* (X_7) is responsible for most of the improvement between Tables 7.2 and 7.3. Table 7.4 shows a model from which *female* (X_5) and *kids* (X_6) have been dropped. χ^2 tests can answer the following questions:

1. Does the model with 5 predictors (Table 7.4) fit significantly worse than the 7-predictor model (Table 7.3)? The answer is no:

$$\chi^2 = -2[-71.326227 - (-70.524689)]$$
$$= 1.60$$

With two degrees of freedom, $P > .30$.

2. Does the 5-predictor model (Table 7.4) fit significantly better than the 4-predictor model (Table 7.2)? The answer is yes:

$$\chi^2 = -2[-74.690816 - (-71.326227)]$$
$$= 6.73$$

With one degree of freedom, $P < .01$.

We therefore need the interaction term *nodad*, but not separate effects for *female* and *kids*, in the model.

Testing hypotheses in this manner is somewhat clumsy, since for each test we estimate two different models and calculate minus twice the difference in log likelihoods by hand. A shortcut utilizes the *t*-statistics usually printed with logit output.

These t-statistics resemble their OLS counterparts:

$$t = \frac{b_k}{\mathrm{SE}_{b_k}} \qquad [7.13]$$

Coefficient standard errors (SE_{b_k}) are calculated as part of the maximum likelihood estimation process.[5] If $\beta_k = 0$, the ratio b_k/SE_{b_k} is asymptotically normally distributed. With large samples, results from a t-test and a corresponding χ^2 test ([7.12]) should be similar. Since the t-test is only an approximation, however, their results may diverge with small samples. In Table 7.4 t-statistics indicate that all five X variables have significant effects, consistent with our earlier χ^2 conclusions.

Confidence intervals for logit coefficients also resemble their OLS counterparts:

$$b_k \pm t(\mathrm{SE}_{b_k}) \qquad [7.14]$$

For example, to form a 95% confidence interval for the coefficient on education (X_2) in Table 7.4, use $t_{n-K} = 1.98$:

$$-.197 \pm 1.98(.093)$$

$$-.381 \leq \beta_2 \leq -.013$$

Other things being equal, each additional year of education:

1. subtracts between .381 and .013 from the log odds of favoring school closings;
2. multiplies the odds by $e^{-.381} = .68$ to $e^{-.013} = .99$.

That is, with each year of education the odds of favoring school closing decline by somewhere between 32% and 1%.

Contamination (X_3) is a dummy variable, with 1 indicating contamination of the respondent's own property or water. A 95% confidence interval for its coefficient is:

$$1.30 \pm 1.98(.477)$$

$$.356 \leq \beta_3 \leq 2.244$$

Being personally affected by the chemical contamination increases the odds of favoring school closing by a factor of $e^{.356} = 1.43$ to $e^{2.244} = 9.43$.

Interpretation

The logit equation from Table 7.4 is approximately

$$\hat{L}_i = 2.18 - .04X_{i1} - .2X_{i2} + 1.3X_{i3} + 2.28X_{i4} - 1.73X_{i7} \qquad [7.15]$$

where X_1 is years of residency, X_2 is education, X_3 is contamination, X_4 is Health and Safety Committee, and X_7 is a dummy variable for male nonparents. $\hat{L}$ represents the predicted log odds of favoring school closings.

We could interpret each coefficient in [7.15] with reference to the following:

1. logit or log odds ($\hat{L}$);
2. odds ($\hat{\mathcal{O}} = e^{\hat{L}}$) or ratios of odds ($\hat{\Omega} = \hat{\mathcal{O}}_1/\hat{\mathcal{O}}_0 = e^b$); or
3. probabilities ($\hat{P} = 1/(1 + e^{-\hat{L}})$).

The logit interpretation is easily stated but not so easily understood. For example, the coefficient on years lived in town is $b_1 = -.04$. For every additional year of residence, other things being equal, the predicted log odds decline by $-.04$.

Odds allow a more intuitive interpretation, especially with dichotomous X variables. The coefficient on X_7, a dummy variable indicating males without young children, is $b_7 = -1.73$. Being a male nonparent therefore multiplies the odds by $e^{-1.73} = .18$. For example, if all other X variables equal their means, predicted odds that a male parent or a female favors school closing are about .89 to 1 (found by substituting means and $X_7 = 0$ into Equation [7.15]). Predicted odds for a male nonparent are less than one-fifth as high: $.89 \times .18 = .16$. Relatively few male nonparents saw a need for closing the contaminated schools.

An *odds ratio* is a ratio of the odds at two different values of X:

$$\hat{\Omega} = \frac{\hat{\mathcal{O}}_1}{\hat{\mathcal{O}}_0} \qquad [7.16]$$

For dummy X variables, the odds ratio equals the antilogarithm (e to the power) of the logit coefficient:

$$\hat{\Omega} = e^b \qquad [7.17]$$

Odds ratios are often used comparatively, to describe the strength of an effect. They provide another way to interpret logit coefficients.

With other variables at their means, the odds favoring school closing are .16 when $X_7 = 1$ (male nonparents) and .89 when $X_7 = 0$. The predicted odds ratio is therefore

$$\hat{\Omega} = \frac{\hat{\mathcal{O}}_1}{\hat{\mathcal{O}}_0} = \frac{.16}{.89} = .18$$

We could also obtain the odds ratio from the coefficient on X_7:

$$\hat{\Omega} = e^{b_7} = e^{-1.73} = .18$$

This ratio does not change if we choose other values for X_1–X_4.

The coefficient on contamination (X_3) is $b_3 = 1.30$, so the corresponding odds ratio is $\hat{\Omega} = e^{1.30} = 3.67$. The odds of favoring school closing are 3.67 times higher if the respondent's own property or water is contaminated (that is, if $X_3 = 1$). The coefficient on hsc (X_4) is $b_4 = 2.28$, so the odds ratio is $e^{2.28} = 9.78$. The odds of favoring school closing are 9.78 times higher for Health and Safety Committee members ($X_4 = 1$) than for nonmembers.

If X is a measurement variable, then e^b describes the effect of a one-unit change. A larger or smaller change might be substantively more meaningful. For example, the odds of favoring school closing are multiplied by $e^{-.20} = .82$ with each additional 1 year of education. With each 10 years of education, the odds are multiplied by $(e^{-.20})^{10} = e^{-2} = .14$.

If Y and X_k are unrelated, the coefficient on X_k equals 0 and the odds ratio $e^0 = 1$. The stronger the relation, the farther the odds ratio will be from 1.

Alternatively, we can describe effects in terms of probabilities rather than odds. Conditional effect plots provide a graphical view of relationships between probabilities and X variables.

Figure 7.5 shows conditional effects of years of residency (X_1). The following predicted logit is for respondents with average education ($\bar{X}_2 = 12.95$) and no contamination ($X_3 = 0$) who are not Health and Safety Committee members ($X_4 = 0$), female, or parents ($X_7 = 1$):

$$
\begin{aligned}
\hat{L}_i &= 2.18 - .04X_{i1} - .2(12.95) + 1.3(0) + 2.28(0) - 1.73(1) \\
&= -2.14 - .04X_{i1}
\end{aligned}
$$

A linear relationship between X_1 and $\hat{L}$ implies a curvilinear relationship between X_1 and predicted probabilities ($\hat{P} = 1/(1 + e^{-\hat{L}})$), graphed at bottom in Figure 7.5.

The middle curve in Figure 7.5 shows this relation for respondents who are average in all other respects:

$$
\begin{aligned}
\hat{L}_i &= 2.18 - .04X_{i1} - .2(12.95) + 1.3(.28) + 2.28(.31) - 1.73(.17) \\
&= .387 - .04X_{i1}
\end{aligned}
$$

The top curve shows the relation for respondents who have average education, were contaminated, are Health and Safety Committee members, and are female or parents:

$$
\begin{aligned}
\hat{L}_i &= 2.18 - .04X_{i1} - .2(12.95) + 1.3(1) + 2.28(1) - 1.73(0) \\
&= 3.17 - .04X_{i1}
\end{aligned}
$$

Notice that differences in the intercepts produce differences in the steepness of the corresponding probability curves. Thus the effects of X_1 change with the level of X_2, $X_3, \ldots$.[6]

Figure 7.6 shows conditional effects of the dummy variable X_3 (contamination). Since X_3 has only two possible values, its effects cannot be a smooth curve. Like any logit effects, they depend on the values of other variables. The nearly horizontal

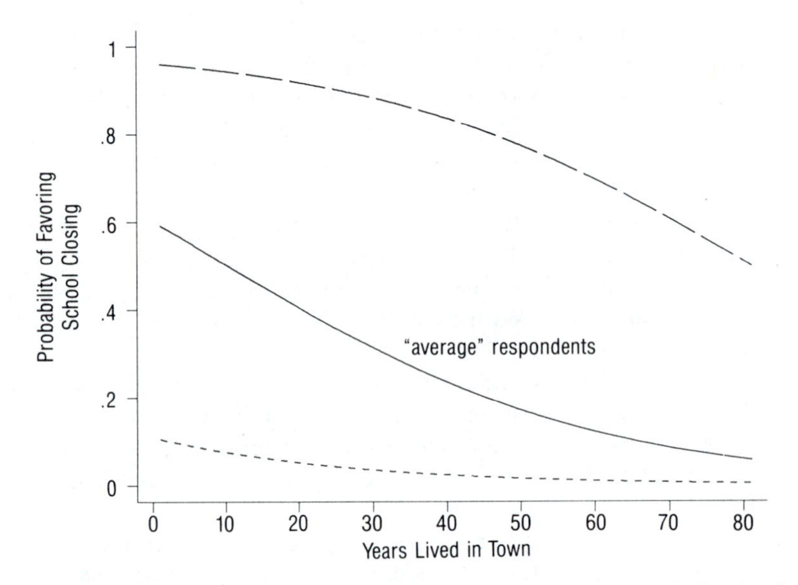

Figure 7.5 Conditional effects of years lived in town, at proclosing (top), average, and anticlosing levels of other X variables.

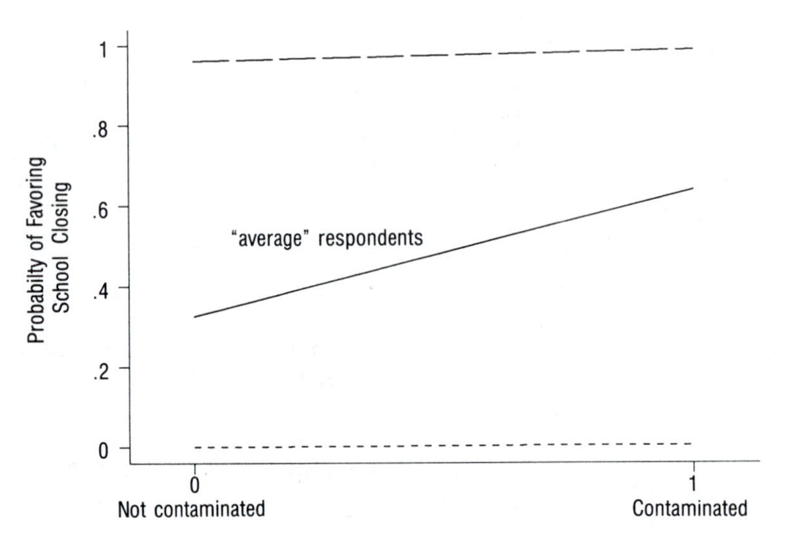

Figure 7.6 Conditional effects of contamination, at proclosing, average, and anticlosing levels of other X variables.

lines at top and bottom show that contamination has little effect when the other X variables are at their proclosing (top) or anticlosing (bottom) extremes. Logit curves are least steep near probabilities of 1 or 0. When the other variables in this example equal their means, resulting in probabilities nearer to .5, a much larger conditional effect appears (middle line).

With OLS the coefficient of determination, R^2 or R_a^2, measures overall fit. Several analogous measures have been proposed for logit models. Aldrich and Nelson (1984) suggest

$$\text{pseudo } R^2 = \frac{\chi^2}{\chi^2 + n} \qquad [7.18]$$

where n is sample size and χ^2 is the test statistic for the null hypothesis that all coefficients but the intercept are zero. For Table 7.4 this gives

$$\text{pseudo } R^2 = \frac{66.56}{66.56 + 153} = .303$$

McKelvey and Zavoina (1976) and Hosmer and Lemeshow (1989) propose other R^2 analogs. Each pseudo R^2 has advantages, but there is no general agreement about which one is best. None supports a straightforward explained-variance interpretation, as does true R^2.

Statistical Problems

Multicollinearity poses similar problems for OLS and logit regression. It leads to unreliable coefficient estimates and large standard errors; at the extreme, perfect multicollinearity makes estimation impossible. Methods for diagnosing multicollinearity include the following:

1. Examine the matrix of correlations among X variables. This method is fallible, because multicollinearity can occur even without high intervariable correlations.
2. Examine the matrix of correlations among estimated coefficients.[7] High correlations indicate a problem but do not necessarily identify its source.
3. Regress each X variable on all the other X variables. Low tolerance $(1 - R_k^2)$ indicates a potential problem with that variable.

To cope with collinearity, consider collecting more data or dropping or combining variables. Alternatively, χ^2 tests ([7.12]) can test the coefficients on a set of highly correlated variables together, when individual effects cannot be disentangled.

High *discrimination*, like multicollinearity, inflates standard errors or makes estimation impossible. Discrimination (also termed *separation*) refers to our ability to predict Y. Table 7.5 illustrates this problem, using data from another community threatened by toxic-waste contamination. People were asked whether water-quality

Table 7.5 Cross-tabulation and logit analysis with perfect one-way discrimination

Standards Made Less Strict?	Female/Young Children No	Yes	Total
no	202	79	281
yes	44	0	44
Total	246	79	325

. logit less mother

Note: mother $\neq 0$ predicts failure perfectly
mother dropped and 79 obs not used

Iteration 0: Log Likelihood = −115.53714

Logit Estimates   Number of obs = 246
                                                                                   chi2(0) = −0.00
Log Likelihood = −115.53714   Prob > chi2 = .

Variable	Coefficient	Std. Error	t	Prob > $\|t\|$	Mean
less					.1788618
_cons	−1.524078	.1663664	−9.161	0.000	1

standards should be made less strict; mildly contaminated water could then be declared drinkable, saving the cost of a cleanup. This proposal was debated in local papers.

Table 7.5 (top) cross-tabulates water-quality opinion by whether the respondent is a woman with young children. None of the mothers in this sample felt that water-quality standards should be made less strict. The odds favoring less strict standards are $\mathcal{O}_0 = 44/202 = .218$ for nonmothers and $\mathcal{O}_1 = 0/79 = 0$ for mothers. The corresponding odds ratio is

$$\Omega = \frac{\mathcal{O}_1}{\mathcal{O}_0} = \frac{0}{.218} = 0$$

so the logit coefficient on X must be $b = \log(\Omega) = \log(0)$, which equals negative infinity.[8] Its standard error will be infinite too.

Some computer programs try to estimate these values anyway, obtaining high numbers before giving up or crashing. Other programs detect the problem and complain, as at bottom in Table 7.5. This particular program (*Stata*) resolved the problem by dropping all mothers (79 cases) from analysis.

High discrimination can take several forms. Table 7.5 illustrates *oneway discrimination by a dummy variable*, in which $X = 1$ perfectly predicts $Y = 0$, but $X = 0$ does not perfectly predict $Y = 1$. That is, all mothers oppose lower standards, but not all nonmothers favor them. If they did, we would have *twoway discrimination by a dummy variable.*

We could also encounter one- or twoway discrimination by a measurement variable: for example, age < 50 might perfectly predict $Y = 0$. Like multicollinearity, high discrimination is a matter of degree. Problems (untrustworthy estimates and increasing standard errors) worsen as we approach perfect discrimination.

Curvilinearity is another potential problem. With measurement Y variables we check for curvilinearity by looking at scatterplots. Scatterplots of dichotomous Y variables are less informative. To check the assumption that logits are linearly related to an X variable, we could follow these steps:

1. Group the X variable.
2. Within each group, find the mean of Y and express this as a logit.
3. Graph the mean logits against grouped X.

Hosmer and Lemeshow (1989) describe alternative procedures. Unfortunately, linearity checks for logit regression tend to be awkward and time consuming (requiring the arbitrary grouping of X, for example) and so are not popular.

If an X variable is badly skewed, seek a symmetrizing transformation before doing the logit analysis. Conditional effect plots display the consequences of nonlinear transformations.

One reason for transformation is to reduce the leverage of outlying cases. The next section looks at direct measures of influence.

Influence Statistics for Logit Regression

Chapter 4 described OLS influence statistics. Analogous statistics exist for logit regression. Typically these focus not on individual cases but on X *patterns*, or combinations of X values. Two or more cases have the same X pattern if they have identical values on all X variables. We will use J to denote the number of unique X patterns in the sample ($J \leq n$) and m_j to denote the number of cases with the jth X pattern. If no two cases have the same X pattern, $J = n$ and $m_j = 1$ for all j. All cases with pattern j have the same predicted probability that $Y = 1$, $\hat{P}_j$. Table 7.6 provides a glossary for this notation.

Table 7.6 Symbols used in logit influence analysis

J	Number of unique X patterns in the data ($J \leq n$)
m_j	Number of cases with X pattern j ($m \geq 1$)
$\hat{P}_j$	Predicted probability that $Y = 1$, for cases with pattern j
Y_j	Sum of Y values, for all cases with pattern j (equivalently, number of cases with X pattern j and $Y = 1$)
r_j	Pearson residual for jth X pattern, [7.19]
χ^2_P	Pearson χ^2 statistic, [7.20]
d_j	Deviance residual for jth X pattern, [7.21]
χ^2_D	Deviance, [7.22]
h_i	Leverage of the ith case, from [7.26]
h_j	Leverage of the jth X pattern, [7.28]

Two kinds of standardized residuals may be calculated: *Pearson* and *deviance residuals*. Sums of squared Pearson or deviance residuals test whether a given logit model is significantly worse than a perfect-fit (*saturated*) model.

The *Pearson residual* is

$$r_j = \frac{Y_j - m_j\hat{P}_j}{\sqrt{m_j\hat{P}_j(1 - \hat{P}_j)}} \quad [7.19]$$

The *Pearson* χ^2 *statistic* is the sum of squared Pearson residuals:

$$\chi^2_P = \sum_{j=1}^{J} r_j^2 \quad [7.20]$$

The *deviance residual* has the same sign as $Y_j - m_j\hat{P}_j$:

$$d_j = \pm\left\{2\left[Y_j\,log_e\left(\frac{Y_j}{m_j\hat{P}_j}\right) + (m_j - Y_j)\log_e\left(\frac{m_j - Y_j}{m_j(1 - \hat{P}_j)}\right)\right]\right\}^{1/2} \quad [7.21]$$

In evaluating [7.21], drop the first part if $Y_j = 0$, and drop the second part if $Y_j = m_j$.
The *deviance* equals the sum of squared deviance residuals:

$$\chi^2_D = \sum_{j=1}^{J} d_j^2 \quad [7.22]$$

Both [7.20] and [7.22] resemble χ^2 tests of the null hypothesis of no difference between the estimated model and a *saturated model* with J parameters (one parameter for each combination of X values).[9]

The *leverage* of the ith case, h_i, is the ith diagonal element of the *hat matrix*. The leverage of the jth X pattern, h_j, equals the leverage of any one jth-pattern case times the number of times the pattern occurs (see [7.26]–[7.29]). Three influence statistics derive from h_j:

ΔB_j, analogous to Cook's D, measures the standardized change in estimated parameters (b_k) that results from deleting all cases with the jth X pattern:

$$\Delta B_j = \frac{r_j^2 h_j}{(1 - h_j)^2} \quad [7.23]$$

A large value of ΔB_j indicates that the jth pattern exerts substantial influence. As with Cook's D, we might view $\Delta B \geq 1$ as "large."

$\Delta\chi^2_{\mathrm{P}(j)}$ measures decrease in Pearson χ^2 ([7.20]) that results from deleting all cases with the jth X pattern:

$$\Delta\chi^2_{\mathrm{P}(j)} = \frac{r_j^2}{(1 - h_j)} \qquad [7.24]$$

$\Delta\chi^2_{\mathrm{D}(j)}$ measures the change in deviance ([7.22]) that results from deleting all cases with the jth X pattern:

$$\Delta\chi^2_{\mathrm{D}(j)} = \frac{d_j^2}{(1 - h_j)} \qquad [7.25a]$$

Equivalently,

$$\Delta\chi^2_{\mathrm{D}(j)} = -2[\log_e \mathscr{L}_K - \log_e \mathscr{L}_{K(j)}] \qquad [7.25b]$$

where $\mathscr{L}_K$ is the likelihood of a model with K parameters, estimated from the full sample, and $\mathscr{L}_{K(j)}$ is the likelihood for the same model estimated after deleting all cases with the jth X pattern.

$\Delta\chi^2_{\mathrm{P}(j)}$ and $\Delta\chi^2_{\mathrm{D}(j)}$ measure how poorly the model fits the jth pattern. Large values indicate that the model would fit the data much better if this pattern were deleted. As a rule of thumb, values greater than 4 indicate a "significant" change (because the distribution is asymptotically χ^2).

For OLS, leverage (h_i) equals diagonal elements of the hat matrix $\mathbf{H}$:

$$\mathbf{H} = \mathbf{X}(\mathbf{X}'\mathbf{X})^{-1}\mathbf{X}'$$

Logit leverage statistics are diagonal elements of

$$\mathbf{H} = \mathbf{X}^*(\mathbf{X}^{*\prime}\mathbf{X}^*)^{-1}\mathbf{X}^{*\prime} \qquad [7.26]$$

where $\mathbf{X}^* = \mathbf{W}^{1/2}\mathbf{X}$. $\mathbf{W}$ is an $n \times n$ matrix with weights

$$w_i = \hat{P}_i(1 - \hat{P}_i) \qquad [7.27]$$

on the main diagonal, zeros elsewhere. The ith diagonal element of $\mathbf{H}$, h_i, is the leverage of the ith case. If there are m_j cases with the same X pattern as case i, the leverage of those cases together is

$$h_j = m_j h_i \qquad [7.28]$$

Equation [7.26] is a weighted least squares hat matrix, with weights based on predicted probabilities ($\hat{P}_i$). We could obtain this hat matrix, and the logit regression coefficients themselves, by a WLS regression using *pseudovalues* (z_i) for the Y variable. Calculate pseudovalues:

$$z_i = \log_e\left(\frac{\hat{P}_i}{1-\hat{P}_i}\right) + \frac{Y_i - \hat{P}_i}{\hat{P}_i(1-\hat{P}_i)} \qquad [7.29]$$

Then regress z on all X variables, weighting by $\mathbf{W}$.[10]

Diagnostic Graphs

$\Delta\chi^2_{\text{P}}$, $\Delta\chi^2_{\text{D}}$, and ΔB have different values for each X pattern in the data. Graphs provide a quick overview of this information. Hosmer and Lemeshow (1989) suggest three diagnostic plots:

1. change in Pearson χ^2 versus predicted probability ($\Delta\chi^2_{\text{P}}$ versus $\hat{P}$);
2. change in deviance versus predicted probability ($\Delta\chi^2_{\text{D}}$ versus $\hat{P}$); and
3. influence versus predicted probability (ΔB versus $\hat{P}$)

As a further refinement, make plotting-symbol sizes in plots 1 and 2 proportional to ΔB.

Figure 7.7 shows a $\Delta\chi^2_{\text{P}}$-versus-$\hat{P}$ plot for the analysis of Table 7.4. Most data points align with one of two curves:

Going down, left to right, are the points for X patterns with $Y_j = 1$ and $m_j = 1$.
Going up, left to right, are the points for X patterns with $Y_j = 0$ and $m_j = 1$.

In these data we have 153 cases. There are 132 distinct X patterns: 115 that occur once ($m_j = 1$), 13 that occur twice ($m_j = 2$), and 4 that occur three times ($m_j = 3$). The 115 unique patterns fall along one of the two curves. Other patterns may be on or off the curves.

Two X patterns stand out as poorly fit: number 131, at upper left, and number 3, at upper right.[11] Table 7.7 lists what we know about these patterns. Each occurs with only a single case ($m_{131} = m_3 = 1$). Pattern 131 is the combination 68 years residency, 12th-grade education, not contaminated, not a Health and Safety Committee member, and not a male nonparent. According to our model, a person with these characteristics is unlikely to favor school closing: the predicted probability is $\hat{P}_{131} = .05$. This person nonetheless does favor school closing ($Y_{131} = 1$), so our model fits her poorly. *Any such case with $Y = 1$ despite low $\hat{P}$ will fall at upper left in a $\Delta\chi^2_{\text{P}}$-versus-$\hat{P}$ plot.*

Conversely, any case with $Y = 0$ despite high $\hat{P}$ will be toward upper right. Pattern 3 describes a person who should be very *likely* to favor school closing ($\hat{P}_3 = .97$) but nonetheless does not ($Y_3 = 0$). Cases that are likely to favor closing, and *do* favor closing ($\hat{P}$ is high, $Y = 1$), plot along bottom right. Cases that are

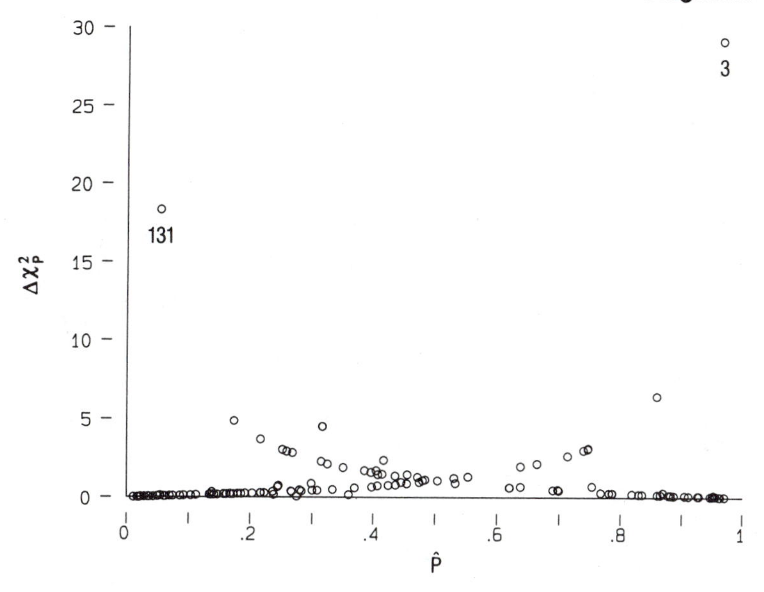

Figure 7.7 Poorness-of-fit statistic $\Delta\chi^2_{\hat{P}}$ versus predicted probability of favoring closed schools; X patterns 131 and 3 are poorly fit (high $\Delta\chi^2_{\hat{P}}$ values).

Table 7.7 Two poorly fit X patterns (Figure 7.7)

	Pattern 131	Pattern 3
Y: favor closing schools	1	0
$\hat{P}$: predicted $P(Y = 1)$	.05	.97
X_1: years resident	68	1
X_2: education	12	12
X_3: contaminated	0	1
X_4: H & S Committee	0	1
X_7: male nonparent	0	0
m: # of cases with this X pattern	1	1
$\Delta\chi^2_{\hat{P}}$	18.34	29.20
($\Delta\chi^2_{\hat{P}}$ rank)	(2)	(1)
$\Delta\chi^2_{\hat{D}}$	6.07	6.89
($\Delta\chi^2_{\hat{D}}$ rank)	(2)	(1)
ΔB	.66	.42
(ΔB rank)	(1)	(2)

unlikely to favor closing, and *do not* favor closing ($\hat{P}$ is low, $Y = 0$), plot along bottom left.

Figure 7.8 graphs $\Delta\chi^2_{\hat{D}}$ versus $\hat{P}$. The two prominent curves have the same explanation as in Figure 7.7. The $\Delta\chi^2_{\hat{D}}$ statistic ranges less than $\Delta\chi^2_{\hat{P}}$ (about 0–7 instead of 0–36), but the two worst-fit patterns are again numbers 131 and 3. We now see that patterns 27, 62, and 115 are also poorly fit. Patterns 27 and 62 both occur twice in the raw data ($m_{27} = m_{62} = 2$) and do not fall on either curve. Their position toward upper left indicates that these people favor school closing,

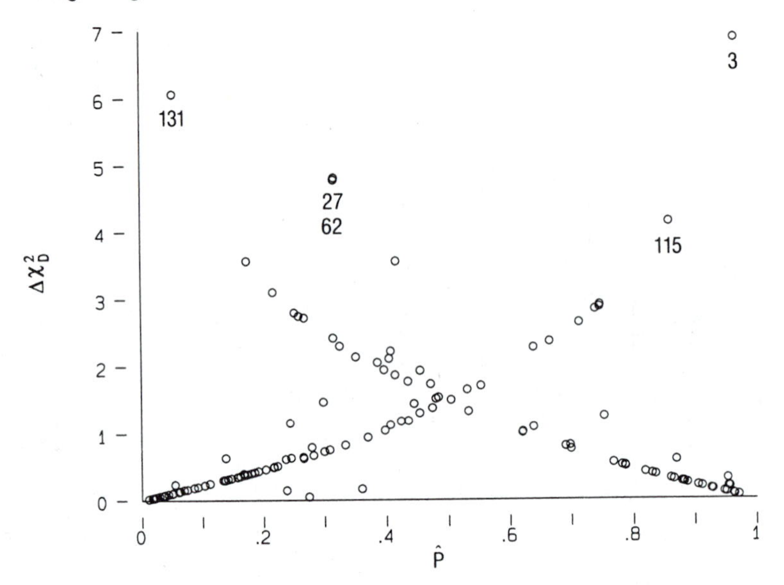

Figure 7.8 Poorness-of-fit statistic $\Delta\chi^2_{\mathrm{D}}$ versus predicted probability of favoring closed schools; X patterns 131, 3, 27, 62, and 115 are poorly fit (high $\Delta\chi^2_{\mathrm{D}}$ values).

despite a relatively low predicted probability ($Y = 1, \hat{P} \approx .32$). The opposite goes for pattern 115, which occurs once ($Y_{115} = 0$, $\hat{P}_{115} = .86$).

Figure 7.9 graphs the influence statistic ΔB against predicted probability. Patterns 131 and 3 exert the most influence, followed by 115, 44, and 94. No ΔB values are numerically large ($\Delta B \geq 1$), however. ΔB, like Cook's D, measures influence on all the model's coefficients. An X pattern could strongly influence one coefficient but still not have a large ΔB. Consequently, we should investigate further any patterns with ΔB values noticeably higher than the rest of the data—such as 131, 3, and 115 in Figure 7.9.

Table 7.8 shows how the logit equation changes when these high-ΔB X patterns are deleted. The shifts are not dramatic, but some coefficients change by as much as 25% with the deletion of a single case, .7% of the data. Remember that, unlike linear regression, logit regression intercepts modify the steepness of X-variable effects. This gives added importance to shifts in intercept. Deleting pattern 3, which had the highest $\Delta\chi^2_{\mathrm{D}}$ and $\Delta\chi^2_{\mathrm{P}}$ values, causes the greatest change in the log likelihood. $\Delta\chi^2_{\mathrm{D}}$ approximately equals minus two times the shifts in log likelihoods shown in Table 7.8.

To combine information about fit and influence, graph $\Delta\chi^2_{\mathrm{P}}$ or $\Delta\chi^2_{\mathrm{D}}$ against $\hat{P}$, with plotting symbols sized proportional to ΔB (Figure 7.10).[12] Patterns 131 and 3 show up as both poorly fit (vertically high) and influential (large circles). In general, high, large circles in graphs like Figure 7.10 deserve close scrutiny. They represent cases that are inconsistent with the rest of the data but that disproportionately affect the model.

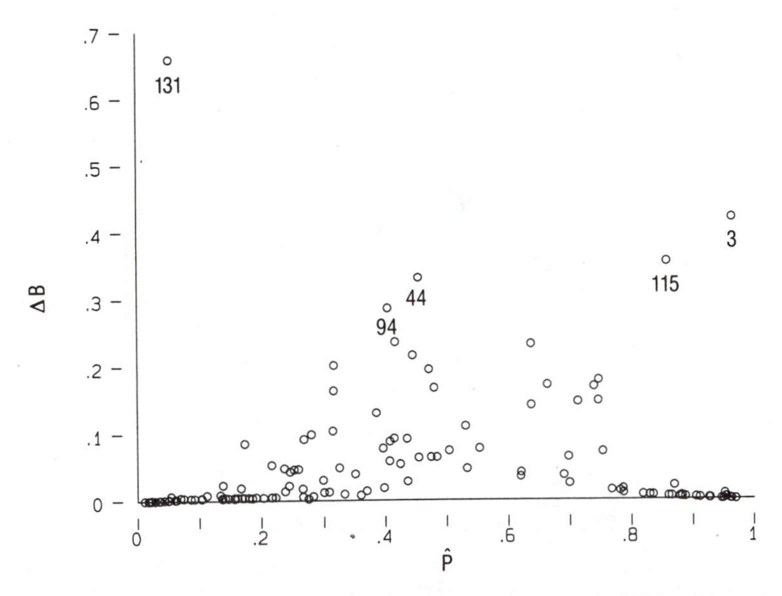

Figure 7.9 Influence statistic ΔB versus predicted probability of favoring closed schools; patterns 131, 3, 115, 44, and 94 are most influential (high ΔB values).

Table 7.8 Consequences of deleting some influential cases

logit model estimated with full sample ($n = 153$):

$$\hat{L} = 2.18 - .04(X_1) - .20(X_2) + 1.30(X_3) + 2.28(X_4) - 1.73(X_7)$$

$\log_e \mathcal{L} = -71.36$

with X pattern 131 deleted ($n = 152$):

$$\hat{L} = 2.53 - .05(X_1) - .21(X_2) + 1.38(X_3) + 2.35(X_4) - 1.65(X_7)$$

$\log_e \mathcal{L} = -68.06$

with X pattern 3 deleted ($n = 152$):

$$\hat{L} = 2.58 - .04(X_1) - .22(X_2) + 1.49(X_3) + 2.49(X_4) - 1.89(X_7)$$

$\log_e \mathcal{L} = -67.71$

with X pattern 115 deleted ($n = 152$):

$$\hat{L} = 2.18 - .04(X_1) - .20(X_2) + 1.45(X_3) + 2.46(X_4) - 1.91(X_7)$$

$\log_e \mathcal{L} = -69.19$

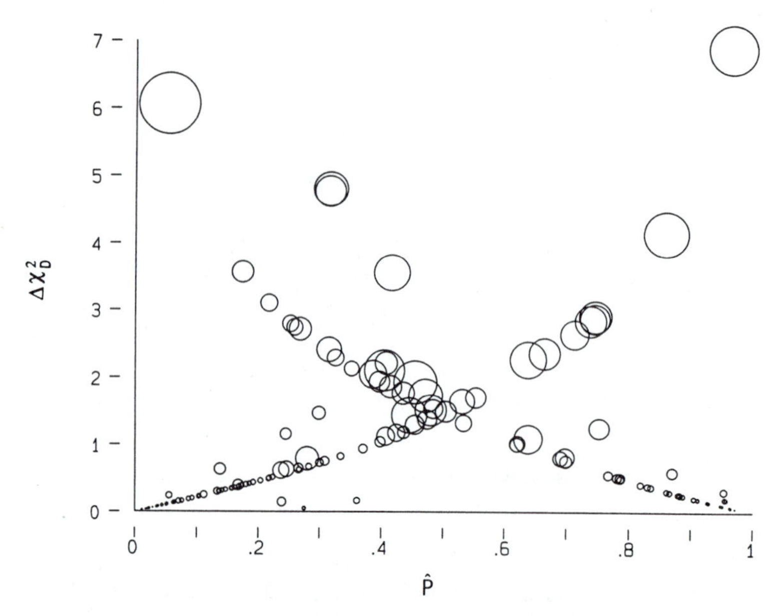

Figure 7.10 $\Delta\chi^2_D$ versus $\hat{P}$ with symbols proportional to ΔB; large, high circles indicate influential, poorly fit X patterns

Conclusion

Linear regression methods work poorly for dichotomous Y variables for the following reasons:

1. With a dichotomous Y variable, errors cannot be Gaussian or have constant variance.
2. Linear predictions can exceed $\{0, 1\}$ boundaries. The true relationship between probabilities and the X variables must be nonlinear.

Logit regression resolves these problems by modeling predicted probabilities as an S-shaped function (the cumulative logistic distribution) of X values.

Likelihood-ratio χ^2 statistics test nested logit models, serving purposes like those of F statistics in OLS. Approximate t tests and confidence intervals for individual logit coefficients resemble their OLS counterparts. The OLS R^2 has no clear logit counterpart, however.

Logit coefficients support interpretations in terms of log odds, odds, odds ratios, or probabilities. Conditional effect plots help to visualize what a model implies about probabilities.

Multicollinearity, nonlinearity, and influential cases undercut logit regression, as they do OLS. Too-high discrimination also creates problems. As with any analysis, careful diagnostic work improves the credibility of our results.

Exercises

Military pilots sometimes black out when their brains are deprived of oxygen due to G-forces during violent maneuvers, a phenomenon called G-induced loss of consciousness (G-LOC). Glaister and Miller (1990) produced similar symptoms by exposing volunteers' lower bodies to negative air pressure, likewise decreasing oxygen to the brain. Their research focused on changes in brain functioning, but they noted that subjects' tolerance varied widely: one lasted 18 minutes without symptoms, but another lost consciousness after only 6 minutes. The following table lists eight subjects by age and by whether they showed syncopal or blackout-related signs (pallor, sweating, slow heartbeat, unconsciousness). Exercises 1–2 relate to these data.

Age and syncopal signs in subjects exposed to negative pressure

Subject	Age	Syncopal Signs (0 = no, 1 = yes)
1 JW	39	0
2 JM	42	1
3 DT	20	0
4 LK	37	1
5 JK	20	1
6 MK	21	0
7 FP	41	1
8 DG	52	1

1. Conduct logit regression of syncopal signs on age, and discuss the results. Since the coefficient on age is not significant, should we conclude that age has no effect?
2. Graph the predicted probability of syncopal signs against age, and describe what this graph shows.

Shultz (1987) conducted a survey in which he asked residents of one city whether they would be willing to pay for a certain groundwater-protection plan. Their answers (0 = no, 1 = yes) became the Y variable in this logit regression model:

$$\hat{L}_i = .13 + .04X_{i1} - .03X_{i2} + .00002X_{i3} - .006X_{i4}$$

where

$\hat{L}$ is log odds of a "yes" answer ($\bar{Y} = .31$);
X_1 is respondent's land value in thousands of dollars ($\bar{X}_1 = 10.42$);
X_2 is respondent's age in years ($\bar{X}_2 = 52.02$);
X_3 is household net income in dollars ($\bar{X}_3 = 36{,}533$);
X_4 is bid value in dollars ($\bar{X}_4 = 214.9$).

All coefficients but b_1 are significant at $\alpha = .01$. The researcher randomly assigned to each respondent a bid (X_4) from $1 to $500 and asked whether each would be willing to pay this amount. This approach, called willingness to pay (WTP) methodology, is a common economic application of logit analysis.

Exercises 3–5 refer to this analysis.

3. Interpret these coefficients in terms of log odds and odds: $b_1 = .04$; $b_2 = -.03$.
4. With other X variables at their means, draw a conditional effect plot for income. Describe what it shows.
5. Draw a conditional effect plot for bid, with two curves: one at income = $5000 and another at income = $80,000, both with other X variables at their means. Describe what this plot shows.

The next table contains data from a study of pesticide residues in human milk, conducted in Western Australia in 1979–80 (Stacey, Perriman, and Whitney, 1985). Earlier research discovered surprisingly high pesticide levels; Stacey et al. hoped to find that levels had decreased due to stronger government controls over the use of pesticides on food crops. They did find decreases for several types of pesticides. Levels of dieldrin, however, had substantially increased. These data help to explain why.

For 45 donors, we have information on: age in years; whether they lived in a new suburb (0 = old, 1 = new); whether their house was treated for termites within the past three years (0 = no, 1 = yes, two missing values); and whether their milk contained above-average (more than .009 parts per million) levels of the pesticide dieldrin. Termites are a common problem in Western Australia, and dieldrin is often used to control them. By law new houses must be pretreated for termites.

Dieldrin residues in human milk from 45 donors

Donor	Age	New Suburb*	House Treated for Termites*	Dieldrin > .009 ppm*
1	33	1	0	1
2	34	0	1	1
3	29	0	0	0
4	28	1	.	0
5	29	1	1	1
6	27	1	1	1
7	27	1	1	1
8	27	1	0	1
9	31	1	0	0
10	27	0	0	0
11	32	0	1	1
12	28	0	1	0
13	25	1	1	0
14	27	1	1	1
15	26	0	0	0
16	28	0	1	1
17	31	1	0	1
18	32	1	0	0

Donor	Age	New Suburb*	House Treated for Termites*	Dieldrin > .009 ppm*
19	32	0	1	1
20	33	0	1	1
21	21	0	1	1
22	32	0	1	0
23	36	0	.	0
24	24	0	0	0
25	34	0	1	1
26	28	0	1	0
27	26	0	0	0
28	34	0	0	0
29	29	0	0	0
30	30	0	0	0
31	28	0	1	1
32	32	0	0	0
33	25	1	1	0
34	29	0	1	0
35	27	0	1	0
36	33	0	0	0
37	37	0	1	0
38	28	1	0	0
39	30	0	0	0
40	27	0	0	0
41	23	0	0	0
42	27	0	1	0
43	25	0	1	0
44	36	0	1	1
45	23	0	1	0

* 0 = no, 1 = yes

6. Perform a logit regression of dieldrin level on age, new suburb, and termites. Which predictors are significant at $\alpha = .05$? At $\alpha = .10$? What *P*-value do we obtain regarding the null hypothesis that all three coefficients are zero?
7. Interpret each of your coefficients in terms of odds.
8. What probability of high dieldrin do we predict for a woman of average age, living in an old-suburb house not treated for termites? Living in a new-suburb house recently treated for termites?
9. Construct a conditional effect plot for the relation between age and probability of high dieldrin level, with separate curves for:
   a. old suburb, house not recently treated for termites
   b. new suburb, house recently treated for termites.
   Briefly describe the results.
10. Construct three diagnostic graphs:
    a. ΔB versus $\hat{P}$ (like Figure 7.9)
    b. $\Delta\chi^2_{\mathrm{P}}$ versus $\hat{P}$, weighted by ΔB
    c. $\Delta\chi^2_{\mathrm{D}}$ versus $\hat{P}$, weighted by ΔB (like Figure 7.10)

11. Two X patterns stand out as poorly fit and influential in your diagnostic graphs of Exercise 10. Identify these patterns and the corresponding donors. How are they unusual?

Notes

1. Given model $E[Y_i] = \beta_0 + \beta_1 X_{i1}$ with dichotomous Y, and errors $\varepsilon_i = Y_i - E[Y_i]$:

$$\text{Var}[\varepsilon_i] = E[Y_i](1 - E[Y_i])$$

which reaches a maximum at $E[Y_i] = .5$

2. Many other S-shaped functions share these properties and so might also be used as models. For example, *probit regression* is based on the S-shaped cumulative normal distribution. In practice, probit and logit analyses tend to reach similar conclusions.

   Practical arguments favor the logit over alternative S-curves. Computation and interpretation are straightforward, and logit models extend readily to polytomous Y variables.

   See Aldrich and Nelson (1984) for a theoretical justification of logit (and probit) models, based on an economic "rational choice" theory.

3. Logit curves are steepest at $\hat{P} = .5$, or $\hat{L} = 0$. We can calculate the value of X_1 that corresponds to the steepest part of the curve. For example, given a bivariate equation

$$\hat{L} = \beta_0 + \beta_1 X_1$$

and setting $\hat{L} = 0$:

$$0 = \beta_0 + \beta_1 X_1$$

We solve for X_1:

$$X_1 = -\beta_0/\beta_1$$

Thus the effect of an additional year's residency ([7.5] or Figure 7.4) is greatest at $X_1 = -.406/(-.041) = 9.9$ years.

   By similar reasoning, given a three-variable equation

$$\hat{L} = \beta_0 + \beta_1 X_1 + \beta_2 X_2$$

the effect of X_1 is greatest when

$$X_1 = (-\beta_0 - \beta_2(X_2))/\beta_1$$

We cannot solve for X_1 without specifying a value for X_2. The location of the curve, and hence of its steepest part, shifts with the value of X_2. For descriptive purposes, we might choose to substitute the median or mean of X_2 (and of X_3, $X_4, \ldots$ if necessary) into such equations.

4. This criterion is arbitrary and can be changed by the analyst.
5. The *information matrix* is a matrix of negatives of the second partial derivatives of the log likelihood function. Variances and covariances of logit coefficients are estimated from the inverse of this matrix. The standard errors printed with logit output are square roots of the diagonal elements of the inverse information matrix.
6. This illustrates a principle mentioned earlier: when the left-hand-side variable is a nonlinear transformation of the variable of substantive interest, all X-variable effects are implicitly interactions.
7. The correlation matrix of the coefficients is a standardization of the inverse of the information matrix (Note 5).
8. For logit analysis with a single dummy X variable, corresponding to a 2×2 cross-tabulation, the coefficient on X will be the natural log of the observed odds ratio:

$$b_1 = \log_e(\mathcal{O}_1/\mathcal{O}_0) = \log_e \Omega$$

The intercept is the log of the observed odds when $X = 0$:

$$b_0 = \log_e \mathcal{O}_0$$

9. See Hosmer and Lemeshow (1989) for further discussion of standardized residuals and other diagnostic statistics.
10. The method for performing WLS "by hand" (page 189 in Chapter 6) works here if your computer program does not calculate logit or WLS diagnostics. Use [7.27] and [7.29]: regress $z_i\sqrt{w_i}$ on $\sqrt{w_i}$ and all $X_{ik}\sqrt{w_i}$, suppressing the constant. The resulting OLS coefficient estimates are the same as those obtained from a corresponding WLS or logit analysis.
11. Pattern numbers do not necessarily correspond to raw-data case ID numbers.
12. Graphs like Figure 7.10 ($\Delta\chi^2_D$ or $\Delta\chi^2_P$ vs. $\hat{P}$, with points proportional to ΔB) are roughly analogous to an OLS e versus $\hat{Y}$ plot, with points proportional to Cook's D (two examples appear in Figure 5.7).

8 Principal Components and Factor Analysis

Principal components and *factor analysis* are methods for data reduction. They seek a few underlying dimensions that account for patterns of variation among the observed variables. Underlying dimensions imply ways to combine variables, simplifying subsequent analysis. For example, a few combined variables could replace many original variables in a regression. Advantages of this approach include more parsimonious models, improved measurement of indirectly observed concepts, new graphical displays, and the avoidance of multicollinearity.

Principal components and factor analysis thus work well with regression, which in some ways they resemble. Linear relations among pairs of measurement variables, as seen in scatterplots and summarized by correlation (or covariance) matrices, lie at the heart of all three methods. Principal components and factor analysis obtain the regression of observed variables on a set of underlying dimensions called *components* or *factors*. They also provide estimates of values on these dimensions.

Factor analysis estimates parameters of regression-like linear models, including error terms and sometimes significance tests. Principal components involves a comparatively simple transformation of the data, rather than a statistical model like factor analysis. Despite basic differences, however, principal components and factor analysis have much in common.

Introduction to Components and Factor Analysis

Factor analysis models each of K observed variables Z_k as a linear function of J unobserved factors F_j ($J < K$), plus a residual or error term (u_k) unique to that variable:

$$Z_k = \ell_{k1}F_1 + \ell_{k2}F_2 + \ell_{k3}F_3 + \cdots + \ell_{kJ}F_J + u_k \quad [8.1]$$

The F_j are sometimes called *common factors* and the u_k *unique factors*. The ℓ_{kj} are standardized regression coefficients termed *factor loadings*. The first subscript (k) denotes the variable, and the second subscript (j) denotes the factor. To avoid triple subscripting, this chapter omits case indicators i.

Figure 8.1 schematically diagrams an example of a factor model with two factors (F_1 and F_2) and five observed variables ($Z_1 - Z_5$). Straight arrows indicate causality; for example, the unobserved variable F_1 causes some of the variation observed in Z_2. The remaining variation in Z_2 we attribute to the unique factor u_2. The factor loading ℓ_{21} equals the standardized regression of variable Z_2 on factor F_1. For simplicity, the diagram omits arrows for near-zero loadings like that of Z_1 on F_2. The factors themselves may be:

uncorrelated, if *not rotated* or *orthogonally rotated* (terms explained later); or
correlated, if *obliquely rotated.*

A curved, double-headed arrow indicates correlation without causal implications. The graphical conventions of Figure 8.1 derive from an approach called *structural equation modeling* (Jöreskog and Sörbom, 1979).

Unlike factor analysis, principal components is not model based but, rather, involves a straightforward mathematical transformation. Data on K observed variables Z_k can be reexpressed as data on K principal components:

$$Z_k = \ell_{k1}F_1 + \ell_{k2}F_2 + \ell_{k3}F_3 + \cdots + \ell_{kK}F_K \quad [8.2]$$

(The same notation (F_j, ℓ_{kj}, and so on) will represent parallel elements of principal components and factor analysis.) Equations [8.1] and [8.2] differ in that factor analysis ([8.1]) allows for an error term; J factors do not explain all of K variables' variance (if $J < K$). K principal components do explain all the variance of K variables.

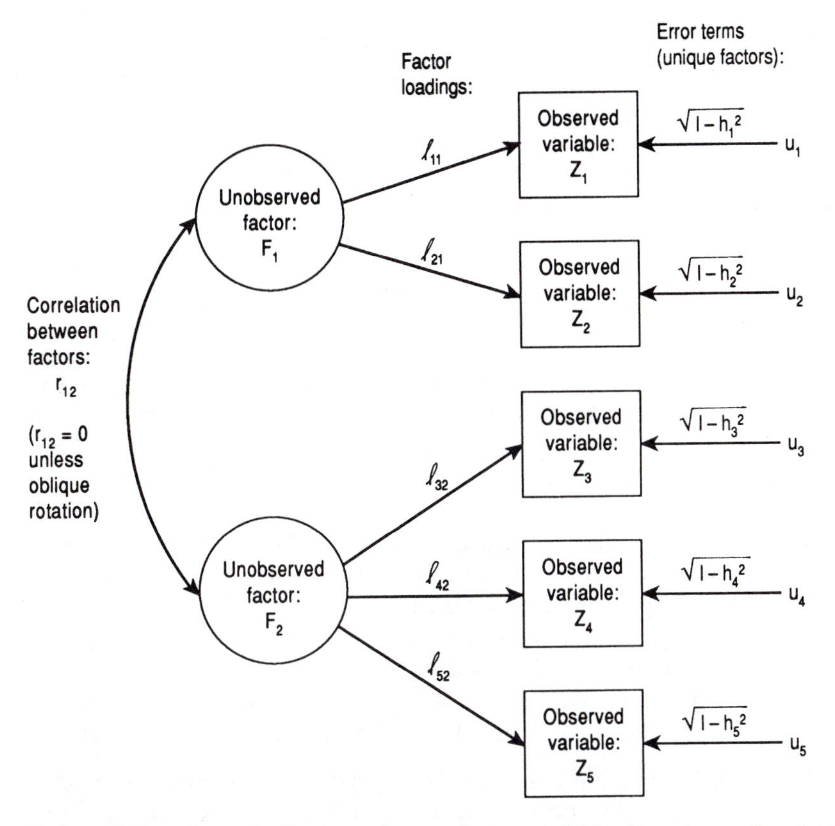

Figure 8.1 Schematic diagram of a two-factor model for five observed variables. Loadings of Z_1 and Z_2 on F_2, and of Z_3, Z_4, and Z_5 on F_1, are near zero and not shown.

In matrix notation the factor model for K variables and J factors ($J < K$) is

$$\underset{n \times K}{\mathbf{Z}} = \underset{n \times J}{\mathbf{F}}\underset{J \times K}{\mathbf{L}'} + \underset{n \times K}{\mathbf{U}} \quad [8.3]$$

where **Z** is a matrix of n observations on **K** variables, **F** is a matrix of values on J unobserved factors, **L** is a $K \times J$ matrix of K loadings on each of the J factors, and **U** is a matrix of residuals or errors.

Principal components requires no residual; K uncorrelated principal components reconstruct K variables:

$$\underset{n \times K}{\mathbf{Z}} = \underset{n \times K}{\mathbf{F}}\underset{K \times K}{\mathbf{L}'} \quad [8.4]$$

Principal components, like factor analysis, can be used for data reduction. This possibility arises when fewer than K components explain *most* of the variance of K variables. If we retain only the J largest components ($J < K$), disregarding the rest, we have

$$Z_k = \ell_{k1}F_1 + \ell_{k2}F_2 + \ell_{k3}F_3 + \cdots + \ell_{kJ}F_J + v_k \qquad [8.5]$$

where the discarded components constitute a residual with small variance:

$$v_k = \ell_{k,J+1}F_{J+1} + \ell_{k,J+2}F_{J+2} + \cdots + \ell_{kK}F_K \qquad [8.6]$$

Equations [8.5]–[8.6] are sometimes termed a *principal components factor* model, emphasizing the resemblance to a true factor analysis model, [8.1] (but also inviting confusion). Since residuals for each of the K variables derive from the same discarded components ($F_{J+1} \cdots F_K$), the v_k residuals of [8.5]–[8.6] cannot be uncorrelated—unlike the unique factors or u_k residuals of [8.1].

A schematic diagram of principal components would show, for example, five components (F_1–F_5) explaining variation in five variables (Z_1–Z_5), with no residual or error terms. Discarding minor components (a "principal components factor analysis") implies a diagram much like Figure 8.1, but with the addition of correlations among the residuals (v_k).

Principal components attempts to explain the observed variables' variance, whereas factor analysis attempts to explain their intercorrelations.

Use principal components to generate composite variables that reproduce the maximum possible variance of the observed variables—for example, one or two composites to stand in for many variables in a regression or graph.

Use factor analysis to model relationships between observed variables and unobserved latent variables, and to obtain estimates of latent-variable values.

In practice, the choice between principal components and factor analysis is often unclear. Many research questions could be addressed by either, so it seems a matter of taste. Principal components appeals more to a "data analysis" perspective, whereas factor analysis fits better with a "model building" approach. Some analysts try both methods and then pick the one that gives the most satisfactory results.

To further blur the distinction, statistical software typically offers principal components as one option under a factor procedure. Factor analysis terms (such as "factors," "factor loadings," and "factor scores") are then applied to results from either method. Where it is not too misleading, I follow this convention.

The first part of this chapter illustrates principal components analysis. We move on to factor analysis itself after establishing some basic ideas.

A Principal Components Analysis

Table 8.1 presents data on runoff, precipitation, glacierization, and area for 19 mountain basins (seen earlier in Exercise 10 of Chapter 4). For this table, all

Table 8.1 Characteristics of 19 mountain basins (based on data from Hicks et al., 1990); all variables expressed as standardized base 10 logs

Basin	Runoff Z_1	Precipitation Z_2	Glacierization Z_3	Area Z_4
1 Ivory	1.306	1.253	1.975	−1.744
2 Cropp	1.321	1.468	.167	−1.613
3 Upper Waitangitoana	.861	1.030	−.608	−.281
4 Hokitika	1.094	1.374	−.119	.336
5 Haast	.747	.873	−.119	.769
6 Little Hopwood Burn	−.505	−.664	−.608	−.894
7 Shotover	−.767	−1.033	−.119	.795
8 Arrow	−1.599	−1.543	−.608	.105
9 Manuherikia	−2.605	−1.926	−.608	.357
10 Karamea	.103	−.049	−.608	.821
11 Buller A	−.374	−.482	−.608	1.377
12 Buller B	−.181	−.373	−.608	1.511
13 Inangahua A	−.328	−.427	−.608	.171
14 Inangahua B	−.088	−.179	−.608	.761
15 Grey	−.066	−.179	−.119	.581
16 Butchers Creek	−.081	−.225	−.608	−1.482
17 Cleddau	.823	.973	−.119	.003
18 Hooker	.812	.873	2.136	−.163
19 Tsidjiore Nouve	−.474	−.766	2.397	−1.408

variables were transformed by first taking base 10 logarithms (to improve linearity and pull in outliers) and then standardizing to have means of 0 and standard deviations of 1. Standardized variables, denoted algebraically by Z_1, Z_2, and so forth, appear throughout this chapter.[1]

Exercise 10 established that both precipitation and runoff predict sediment yield, but we cannot reliably estimate their separate effects due to collinearity. Precipitation (water inflow) and runoff (water outflow) vary so closely together (Figure 8.2) that we might reasonably view them as two different measures of the same thing: water flow. The possibility that several different variables measure the same underlying concept can be explored by principal components or factor analysis.

Table 8.2 gives a correlation matrix for these variables. Runoff and precipitation correlate strongly ($r = .97$); glacierization and area correlate moderately ($r = -.51$). Other correlations are weaker. Figure 8.3, a scatterplot matrix with exploratory band regression curves (Chapter 5), confirms these relations graphically.

Principal components and factor analysis look for subsets of variables correlated among themselves and uncorrelated (or correlated more weakly) with other variables. Runoff and precipitation form one such subset in Table 8.2; glacierization and area form a second, less well-defined, subset.

Table 8.3 presents results from a principal components analysis. It begins by deriving four principal components, which the output labels as factors 1–4. Four components explain 100% of the combined variance of the original four variables. The first two components, however, explain 87% of the combined variance; the second two contribute relatively little. We could reconstruct most of the information of the four original variables from just the first two components.

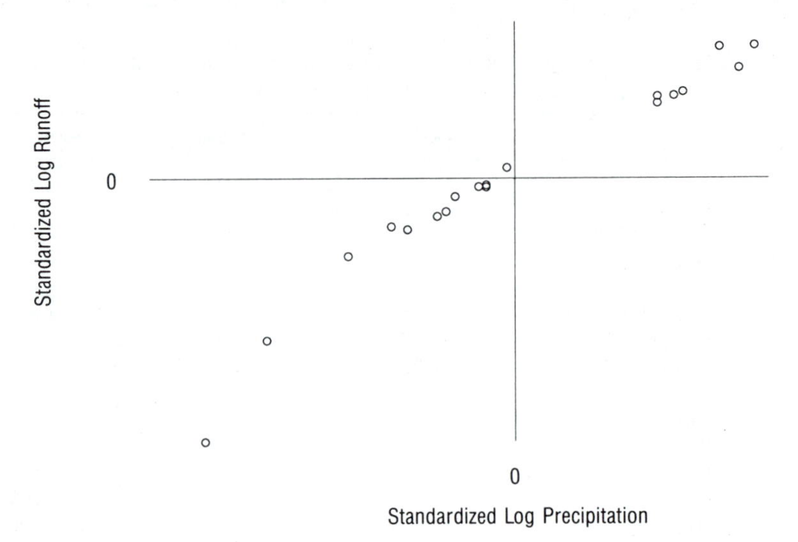

Figure 8.2 Mountain basin runoff versus precipitation; both variables as standardized base 10 logarithms.

Table 8.2 Correlations among mountain basin variables

	Runoff	Precipitation	Glacierization	Area
Runoff	1.00			
Precipitation	.97	1.00		
Glacierization	.34	.30	1.00	
Area	−.29	.28	−.51	1.00

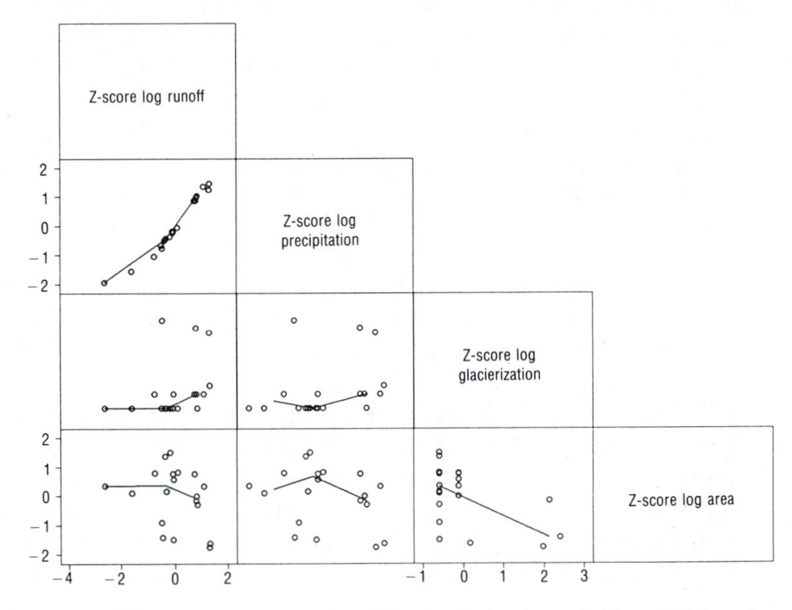

Figure 8.3 Scatterplot matrix of mountain basin variables, with band regression lines.

Table 8.3 Principal components analysis of mountain basin variables

Factor	(principal component factors; 2 factors retained) Eigenvalue	Difference	Proportion	Cumulative
1	2.39152	1.29575	0.5979	0.5979
2	1.09578	0.60839	0.2739	0.8718
3	0.48739	0.46207	0.1218	0.9937
4	0.02532		0.0063	1.0000

Variable	Factor Loadings 1	2	Uniqueness
runoff	0.90586	0.40821	0.01278
precip	0.89510	0.43040	0.01356
glacier	0.63697	−0.58522	0.25178
area	−0.60333	0.63357	0.23458

Eigenvalues are variances of the original components (see Appendix 3 for a definition). The first principal component (F_1) has the highest eigenvalue or variance (λ_1), the second component (F_2) has the second highest eigenvalue (λ_2), and so on. When we derive K components from a set of K variables, the sum of eigenvalues equals the number of variables:

$$\lambda_1 + \lambda_2 + \cdots + \lambda_K = K \qquad [8.7]$$

In Table 8.3:

$$2.392 + 1.096 + 0.487 + 0.025 = 4$$

Standardized variables have variances of 1, so the number of variables also equals the total variance of all variables. The proportion of this total variance explained by the jth component is therefore

$$\frac{\lambda_j}{K} \qquad [8.8]$$

For example, component 1 (labeled factor 1) in Table 8.3 explains 59.79% of the total variance:

$$\frac{2.3915}{4} = .5979$$

Components 1 and 2 together explain 87.18% of the total variance:

$$\frac{2.3915 + 1.0958}{4} = .8718$$

By construction, principal components are uncorrelated with each other.

We can reexpress the information of K variables, $Z_1, Z_2, Z_3, \ldots, Z_K$, in terms of K *principal components*, $F_1, F_2, F_3, \ldots, F_K$. The first principal component, F_1, is that linear combination of original variables having the largest sample variance (λ_1):

$$F_1 = a_{11}Z_1 + a_{21}Z_2 + a_{31}Z_3 + \cdots + a_{K1}Z_K \qquad [8.9]$$

given the constraint $\sum_{k=1}^{K} a_{k1}^2 = 1$. (Without such a constraint, variances can be made arbitrarily large by increasing the magnitudes of the a_{kj} coefficients.)

The second principal component, F_2, is that linear combination *uncorrelated with* F_1 having the largest variance (λ_2):

$$F_2 = a_{12}Z_1 + a_{22}Z_2 + a_{32}Z_3 + \cdots + a_{K2}Z_K \qquad [8.10]$$

given the constraint $\sum_{k=1}^{K} a_{k2}^2 = 1$.

The third principal component is that linear combination uncorrelated with F_1 and F_2 having the largest variance (λ_3), and so forth. The a_{kj} in these equations represent coefficients from the regression of the jth component on the kth variable.

Principal components can be stated in matrix terms: the first principal component, F_1, is that linear combination

$$\underset{n \times 1}{\mathbf{F}_1} = \underset{n \times K}{\mathbf{Z}}\underset{K \times 1}{\mathbf{a}_1} \qquad [8.11]$$

having the largest variance, subject to the constraint

$$\underset{1 \times K}{\mathbf{a}'_1}\underset{K \times 1}{\mathbf{a}_1} = 1 \qquad [8.12]$$

The solution is that the column vector $\mathbf{a}_1$, where

$$\mathbf{a}_1 = [a_{11} \quad a_{12} \quad a_{13} \quad \cdots \quad a_{1K}]$$

equals the eigenvector associated with the largest eigenvalue (λ_1) of the correlation matrix $\mathbf{R}$ (or, less commonly, of the covariance matrix $\mathbf{C}$).

The jth principal component, $\mathbf{F}_j$, is that linear combination

$$\mathbf{F}_j = \mathbf{Z}\mathbf{a}_j \qquad [8.13]$$

where $\mathbf{a}_j$ is the eigenvector associated with the jth-largest eigenvalue, λ_j. This ensures that the sum of squared coefficients for each component equals 1:

$$\mathbf{a}'_j\mathbf{a}_j = 1 \qquad \text{for all } j \qquad [8.14]$$

and each component is uncorrelated with all previous components:

$$\mathbf{a}'_j\mathbf{a}_m = 0 \qquad \text{for all } j < m \qquad [8.15]$$

Equations [8.9]–[8.15] define unstandardized principal components ($\text{Var}[F_j] = \lambda_j$). For some purposes, such as factor loadings, it is more convenient to work with standardized components (defined in the same way but then linearly transformed so that $\text{Var}[F_j] = 1$).

Factor loadings are standardized coefficients in the *regression of variables on components* (*or factors*). Each observed variable can be expressed as a linear function of K uncorrelated principal components:

$$Z_k = \ell_{k1}F_1 + \ell_{k2}F_2 + \cdots + \ell_{kK}F_K \qquad [8.16]$$

Here ℓ_{k1} is the loading of variable Z_k on *standardized* component F_1, and so forth. Factor loadings reflect the strength of relations between variables and components.

The loading of the kth variable on the jth principal component is

$$\ell_{kj} = a_{kj}\sqrt{\lambda_j} \qquad [8.17]$$

where a_{kj} is the kth element of the jth eigenvector and λ_j is the corresponding eigenvalue. Multiplying by $\sqrt{\lambda_j}$ in [8.17] standardizes the loadings.

For variable Z_1, given four (standardized) components, [8.16] becomes:

$$Z_1 = \ell_{11}F_1 + \ell_{12}F_2 + \ell_{13}F_3 + \ell_{14}F_4$$

If, as in Table 8.3, we find the last two components unimportant, we might simplify:

$$Z_1 = \ell_{11}F_1 + \ell_{12}F_2 + v_1$$

where residual $v_1 = \ell_{13}F_3 + \ell_{14}F_4$. Table 8.3 supplies this equation for standardized log runoff (Z_1):

$$Z_1 = .90586F_1 + .40821F_2 + v_1$$

The *uniqueness* of each variable equals the proportion of its variance not explained by the retained components or factors. Only 1.279% of the variance of runoff, but 25.178% of the variance of glacierization, is not explained by the first two components. *Communality* equals the proportion of a variable's variance shared with the components:

$$\text{communality} = 1 - \text{uniqueness}$$

The communality of variable Z_k, (h_k^2), equals the proportion of Z_k's variance explained by the retained factors.

$$h_k^2 = 1 - \text{uniqueness}$$

$$= \sum_{j=1}^{J} \ell_{kj}^2 \qquad [8.18]$$

where the ℓ_{kj} are unrotated loadings of Z_k on the retained factors F_1–F_J.

How Many Components?

K noncollinear variables yield K principal components, but this achieves no simplification. When the first few components account for most of the variation, however, we can concentrate on them and disregard remaining components. For example, we lose little (and gain simplicity) by omitting the last two components in Table 8.3 from further analyses. The computer program did this automatically and printed loadings only for components 1 and 2.

How do we judge which components are unimportant and which should be kept? Eigenvalues provide one criterion. A common rule of thumb, applied in Table 8.3, is to disregard principal components with eigenvalues less than 1. Since each standardized variable has a variance of 1, a component with $\lambda < 1$ accounts for less than a single variable's variation—and hence is useless for data reduction.

Scree graphs (Figure 8.4) plot eigenvalues against component or factor number. Look for a point at which eigenvalues stop falling steeply and begin to level off. ("Scree" is the geologist's term for a slope of small rock fragments fallen to the base of a steeper mountainside.) In Figure 8.4 we see that leveling off (like that of a scree slope) begins after component 2. Components 3 and 4 account for relatively little additional variance, reinforcing the conclusion that only the first two components matter.

Although it is subjective, scree graph inspection provides useful guidance. The eigenvalue-1 rule, by itself, could lead to arbitrary distinctions between components of similar importance. For example, it might lead us to keep one component with $\lambda = 1.01$ but ignore the next one with $\lambda = .99$. Scree graphs may suggest more natural cutoffs.

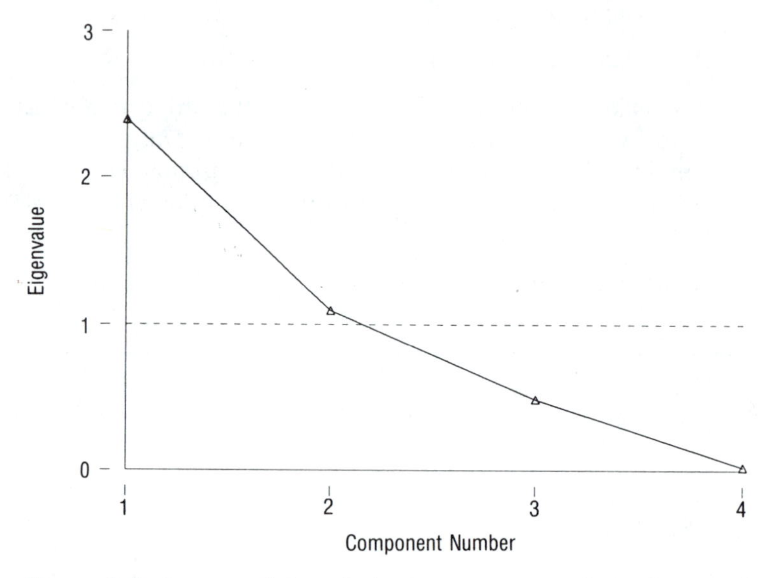

Figure 8.4 Scree graph for principal components analysis of mountain basin variables.

Another criterion for keeping components is whether they are substantively meaningful or interpretable. An uninterpretable component may have limited analytical use despite a large eigenvalue.

Sometimes we retain just a single component, representing the "best possible" (maximum variance) linear combination of the observed variables. Then we skip the next step (rotation) and proceed directly to factor scores.

Rotation

The two components or factors retained at bottom in Table 8.3 fit the data well but are hard to interpret. All four variables load on both factors. Judging from the loadings' signs, factor 1 represents a dimension that ranges from high-runoff, high-precipitation, high-glacierization, small-area basins, at one extreme, to low-runoff, low-precipitation, little-glacierization, and large-area basins at the other. Factor 2 makes even less intuitive sense: it ranges from high runoff, high precipitation, little glacierization, and large area to low runoff, low precipitation, high glacierization, and large area. By definition, these factors are uncorrelated with each other.

Rotation seeks more interpretable factors. We obtain new factors that fit the data equally well but have simpler structure than the initial factors. A "simple structure" might mean that each variable loads strongly (either positively or negatively) on only one factor, and near zero on the other factors. For example:

	Loadings after Rotation	
	Factor 1	**Factor 2**
Variable Z_1	**near ±1**	near 0
Variable Z_2	**near ±1**	near 0
Variable Z_3	near 0	**near ±1**
Variable Z_4	near 0	**near ±1**

Table 8.4 lists *orthogonally rotated* loadings for the mountain basin factors. Compared with the initial loadings in Table 8.3, the rotated loadings come closer to a "simple structure" ideal. Runoff and precipitation load mostly on factor 1 and

Table 8.4 Orthogonally rotated factor loadings

	(varimax rotation) Rotation Factor Loadings		
Variable	**1**	**2**	**Uniqueness**
runoff	**−0.98881**	0.09735	0.01278
precip	**−0.99053**	0.07275	0.01356
glacier	−0.26074	**0.82476**	0.25178
area	0.20749	**−0.84992**	0.23458

very little on factor 2. Glacierization and area load mainly on factor 2. These rotated factors are easier to interpret. Since the factor-1 loadings of precipitation and runoff are negative, a basin high on factor 1 has low water flow. We might call this factor a "dryness" dimension. Factor 2 is harder to name but ranges from small, glacier-filled basins to large, unglacierized ones. Orthogonal rotation ensures that these dimensions, like the original components, are uncorrelated.

To interpret factors, study rotated loading matrices like Table 8.4. Mark the highest loadings in each row. Interpretation follows from which variables load highly (and with what sign) on a given factor.

Rotation of J factors involves postmultiplying the loading matrix **L** by a $J \times J$ transformation matrix **M**:

$$\mathbf{L}^* = \mathbf{L}\mathbf{M} \qquad [8.19]$$

Properties of the rotated loading matrix $\mathbf{L}^*$ (simple structure, correlated or uncorrelated factors) depend on the choice of **M**.

Figure 8.5 graphs the four variables' loadings on unrotated factors 1 and 2 (from Table 8.3). Area (Z_4), for instance, plots at upper left with coordinates (−0.60333,

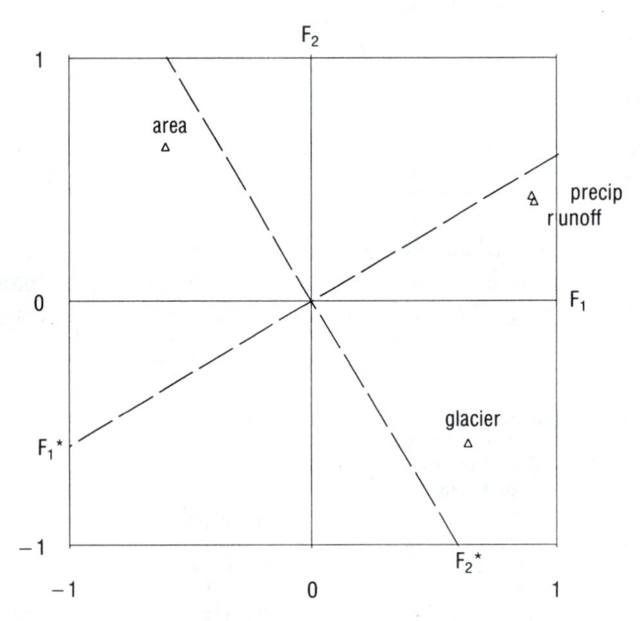

Figure 8.5 Mountain basin variables' loadings on unrotated principal components F_1 and F_2, also showing orthogonally rotated dimensions F_1^* and F_2^*.

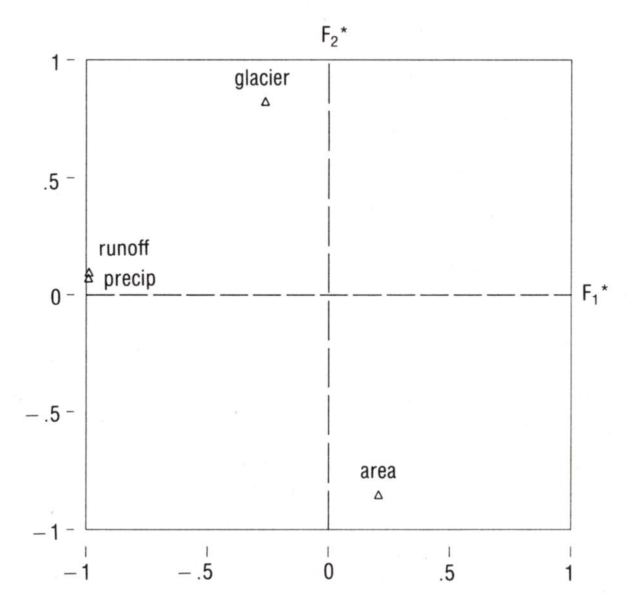

Figure 8.6 Mountain basin variables' loadings on orthogonally rotated principal components F_1^* and F_2^*.

0.63357). Glacierization (Z_3) plots at lower right with coordinates (0.63697, −0.58522). Thus the horizontal axis represents factor 1 (F_1) and the vertical axis factor 2 (F_2). Orthogonal rotation amounts to a clockwise rotation of these axes, like rotating wheel spokes around a hub. F_1 rotates to F_1^*, and F_2 rotates to F_2^*. The angle between the F_1 and F_2 axes remains 90°, reflecting their zero correlation.[2] We seek a new orientation such that each variable is close to one of the axes.

In Figure 8.5, rotation shifts the factor-1 axis to lie near precipitation and runoff. After rotation these two variables load highly on new dimension F_1^* and near zero on new dimension F_2^*. The opposite applies to area and glacierization, which have high loading on F_2^* (area negative, glacierization positive) and lower loadings on F_1^*. Redrawing the graph with the new, rotated axes yields Figure 8.6.

Table 8.4 and Figures 8.5 and 8.6 illustrate *varimax rotation*, a widely used orthogonal rotation method. Varimax maximizes the variance of the squared loadings for each factor. This tends to polarize loadings so that they are either high or low, making it easier to identify factors with specific variables.

Varimax rotation seeks an orthogonal transformation matrix **M** such that

$$\mathbf{L}^* = \mathbf{LM}$$

maximizes

$$\sum_{k=1}^{K} \sum_{j=1}^{J} (\ell_{kj}^{*2} - d_j)^2 \qquad [8.20]$$

where ℓ_{kj} is the loading of the kth variable on the jth factor and d_j is the mean squared loading on the jth factor:

$$d_j = \frac{\sum_{k=1}^{K} \ell_{kj}^2}{K} \qquad [8.21]$$

Equation [8.20] increases as loadings on each factor polarize toward zeros and ones.

Oblique rotation, permitting correlation between factors, seeks even greater polarization—hence simpler structure and more interpretable dimensions. If we view factors as unmeasured variables that underlie the observed data, oblique rotation is more realistic: such variables need not have zero correlation.

Table 8.5 shows results of oblique rotation (using a method called *promax*) applied to the mountain basin data. After oblique rotation, all four variables load highly on one factor and near zero on the other. We could label these moderately correlated factors dryness (F_1^*) and glacierization/area (F_2^*). Geometrically, by relaxing the requirement for right-angle (uncorrelated) axes, we can place them closer to the original variables (Figure 8.7).

Promax rotation begins with the varimax-rotated loadings $\mathbf{L}^*$. We seek a transformation matrix $\mathbf{P}$ with columns that minimize

$$\text{trace}(\mathbf{T} - \mathbf{L}^*\mathbf{P})'(\mathbf{T} - \mathbf{L}^*\mathbf{P}) \qquad [8.22]$$

where trace ( ) denotes the trace operator (Appendix 3). Elements of $\mathbf{T}$ are factor loadings (elements of $\mathbf{L}^*$) raised to an arbitrary power greater than 1 but usually less than 4 (examples in this chapter use 3). The higher the power, the stronger the correlations allowed between factors.

Promax-rotated loadings $\mathbf{L}^{**}$ result from postmultiplication of varimax loadings ($\mathbf{L}^*$) by $\mathbf{P}$ (after scaling $\mathbf{P}$ for unit-variance factors):

$$\mathbf{L}^{**} = \mathbf{L}^*\mathbf{P} \qquad [8.23]$$

Table 8.5 Obliquely rotated loadings for mountain basin factors (compare with Table 8.3 and 8.4)

Variable	(promax rotation) Rotation Factor Loadings 1	2	Uniqueness
runoff	**−0.98901**	0.01338	0.01278
precip	**−0.99742**	−0.01280	0.01356
glacier	−0.04491	**0.84887**	0.25178
area	−0.01657	**−0.88031**	0.23458

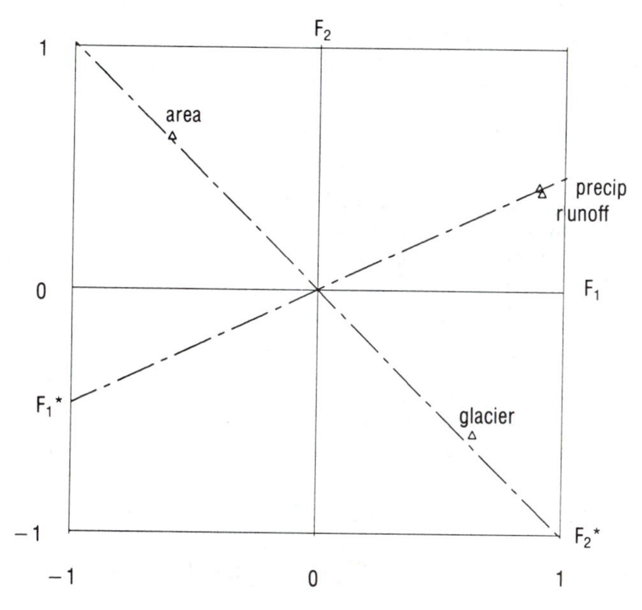

Figure 8.7 Oblique rotation of principal components F_1 and F_2 to F_1^* and F_2^*.

Advantages of oblique rotation include easier interpretation (because loadings are further polarized) and greater realism (because real underlying variables may be correlated). Its disadvantages include arbitrariness and greater complexity.

Rotation makes factor analysis more usable, but it also points out a fundamental uncertainty. Any linear transformation of the initial loadings (that is, any choice of **M** in [8.19]) produces rotated factors that fit the data equally well. There are many methods of orthogonal rotation, and oblique rotation can achieve different results depending on how much interfactor correlation we allow. Thus a rotation outcome like Table 8.4 or Table 8.5 depends on a series of choices by the analyst.

To put our conclusions on sounder footing, we could repeat the analysis a number of different ways. This would allow us to choose the more plausible or substantively satisfying result and would also provide an indication of how stable it is. Sometimes different rotation methods lead to basically similar conclusions, as with Tables 8.4 and 8.5. Then it matters less which method we choose. On the other hand, if different methods disagree, we would be warned that conclusions are arbitrary and unstable.

Factor Scores

Principal components and factor analysis provide a basis for combining variables, forming composites called *factor scores*. Factor scores derive from *factor score coefficients*: coefficients in the *regression of factors on variables*. That is, factor score coefficients for the jth factor are the c_{kj} in

$$\tilde{F}_j = c_{1j}Z_1 + c_{2j}Z_2 + \cdots + c_{Kj}Z_K \qquad [8.24]$$

Factor scores ($\tilde{F}_j$) are estimates of the unknown true values of the factors (F_j).

Table 8.6 Factor score coefficients after oblique rotation

Variable	(based on rotated factors) Scoring Coefficients 1	2
runoff	**−0.50068**	0.00091
precip	**−0.50515**	−0.01666
glacier	−0.01584	**0.56722**
area	−0.01554	**−0.58874**

Table 8.6 lists factor score coefficients based on the obliquely rotated loadings of Table 8.5. Using these coefficients, we calculate scores on factor 1 (dryness):

$$\tilde{F}_1 = -.50068Z_1 - .50515Z_2 - .01584Z_3 - .01554Z_4$$

Similarly, to calculate scores on factor 2 (glacierization/area):

$$\tilde{F}_2 = .00091Z_1 - .01666Z_2 + .56722Z_3 - .58874Z_4$$

A matrix of orthogonal factor scores, $\tilde{\mathbf{F}}$, can be obtained by *regression scoring*:

$$\underset{n \times J}{\tilde{\mathbf{F}}} = \underset{n \times K}{\mathbf{Z}}\underset{K \times K}{\mathbf{R}^{-1}}\underset{K \times J}{\mathbf{L}} \qquad [8.25]$$

where **Z** is the matrix of standardized variables, **R** is their correlation matrix, and **L** is the *unrotated or varimax-rotated* factor-loading matrix.

For oblique factor scores, regression scoring is

$$\tilde{\mathbf{F}} = \mathbf{ZR}^{-1}\mathbf{L}(\mathbf{P}'\mathbf{P})^{-1} \qquad [8.26]$$

where **L** is the *varimax-rotated* factor-loading matrix and **P** is the $J \times J$ scaled oblique transformation matrix (see [8.22]–[8.23]).

These score coefficients define two new variables, listed in Table 8.7. For example, the first basin's score on factor 1 is

$$\tilde{F}_1 = -.50068 \times 1.306 - .50515 \times 1.253 - .01584 \times 1.975 - .01554 \times (-1.744)$$
$$= -1.291$$

Table 8.7 Factor scores based on coefficients from Table 8.6

Basin	Factor 1 Score $\tilde{F}_1$	Factor 2 Score $\tilde{F}_2$
1	−1.291	2.127
2	−1.381	1.021
3	−0.938	−0.196
4	−1.245	−0.288
5	−0.825	−0.534
6	0.611	0.192
7	0.895	−0.519
8	1.588	−0.382
9	2.281	−0.525
10	−0.030	−0.827
11	0.419	−1.148
12	0.265	−1.229
13	0.387	−0.438
14	0.132	−0.790
15	0.116	−0.406
16	0.187	0.532
17	−0.902	−0.085
18	−0.879	1.293
19	0.609	2.201

The negative score indicates that this basin (Ivory) has less dryness (that is, more water flow) than average. Ivory Basin scores above zero on factor 2:

$$\tilde{F}_2 = .00091 \times 1.306 - .01666 \times 1.253 + .56722 \times 1.975 - .58874 \times (-1.744)$$
$$= 2.127$$

indicating above-average glacierization and "smallness" (because the coefficient on area, Z_4, is negative). We can analyze factor scores like other variables, using them in graphs, regression, ANOVA, and so forth.

The principal components analysis suggests that we lose little information, and gain simplicity, by working with the two composite variables of Table 8.7 instead of the four original variables of Table 8.1. Table 8.8 illustrates this. The top part shows regression of log sediment yield (the Y variable from Exercise 10 of Chapter 4) on standardized log runoff, precipitation, glacierization, and area. These four variables explain 79.74% of the variance in log yield. Note symptoms of multicollinearity: despite the high R^2, none of the coefficients are significant. Furthermore, runoff and precipitation, both positively correlated with yield and strongly correlated with each other, here have opposite-sign coefficients.

The bottom of Table 8.8 shows regression of log yield on two factor scores that combine the four original variables. Although this regression uses only two predictors, it fits almost as well as the four-predictor regression at top: $R^2 = .7832$ instead of .7974. The partial regression coefficient on the first factor score is significant at $\alpha = .05$ ($t = -6.348$, $P < .0005$); that on the second is not.

Table 8.8 Regression of log sediment yield on four variables (from Table 8.1) or two factor scores (from Table 8.7)

Source	SS	df	MS	
Model	10.265554	4	2.5663885	Number of obs = 19; $F(4, 14) = 13.77$
Residual	2.60862507	14	.186330362	Prob > F = 0.0001; R-square = 0.7974
Total	12.8741791	18	.715232171	Adj R-square = 0.7395; Root MSE = .43166

Variable	Coefficient	Std. Error	t	Prob > $\|t\|$	Mean
yield					3.200158
runoff	−.1151779	.4575377	−0.252	0.805	−9.02e − 09
precip	.7751374	.4527634	1.712	0.109	4.12e − 09
glacier	.1315507	.1231924	1.068	0.304	−4.31e − 09
area	−.1008783	.1200595	−0.840	0.415	−6.35e − 09
_cons	3.200158	.0990296	32.315	0.000	1

Source	SS	df	MS	
Model	10.0835282	2	5.0417641	Number of obs = 19; $F(2, 16) = 28.91$
Residual	2.79065088	16	.17441568	Prob > F = 0.0000; R-square = 0.7832
Total	12.8741791	18	.715232171	Adj R-square = 0.7561; Root MSE = .41763

Variable	Coefficient	Std. Error	t	Prob > $\|t\|$	Mean
yield					3.200158
F1	−.6634639	.104515	−6.348	0.000	−6.96e − 09
F2	.1890137	.104515	1.808	0.089	3.14e − 09
_cons	3.200158	.0958111	33.401	0.000	1

In this example, regression on factor scores improves parsimony and avoids multicollinearity, with little loss in prediction. Such qualities become more valuable with more variables.

By definition, the first principal component is that linear combination accounting for the largest possible fraction of the observed variables' variance. Scores on the first *unrotated* principal component are therefore a reasonable choice for combining many observed variables into a single composite. A *reliability coefficient* called *theta* (θ) may be calculated for the first principal component:

$$\theta = \left(\frac{K}{K-1}\right)\left(1 - \frac{1}{\lambda_1}\right) \qquad [8.27]$$

where K equals the number of variables included and λ_1 equals the eigenvalue. For the first component in Table 8.3:

$$\theta = \left(\frac{4}{4-1}\right)\left(1 - \frac{1}{2.39152}\right) = .7758$$

Here θ can be viewed as a special case of Cronbach's α. See Carmines and Zeller (1979) for an introduction to reliability coefficients.

Graphical Applications: Detecting Outliers and Clusters

Reducing multivariate information to a few underlying dimensions makes it easier to graph. Figure 8.8 plots factor scores for the glacier basins data from Table 8.7. Such two-dimensional plots of the first two principal components can graphically display much of the information from the original multivariate data.

Principal component graphs may uncover multivariate outliers—cases having an exceptional combination of values on the original variables. One such case appears at upper right in Figure 8.8. It does not appear unusual in univariate plots of any of the four original variables. But seeing how it stands apart in Figure 8.8 should lead us to wonder, what makes this basin so different from the rest?

Figure 8.9 answers the question by replacing Figure 8.8's points with each basin's geographic location. Eighteen basins are in New Zealand; the outlier (Tsidjiore Nouve) turns out to be in Switzerland. Researchers would ordinarily be cautious with such mixed data, knowing that one case originates halfway around the world from the rest. I included the Swiss basin here to illustrate the power of this graphical approach for detecting multivariate outliers, or observations that do not belong.

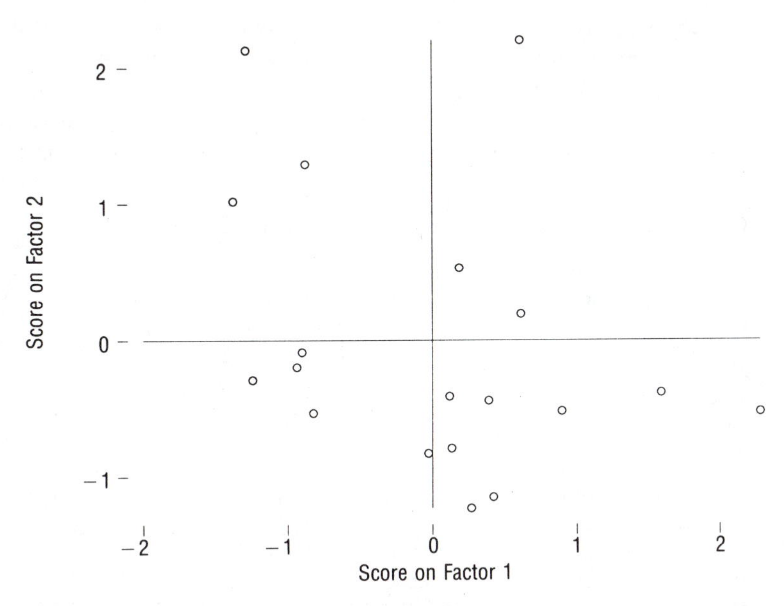

Figure 8.8 Principal components plot of scores on factor (component) F_2 versus factor (component) F_1.

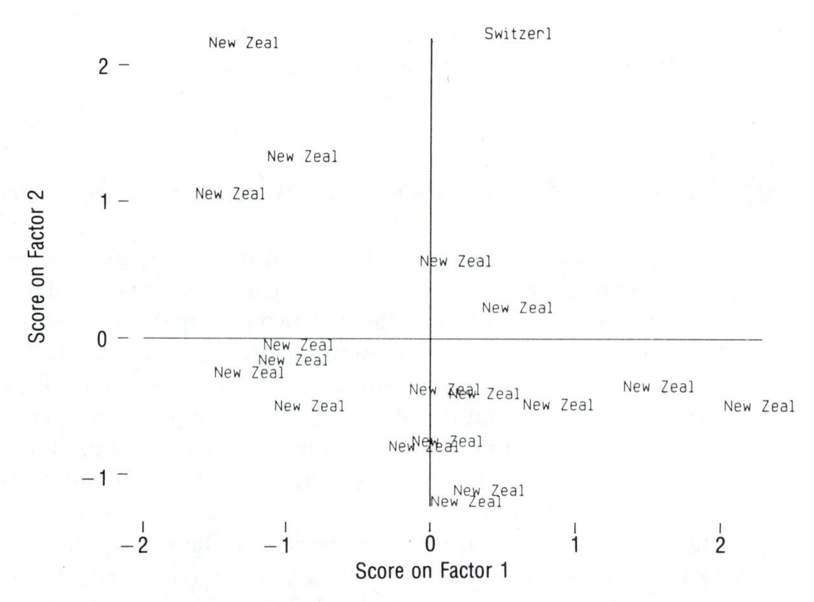

Figure 8.9 Principal components plot identifying mountain basin locations.

Sometimes principal component graphs like Figures 8.8 and 8.9 reveal not just a few outliers, but clusters of cases apart from the rest. Then such graphs provide a tool for *cluster analysis*: identification of subgroups of cases having similar multivariate profiles.

For illustration, consider the planetary data of Table 8.9. You may already know that there are two main types of planets, the inner rocky planets and the outer gas giants. Setting that knowledge aside, we will see how well principal components performs in discovering this typology.

Figure 8.10 shows a scatterplot matrix with all variables (except a dummy for rings) as natural logarithms. Logarithms pull in the servere outliers and make

Table 8.9 Physical characteristics of nine planets[1]

Planet	Mean Dist. from Sun (10^6 km)	Radius At Equator (km)	Mass (kg)	Mean Density (g/m^3)	Number of Known Moons	Rings Present
Mercury	57.9	2,439	3.30*e* + 23	5.42	0	no
Venus	108.2	6,050	4.87*e* + 24	5.25	0	no
Earth	149.6	6,378	5.98*e* + 24	5.52	1	no
Mars	227.9	3,398	6.42*e* + 23	3.94	2	no
Jupiter	778.3	71,900	1.90*e* + 27	1.31	16	yes
Saturn	1,427.0	60,000	5.69*e* + 26	.69	17	yes
Uranus	2,869.6	26,145	8.66*e* + 25	1.19	15	yes
Neptune	4,496.6	24,750	1.03*e* + 26	1.66	8	yes
Pluto	5,900.0	1,550	1.10*e* + 22	1.20	1	no

[1] Data from Beatty, O'Leary, and Chaikin, 1981.

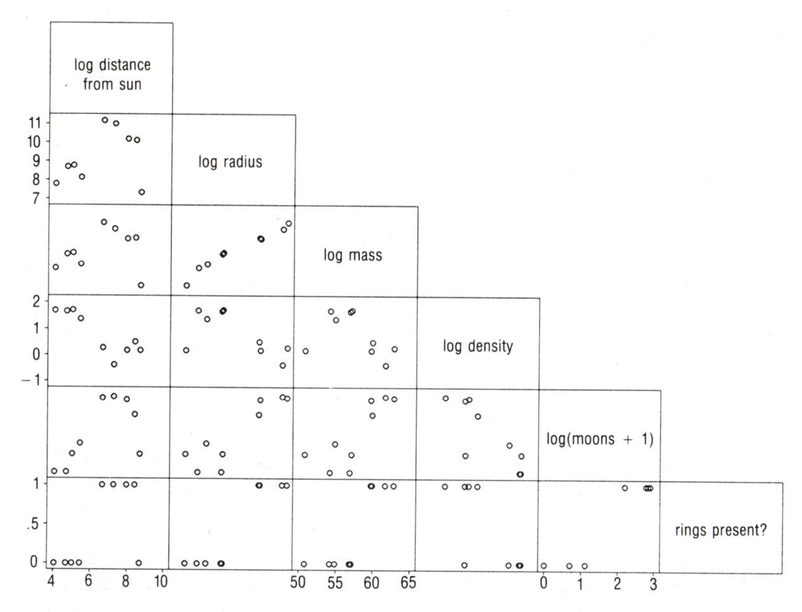

Figure 8.10 Scatterplot matrix for planetary data of Table 8.9.

the basically nonlinear relationships somewhat more linear, at least to a crude approximation.

This linear approximation is good enough that the first two principal components account for 96.5% of the total variation of all six variables (as logarithms). The four remaining components add little and have eigenvalues below .12, as shown in the scree graph in Figure 8.11.

Table 8.10 lists scores on the first two principal components after oblique rotation. Logarithms of radius and mass, and the rings dummy variable, load most heavily (and negatively) on the first component. High $\tilde{F}_1$ scores indicate a small, unringed planet. Distance (positive) and density (negative) load more heavily on the second component. High $\tilde{F}_2$ scores indicate a far-out, low-density world.

The inner rocky planets all possess positive $\tilde{F}_1$ scores and negative $\tilde{F}_2$ scores; the opposite applies to the outer gas giants. Pluto is unique in being above average (positive) in both the "smallness" ($\tilde{F}_1$) and "far-out/low-density" ($\tilde{F}_2$) dimensions. Figure 8.12 displays these scores graphically.

Figure 8.12 reveals three distinct types of planets. (The third type is Pluto, a remote iceball apparently in a class by itself. Pluto actually resembles the large outer-system moons more than it does other planets.) To use principal components for cluster analysis, we look for graphically distinct subgroups of cases. Identifying such subgroups or clusters may be a research goal in itself, or an exploratory step in developing better-fitting models.

Outliers and clusters are related concepts: cases notably different (on the dimensions of interest) from the rest of the data. We can look for both in the same way, by scanning plots of component scores. This approach is particularly effective if the first few components explain most of the variation.

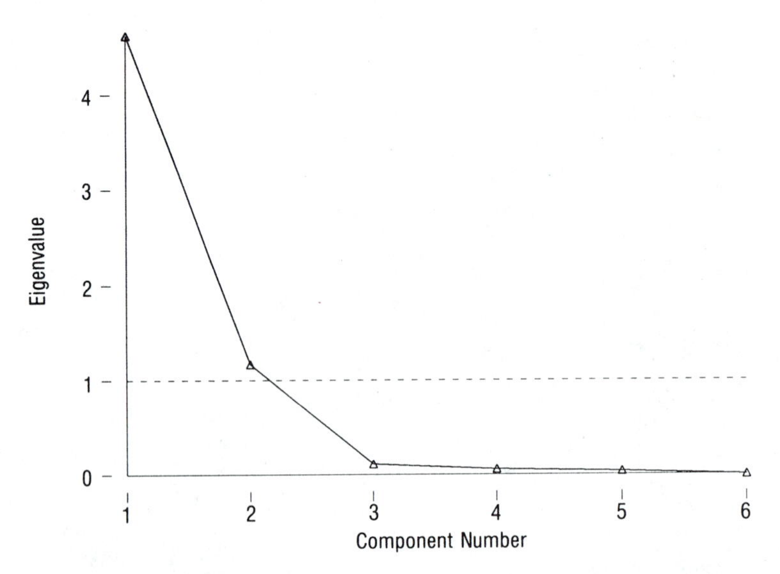

Figure 8.11 Scree graph for principal components of planetary data.

Table 8.10 Factor scores after oblique (promax) rotation of first two principal components

Planet	$\tilde{F}_1$	$\tilde{F}_2$
Mercury	0.85	−1.27
Venus	0.44	−1.18
Earth	0.33	−1.03
Mars	0.66	−0.62
Jupiter	−1.36	0.46
Saturn	−1.16	0.97
Uranus	−0.72	0.95
Neptune	−0.61	0.83
Pluto	1.57	0.89

A cluster of unusual cases can together exert much influence in regression and yet remain undetected by the single-case diagnostic statistics of Chapter 4 or the case-downweighting methods of Chapter 6. Principal component graphs like Figures 8.8 or 8.12 supplement methods for detecting individual outliers.

Principal Factor Analysis

The simplest true factoring procedure, called *principal factor* or *principal axis* analysis, extracts principal components of a modified correlation matrix **R***. Whereas

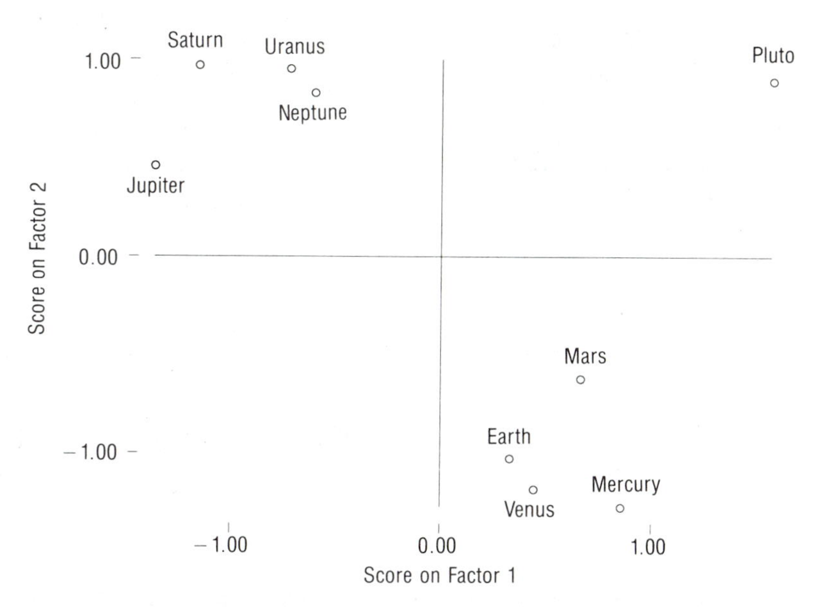

Figure 8.12 Principal components plot identifying three types of planets.

the original correlation matrix **R** has ones on its major diagonal, in **R*** these ones are replaced by estimates of communality (shared variance).[3] Thus principal factor analysis tries not to explain the *variance* of $Z_1, Z_2, \ldots, Z_K$, as principal components does, but rather to explain *correlations* or *covariance* among $Z_1, Z_2, \ldots, Z_K$. Some of the variance of Z_1, Z_2, $\ldots$, Z_K is left unexplained, attributed to an error term or "unique factor" (see [8.1]). A variable's uniqueness might reflect measurement error, or something measured accurately by that variable but not by any others.

A variable's communality (1 − uniqueness) equals the proportion of its variance explained by the retained factors. We can calculate this from the sum of its squared unrotated loadings on the retained factors ([8.18]). Principal factor analysis, however, must estimate communality before obtaining factor loadings. One approach uses R_k^2, the coefficient of determination from regressing Z_k on all other variables, as estimates of communality. That is, we extract principal components of the matrix

$$\mathbf{R}^* = \begin{bmatrix} R_1^2 & r_{12} & r_{13} & \cdots & r_{1K} \\ r_{21} & R_2^2 & r_{23} & \cdots & r_{2K} \\ r_{31} & r_{32} & R_3^2 & \cdots & r_{3K} \\ \cdot & \cdot & \cdot & \cdots & \cdot \\ \cdot & \cdot & \cdot & \cdots & \cdot \\ \cdot & \cdot & \cdot & \cdots & \cdot \\ r_{K1} & r_{K2} & r_{K3} & \cdots & R_K^2 \end{bmatrix}$$

where R_1^2 is the coefficient of determination regressing Z_1 on $Z_2, Z_3, \ldots, Z_K$; r_{12} is the correlation between Z_1 and Z_2. The first principal component of $\mathbf{R}^*$ is the first principal factor, and so forth.

A K-variable correlation matrix $\mathbf{R}$ decomposes to

$$\underset{K \times K}{\mathbf{R}} = \underset{K \times J}{\mathbf{L}}\underset{J \times K}{\mathbf{L}'} + \underset{K \times K}{\mathbf{Q}} \qquad [8.28]$$

where $\mathbf{L}$ is the matrix of loadings on J factors ($J < K$) and $\mathbf{Q}$ is a matrix with uniquenesses on the main diagonal and zeros elsewhere.

Given an estimate of $\mathbf{Q}$, columns of $\mathbf{L}$ equal eigenvectors (scaled by eigenvalues as in [8.17]) of the modified correlation matrix $\mathbf{R}^*$:

$$\mathbf{R}^* = \mathbf{R} - \mathbf{Q} \qquad [8.29]$$

That is, principal factor loadings $\mathbf{L}$ derive from principal components of $\mathbf{R}^*$.

After extraction of these principal factors, Equation [8.18] provides better estimates of communality. We could then place the new estimates of communality on the major diagonal of our modified correlation matrix $\mathbf{R}^*$ and extract new principal factors. This procedure, called *principal factoring with iteration,* continues until there is little change in the communality estimates.

How many factors should we retain? Eigenvalues, proportion of variance explained, scree graphs, and interpretability provide subjective guidance, as they do with principal components. Factor eigenvalues tend to be lower than those of principal components, so some authors recommend a minimum eigenvalue of 0 rather than 1. An eigenvalue-0 criterion is not very selective, however.

With principal components analysis, or principal factoring *without* iteration, we work in four basic steps:

1. extract initial factors or components
2. rotate to simplify structure
3. decide how many factors to retain
4. obtain and use scores for the retained factors, ignoring any lesser unwanted factors

Communality estimates using [8.18] depend upon how many factors we retain. Consequently, when communalities are estimated iteratively, the final factor analysis should include *only the retained factors.* We therefore repeat the factor and rotation steps:

1. (iteratively) extract factors
2. rotate to simplify structure
3. decide how many factors to retain
4. factor again, specifying how many factors to retain
5. rotate again, using these retained factors
6. obtain and use scores for the retained factors

Like principal components, principal factors (with or without iteration) can be orthogonally or obliquely rotated as in [8.20]–[8.23], and scored as in [8.25]–[8.26].

An Example of Principal Factor Analysis

Survey questionnaires, psychological profiles, and aptitude tests often produce data suitable for factor analysis: many different questions intended to measure a smaller number of latent variables. For example, Table 8.11 lists six questions from a survey conducted in Tulsa, Oklahoma (Blocker and Eckberg, 1989). The questions ask respondents to rate the seriousness of some local environmental hazards. This survey produced data on six different, but conceptually related, variables.

Applied to such data, factor analysis serves two purposes:

1. Perhaps people's responses to these six specific questions partly reflect general traits like "concern about pollution" or "fear of natural disasters." Factor analysis helps identify and measure such latent variables.
2. If plausible latent variables emerge, they simplify subsequent analyses. A few factor scores (estimates of latent variables) might replace many individual items.

Table 8.11 Tulsa area survey questions on local problems[1]

How Serious Do You Think These Local Problems Are?	
Z_1 deepwell	Injecting chemical wastes into deep underground wells
Z_2 chandler	The Chandler Park landfill that has been burning underground
Z_3 tornados	Tornados
Z_4 floods	Floods
Z_5 airpol	Pollution in Tulsa's air
Z_6 rivpol	Pollution in Tulsa's rivers and streams

[1] Described in Blocker and Eckberg (1989).

Table 8.12 Correlation matrix for seriousness of Tulsa environmental hazards

	Deepwell	Chandler	Tornados	Floods	Airpol	Rivpol
deepwell	1.0000					
chandler	0.4726	1.0000				
tornados	0.1131	0.2027	.10000			
floods	0.0928	0.0480	0.4052	1.0000		
airpol	0.2805	0.1661	0.1524	0.0712	1.0000	
rivpol	0.3365	0.2587	0.1007	0.1511	0.3861	1.0000

Successful factor analysis requires a pattern in the correlation matrix: subgroups of variables that correlate more strongly with each other than with other subgroups of variables.

Table 8.12 gives a correlation matrix for the six Tulsa survey questions ($n = 199$). These correlations are moderate at best, which is typical of survey data. Some patterns can be discerned. For example, tornados and floods correlate moderately ($r = .4052$) with each other but weakly with other variables. Other moderate correlations occur between deep well injection and the burning Chandler landfill ($r = .4726$) and between air and river pollution ($r = .3861$).

Figure 8.13 shows a scatterplot matrix for the Tulsa data. These variables have only three possible values (1 = not very serious, 2 = somewhat serious, 3 = very serious), so jittering (see Chapter 4, Note 4) was employed in graphing. Scatterplot

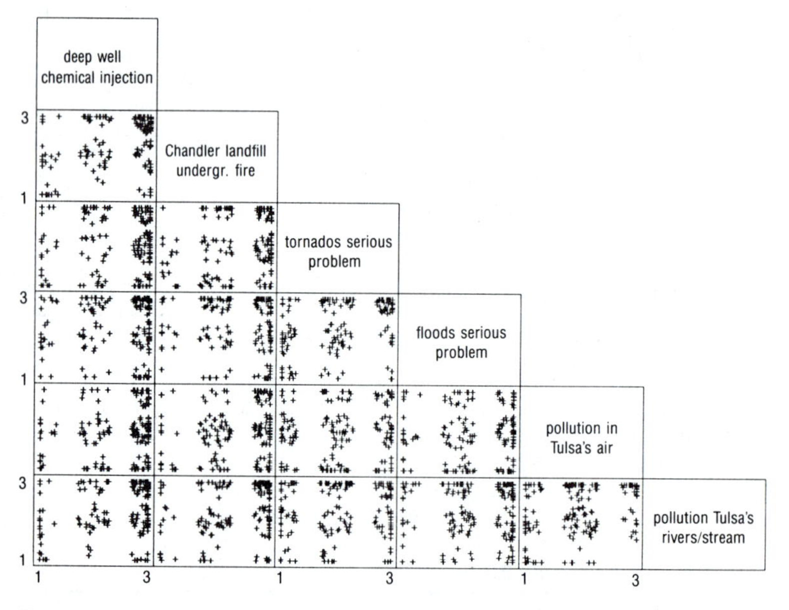

Figure 8.13 Scatterplot matrix for Tulsa survey data (jittered).

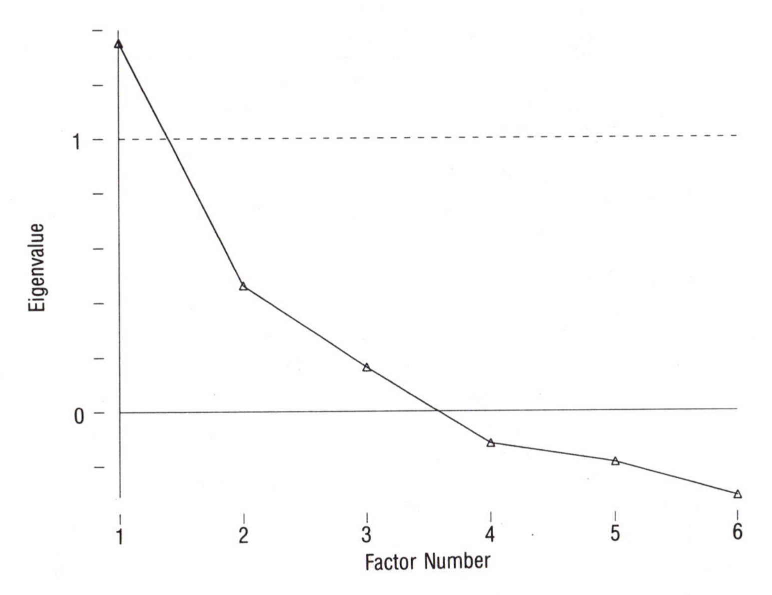

Figure 8.14 Scree graph for principal factors of Tulsa survey data.

matrices generally provide a good screening device before multivariate analysis, but, like other graphs, they illuminate less when the variables have only a few distinct values, as in Figure 8.13.

Principal factor analysis of the Tulsa data yields the scree graph in Figure 8.14. Only three factors have positive eigenvalues. After oblique rotation, the first three factors (Table 8.13) make intuitive sense:

F_1—Perceived seriousness of specific human-made hazards: deep-well waste injection and the burning underground landfill.
F_2—Perceived seriousness of Tulsa's natural hazards: floods and tornados.
F_3—Perceived seriousness of Tulsa's air and water pollution.

Since relations among the original variables are not strong, each has substantial uniqueness, ranging from .59–.73.

This factor analysis generates three new variables: the factor scores $\tilde{F}_1$, $\tilde{F}_2$, and $\tilde{F}_3$. Oblique rotation permits correlations between factors, shown at bottom in Table 8.13. Factors representing perceived seriousness of human-made environmental hazards ($\tilde{F}_1$ and $\tilde{F}_3$) correlate strongly with each other ($r_{13} = .87$) and moderately with perceived seriousness of natural hazards. Figure 8.15 shows the information of Table 8.13 schematically, in the manner of Figure 8.1.[4]

For simplicity and for estimates of the underlying latent variables (perceived seriousness of natural hazards, and so on), we can employ factor scores, instead of the individual opinion variables, in further analysis. After factor-analyzing a larger

Table 8.13 Principal factor analysis of seriousness of Tulsa environmental hazards

(principal factors; 3 factors retained)

Factor	Eigenvalue	Difference	Proportion	Cumulative
1	1.35194	0.88877	0.9929	0.9929
2	0.46317	0.29830	0.3402	1.3331
3	0.16487	0.28236	0.1211	1.4542
4	−0.11749	0.07032	−0.0863	1.3679
5	−0.18781	0.12528	−0.1379	1.2299
6	−0.31309		−0.2299	1.0000

(promax rotation)
Rotated Factor Loadings

Variable	1	2	3	Uniqueness
deepwell	**0.54661**	−0.02946	0.14547	0.59327
chandler	**0.61054**	0.03497	−0.03720	0.64047
tornados	0.06995	**0.54955**	−0.02505	0.68031
floods	−0.06653	**0.54282**	0.03733	0.71092
airpol	0.04270	0.00781	**0.49339**	0.72551
rivpol	0.13743	0.01013	**0.47524**	0.66954

(correlations between factors)
(obs = 199)

	F_1	F_2	F_3
F_1	1.0000		
F_2	0.4957	1.0000	
F_3	0.8732	0.5503	1.0000

set of opinion variables, Blocker and Eckberg examined how factor scores varied with respondent demographic characteristics. They found that mean scores on factors representing local environmental concerns were significantly higher for women than for men: women appeared to take local hazards more seriously. There were no gender differences on factors representing more general, nonlocal environmental concerns.

Figure 8.16 shows distributions of $\tilde{F}_3$, the factor score representing concern about Tulsa's air and water pollution. Both distributions are negatively skewed: most respondents thought pollution was serious, but a few did not.[5] Women more than men tended to assign these problems the highest seriousness. Factor scores $\tilde{F}_1$ and $\tilde{F}_2$ exhibit similar gender differences.

Tested using ANOVA or an equivalent dummy variable regression, the mean $\tilde{F}_3$ score for women is significantly higher than the mean score for men (.128 versus −.166; $P = .004$). The skewed distributions in Figure 8.16 cast doubt on these tests' normality assumption, so it is reassuring to know that robust regression, which does not assume normality, finds a significant difference in robust means ($P = .011$). Once factor scores are defined, we can study them like other variables, using all available analytical techniques.

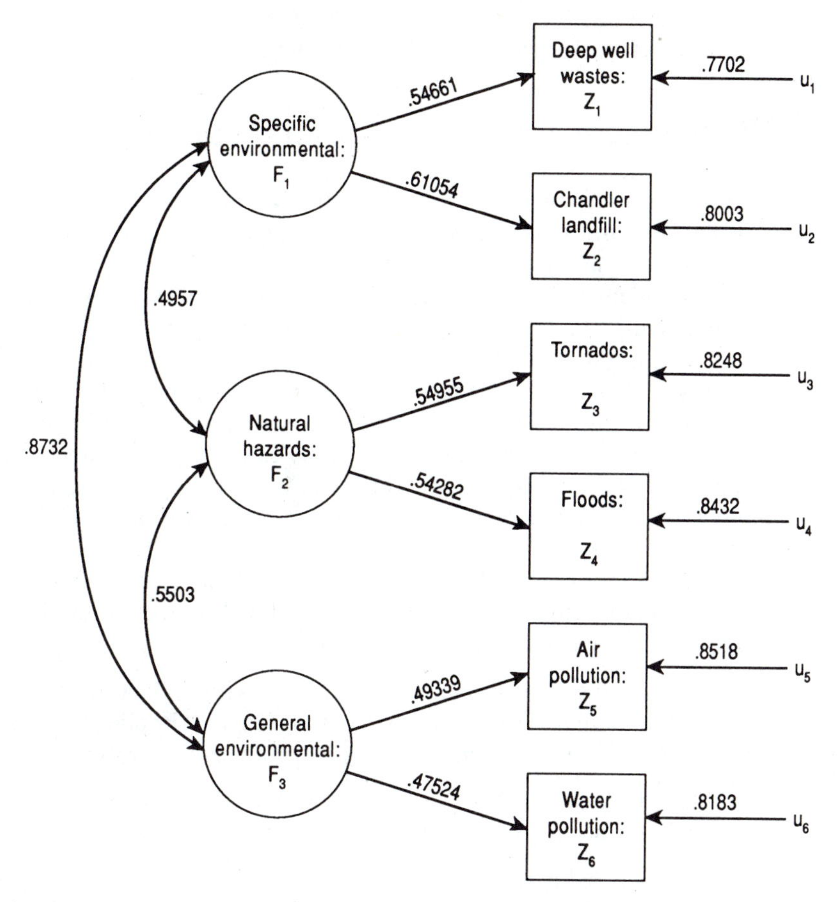

Figure 8.15 Schematic diagram of factor model for perceived seriousness of environmental hazards (Table 8.13); weak factor loadings not shown.

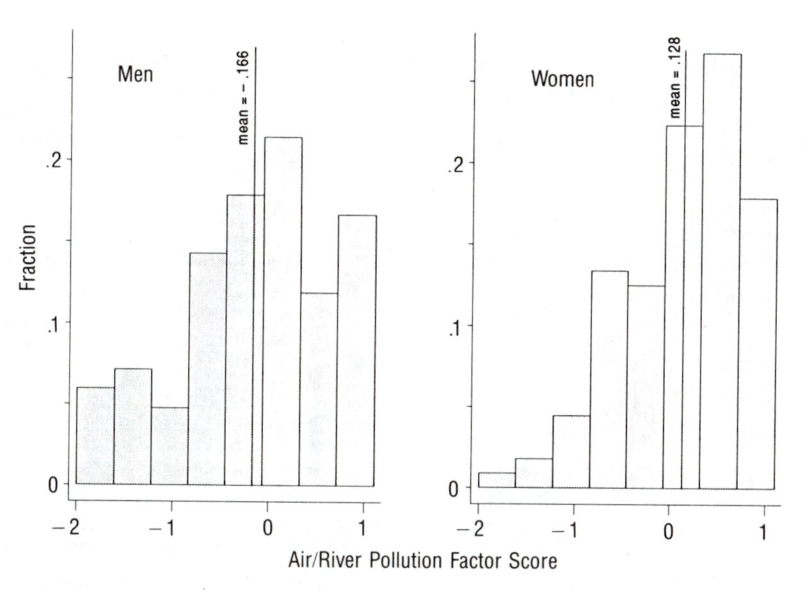

Figure 8.16 Distributions of factor-3 scores for men and women; women tend to express more concern.

Maximum-Likelihood Factor Analysis

Principal components and principal factor analysis lack a well-developed theoretical framework like that of least squares regression. They consequently provide no systematic way to test hypotheses about the number of factors to retain, the size of factor loadings, or the correlations between factors, for example. Such tests are possible using a different approach, based on maximum-likelihood estimation.

There are several alternative methods for maximum-likelihood factor analysis. Many statistical programs offer maximum-likelihood factoring as one option, producing output similar to principal factoring but with a likelihood-ratio χ^2 test for the number of factors. For illustration, Table 8.14 lists nine questions regarding general environmental issues, from the same Tulsa survey described earlier. Are all nine measuring the same concept, "environmentalism," or is there more than one underlying dimension here?

Maximum-likelihood factor analysis (Table 8.15) finds three significant factors. The two reported χ^2 tests both compare the observed correlation matrix with that implied by the estimated factor model:

1. A test of the fitted model versus a model with no factors. Low P-values indicate that the actual correlation matrix resembles that implied by the fitted model more closely than it resembles that implied by a no-factor model (an identity matrix).

Table 8.15 reports that $\chi^2(27) = 223.59$, $P = .0000$ (meaning $P < .00005$): the three-factor model fits significantly better than a no-factor model.

2. A test of the fitted model versus a model with enough factors to perfectly fit the data. Low P-values indicate that the actual correlation matrix does not resemble that implied by the fitted model.

Table 8.14 Opinions regarding general environmental issues from Tulsa area survey (Blocker and Eckberg, 1989); questions not in original order

How Strongly Do You Agree Or Disagree With The Following Statements About Population And Environmental Issues? (1 = strongly disagree, 4 = strongly agree)	
Z_1 taxbabes	To control population, our government should increase taxes on couples who have more than two children.
Z_2 manykids	A married couple should have as many children as they wish, as long as they can adequately provide for them.
Z_3 lessenvt	We should relax our effort to control pollution in order to improve the economy.
Z_4 toocons	There has been too much emphasis on conserving natural resources, and not enough on using them, in recent years.
Z_5 pollburd	Pollution control measures have created unfair burdens on industry.
Z_6 privown	Where natural resources are privately owned, society should have no control over what the owner does with them.
Z_7 shutdown	If an industry cannot control its pollution, the industry should be shut down.
Z_8 punish	Managers of polluting industries should be punished by fines or imprisonment.
Z_9 preserve	Natural resources must be preserved for the future, even if people must do without.

Table 8.15 Maximum-likelihood factor analysis of environmental-issue opinions (see Table 8.14)

(obs = 241)

	(maximum-likelihood factors; 3 factors retained)			
Factor	**Variance**	**Difference**	**Proportion**	**Cumulative**
1	1.09159	−0.42938	0.3398	0.3398
2	1.52096	0.92086	0.4734	0.8132
3	0.60010		0.1868	1.0000

Test: 3 vs. no factors. $\chi^2(27) = 223.59$, Prob > $\chi^2 = 0.0000$
Test: 3 vs. more factors. $\chi^2(12) = 20.20$, Prob > $\chi^2 = 0.0635$

	(promax rotation) Rotated Factor Loadings			
Variable	**1**	**2**	**3**	**Uniqueness**
taxbabes	**0.94368**	0.04826	0.00168	0.10743
manykids	**−0.36004**	0.15241	0.12783	0.85602
lessenvt	0.12017	**0.47898**	−0.11160	0.70257
toocons	0.00568	**0.70413**	0.19670	0.58085
pollburd	0.00180	**0.60911**	−0.12521	0.54973
privown	0.01370	**0.26475**	−0.15846	0.87010
shutdown	−0.06779	0.05460	**0.72057**	0.51620
punish	0.18163	0.01425	**0.51160**	0.69177
preserve	0.07924	−0.11676	**0.20860**	0.91269

Table 8.15 reports that $\chi^2(12) = 20.20$, $P = .0635$. At $\alpha = .05$ the implied and actual correlation matrices do not significantly differ. We need not reject the three-factor model in favor of one with more factors. Specific procedures vary, but most computer programs for maximum-likelihood factor analysis include some version of test 2. The *higher* the *P*-values from such tests, the more confident we can be that our model includes enough factors. For parsimony, we might seek the simplest (fewest-factor) model that obtains $P > .05$.

Apart from these testing capabilities, maximum-likelihood results can be interpreted and used like those of other factoring methods. Factor loadings exceeding $\pm.2$ appear in bold in Table 8.15. They suggest the following interpretations:

Factor 1: "Zero population growth (ZPG) beliefs." Respondents with high factor-1 scores support a tax on large families and disagree that couples should have as many children as they wish.

Factor 2: "Antienvironmentalism." High scores indicate agreement that there should be relaxed pollution controls, that there should be more resource use, that pollution controls are an unfair burden, and that private owners should do what they want.

Factor 3: "Punish polluters." High scores indicate agreement that polluting industries should be shut down, that polluting managers should be fined or jailed, and that we must sacrifice to preserve natural resources.

High uniquenesses reveal that the three factors explain only a fraction of each variable's variance, however.

Factor scores from this analysis provide variables for further study. One unexpected discovery is a weak but significant positive relationship between factor 1 and number of children ($r = .16, p = .02$). Respondents expressing stronger agreement with zero-population-growth beliefs tended themselves to have larger families, as shown in Figure 8.17 (jittered). At first glance we might suspect that this effect is caused by a few large-family outliers, but further investigation suggests otherwise. No individual cases are especially influential; only three have DFBETAS exceeding $\pm.2$, and none are beyond $\pm.3$ (shown in Figure 8.17). Power transformations cannot symmetrize the distribution of number of children, but the relation remains about as strong even if we restrict analysis to families of three children or fewer. It would be interesting to explore whether similar relationships occur in other survey samples.

Maximum-likelihood factor analysis assumes multivariate normality, a K-dimensional generalization of the normal distribution. Since even simple normality is implausible for many kinds of data, the stronger assumption of multivariate normality must be regarded skeptically. Graphical or other normality checks, together with transformations as needed, may make this assumption more defensible.

"Agreement" scales like those in Table 8.14 cannot be normally distributed, since they have only a few values. Therefore Table 8.15's analyses, and particularly the hypothesis tests, are questionable. Maximum-likelihood methods were applied

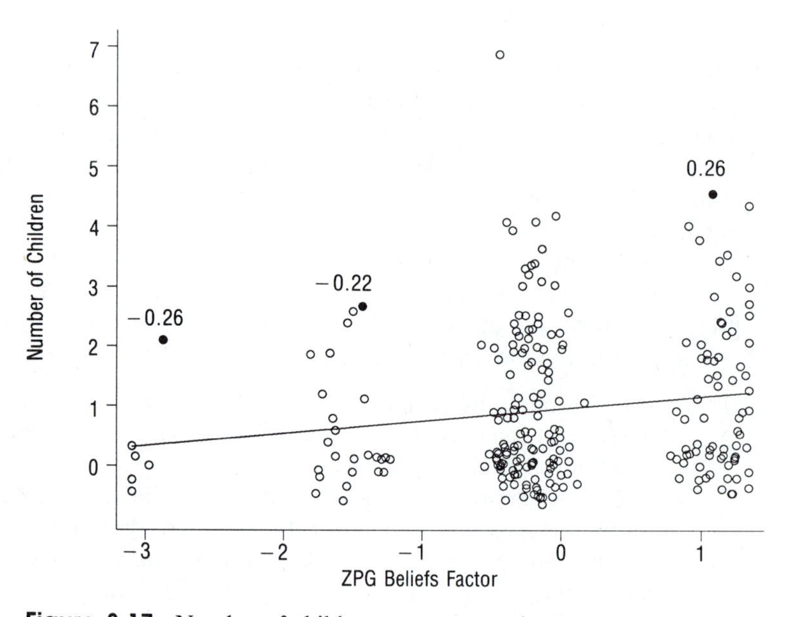

Figure 8.17 Number of children versus scores on "zero population growth" beliefs factor, showing OLS line and three largest DFBETAS.

here for illustration; a more robust method like principal factoring would actually have been more appropriate. Consequences of violating the normality assumption with maximum-likelihood factor analysis (and related methods) are a topic of ongoing research. Caution suggests that we should distrust methods built upon unbelievable assumptions, but in some circumstances the consequences of violating this assumption may not be severe (Bollen, 1989).

Two further problems complicate use of maximum-likelihood factor analysis. First, we often get *improper solutions*: substantively unlikely or impossible results such as variables with 0 uniqueness, negative variances, or correlations exceeding ± 1. Improper solutions might result from too few cases, too few variables per factor, outliers, or nonnormality—the cause can be hard to diagnose. Second, there is a problem with relying on significance tests, which depend upon sample size, to determine the number of factors retained. The number of "significant" factors tends to increase with sample size. In large samples we may end up with more factors than we can interpret or use.

Certain maximum-likelihood methods, notably *LISREL* (for *Li*near *S*tructural *Rel*ations), combine the work of factor analysis and regression into a single step. This simplifies estimation and permits testing of a wide variety of hypotheses. The power and advantages of LISREL-type methods have made them very popular, particularly in social science. See Bollen (1989) for a lucid introduction.

Conclusion

Principal components and factor analysis offer a confusing variety of choices. Table 8.16 outlines some of the main options.

Principal components analysis obtains K underlying dimensions that explain all the variance of K variables. Since variable Z_k can be reconstructed exactly from K principal components $F_1 \cdots F_K$, the principal components "model" requires no error term. Fewer than K components often explain most of the variance of K variables. We may retain only the first few components, treating the remainder as a residual or error term. This approach looks much like factor analysis. For extreme simplification, we retain only the unrotated first principal component.

Factor analysis seeks to explain variables' correlations (or covariance) rather than their variance. The factor analysis model for variable Z_k includes an explicit residual or error term u_k, sometimes called the unique factor. Factor analysis is more appropriate than principal components when our interest lies with a theoretical causal or measurement model.

Principal components and principal factoring produce similar results when applied to strongly correlated data. With strong correlations, communalities approach 1, so principal factoring (basically principal components with communalities replacing ones on the major diagonal of **R**) resembles principal components. Successful data reduction is also more likely with strong correlations: the first few components or factors account for much of the data's structure. On the other hand,

Table 8.16 Summary of factor analysis options

EXTRACTING INITIAL COMPONENTS OR FACTORS

Principal components. A transformation of the data, not model based. Appropriate if goal is to compactly express most of the variance of K variables. Minor components (perhaps all but first component) may be discarded and viewed as residual.

Factor analysis. Estimates parameters of a measurement model with latent (unobserved) variables. Types:

Principal factoring—principal components of a modified correlation matrix $\mathbf{R}^*$, in which communality estimates replace ones on the main diagonal.

principal factoring without iteration—communalities estimated from R_k^2; computationally simple.

principal factoring with iteration—communalities initially estimated from R_k^2, then iteratively from factor loadings. Provides better communality estimates but takes more computing. The final analysis should include only those factors to be retained.

Maximum-likelihood factoring—significance tests regarding number of factors and other hypotheses, assuming multivariate normality.

ROTATION

If we retain more than one factor, rotation simplifies structure and improves interpretability.

Orthogonal rotation (e.g., *varimax*)—maximum polarization given uncorrelated factors.

Oblique rotation (e.g., *promax*)—further polarization by permitting interfactor correlations. The results may be more interpretable and more realistic than uncorrelated factors.

SCORES

Factor scores can be calculated for use in graphs and further analysis, based on rotated or unrotated factors or principal components.

when all correlations approach 0, principal components or factor analysis becomes pointless: we need as many components or factors as variables.

When a correlation matrix divides neatly into subgroups of variables, each containing variables correlated with each other but not with variables of other subgroups, most procedures yield approximately the same factors. If different procedures obtain quite different results, however, this warns that there is no clear structure and solutions are arbitrary.

Correlation and covariance are linear regression statistics. Nonlinearity and influential cases cause the same problems for correlations, and hence for principal components/factor analysis, as they do for regression. Scatterplots should be examined routinely to check for nonlinearity and outliers. Diagnostic checks become even more important with maximum-likelihood factor analysis, which makes stronger assumptions and may be less robust than principal components or principal factors.

If their assumptions are justified, maximum-likelihood factor analysis and related methods support a sophisticated range of hypothesis tests. These tests provide systematic guidance for constructing elaborate multivariate models of measurement and causality.

Exercises

Exercise 7 of Chapter 2 examined the relationship between leaded gasoline sales and lead concentrations in umbilical cord blood from births at a major Boston hospital. The following table presents further data from this research (Rabinowitz, Needleman, Burley, Finch, and Rees, 1984; Rabinowitz, Leviton, Needleman, Bellinger, and Waternaux, 1985). Researchers measured levels of lead in the air and dust from a sample of 30 Boston households over the study period. The table lists monthly average lead levels in micrograms. (μg).

Boston household indoor lead levels (from Rabinowitz et al., 1984)

			Lead in Dust on:		
Month	Year	Lead in Air (μg/m^3)	Furniture (μg/area)	Floor (μg/area)	Windowsill (μg/area)
3	1980	.12	4.5	5.8	38
4	1980	.15	6.5	6.5	24
5	1980	.18	9.3	10.7	30
6	1980	.09	3.6	5.4	70
7	1980	.16	14.5	35.6	105
8	1980	.22	18.1	31.7	62
9	1980	.16	16.0	25.0	87
10	1980	.15	13.8	17.3	72
11	1980	.07	5.2	65.7	172
12	1980	.05	7.5	5.3	106
1	1981	.08	7.6	11.8	30
2	1981	.01	8.5	9.2	28
3	1981	.02	1.4	2.3	14
4	1981	.09	1.6	2.8	20

1. It seems possible that the four lead-level variables reflect one underlying dimension: airborne lead contamination of the household. Use principal components to explore this possibility:
   a. To linearize relationships and reduce outliers, take natural logarithms of floor and windowsill lead levels.
   b. Extract principal components of air, furniture, log floor, and log dust lead, and interpret the four resulting eigenvalues. Applying the eigenvalue-1 rule, how many components should we retain?
   c. Calculate scores for the first component. In what months was overall airborne lead contamination highest? Lowest?
2. Regress umbilical lead level (from Exercise 7 of Chapter 2) on your first principal component scores from Exercise 1c. Interpret the results.
3. Carry out and interpret two diagnostic checks on the regression of Exercise 2:
   a. Durbin-Watson test for autocorrelation. (Why should we worry about autocorrelation here?)
   b. Residual versus predicted Y-plot.

Following are data on 49 South Carolina streams. The zoologists who collected these data (Meffe and Sheldon, 1988) summarized 15 stream-characteristic variables (seven of them listed below) with the first four principal components. They then explored relations between component scores, viewed as measures of habitat structure, and populations of various fish species. Exercises 4–6 refer to these data.

Characteristics of 49 South Carolina streams[6]

Stream	Mean Depth	Mean Width	Mean Velocity	% silt Bottom	% leaves Bottom	% sand Bottom	% mud Bottom
1	4.04	1.41	1.46	35.73	0.00	27.06	40.80
2	3.41	1.55	1.89	45.00	13.94	38.35	7.92
3	3.67	1.82	1.61	45.29	29.67	28.93	5.13
4	3.08	1.69	2.17	37.29	0.00	37.94	28.25
5	2.65	1.44	2.95	21.64	18.63	57.29	0.00
6	1.57	−0.22	2.27	32.77	24.58	46.89	0.00
7	1.95	0.26	1.41	44.26	27.35	31.69	0.00
8	1.96	0.53	1.36	46.66	25.91	26.99	0.00
9	1.84	−0.11	1.90	44.31	15.45	38.12	0.00
10	1.65	−0.11	2.22	27.56	11.97	53.31	17.05
11	3.33	1.31	3.13	7.71	16.64	61.41	16.95
12	3.43	1.59	2.79	20.00	13.44	54.03	22.63
13	3.15	1.55	3.09	21.13	8.13	60.00	18.43
14	3.20	1.90	2.74	26.64	21.05	48.33	16.32
15	1.70	1.03	3.77	0.00	0.00	24.12	0.00
16	2.15	0.69	3.60	19.19	0.00	46.89	0.00
17	2.07	0.83	3.34	20.00	7.92	45.34	0.00
18	3.27	0.59	2.78	0.00	15.45	67.78	15.45
19	2.80	0.83	1.39	33.27	24.20	32.65	9.97
20	2.88	1.06	1.90	33.21	34.57	27.76	0.00
21	1.41	0.69	2.98	21.56	30.66	14.54	0.00
22	2.24	0.79	1.53	27.62	27.97	30.85	10.30
23	2.86	1.10	1.28	44.37	18.53	35.06	0.00
24	3.02	1.22	3.19	21.47	8.53	64.30	10.30
25	2.83	0.83	2.24	27.56	0.00	62.44	0.00
26	3.16	1.93	3.14	19.37	15.45	62.73	9.10
27	1.67	0.53	2.36	41.44	14.42	45.00	0.00
28	2.52	0.79	1.48	36.33	29.73	39.47	0.00
29	3.03	0.79	1.53	43.74	19.46	39.87	0.00
30	2.07	0.41	2.22	28.73	23.34	48.73	0.00
31	2.27	1.31	3.07	25.62	14.89	59.80	0.00
32	3.05	1.03	2.65	30.00	14.18	50.77	0.00
33	2.64	1.13	2.93	19.64	8.91	49.20	0.00
34	2.52	1.19	3.27	20.70	0.00	55.06	10.14
35	3.37	1.16	2.87	24.43	7.92	51.94	9.63
36	3.03	1.31	2.91	19.55	13.94	59.02	9.46
37	3.50	1.22	2.51	18.72	20.96	51.65	22.06
38	2.78	1.48	3.14	15.23	14.06	46.61	0.00
39	3.56	1.57	1.63	26.21	19.28	26.21	45.00
40	2.58	0.10	2.33	28.11	21.56	46.83	0.00
41	1.96	0.00	2.84	20.09	6.80	68.70	0.00
42	2.23	0.34	2.30	25.33	17.15	58.69	0.00

Stream	Mean Depth	Mean Width	Mean Velocity	% silt Bottom	% leaves Bottom	% sand Bottom	% mud Bottom
43	2.26	0.26	2.40	20.70	0.00	69.30	0.00
44	3.28	1.70	2.69	18.43	24.58	30.07	11.54
45	3.77	1.61	2.65	26.92	10.30	39.82	26.71
46	3.10	1.82	3.00	21.30	16.22	56.23	9.80
47	2.80	1.31	3.25	21.13	11.09	60.60	0.00
48	2.90	1.31	2.91	17.66	15.89	63.36	0.00
49	2.72	1.50	3.08	17.15	9.46	65.42	11.54

4. Extract principal components, and construct a scree graph. How many components should we retain?
5. Referring to your principal components output from Exercise 4, confirm the following:
   a. The sum of eigenvalues equals the number of variables (Equation [8.7]).
   b. The proportion of variance explained by the jth component equals its eigenvalue divided by the number of variables (Equation [8.8]).
   c. Factor loadings equal eigenvector elements times the square root of the corresponding eigenvalue (Equation [8.17]).
   d. Squared loadings for each variable sum to its communality (Equation [8.18]).
6. Perform varimax rotation of the retained components from Exercise 4. How would you characterize a stream with a high score on rotated component 1? A low score on rotated component 1? A high score or a low score on rotated component 2?

The discovery of oilfields under the North Sea brought rapid economic development to Scotland's Shetland and Orkney Islands. In other areas, such as the western United States, energy-project booms often disrupted local ways of life. The following data come from a survey assessing how oil development had affected Shetland and Orkney youth (Seyfrit, 1989). This particular set of items concerns feelings about life on the islands:

famconf—One should confide more fully in members of his/her family.
differ—You are out of luck here if you happen to be different.
critical—People are generally critical of others in this community.
home—Home is the most pleasant place in the world.
badname—People here give you a bad name if you insist on being different.
famsac—A person should be willing to sacrifice everything for his/her family.

Answers are coded from 1 (strongly disagree) to 5 (strongly agree).[7] Although 791 students responded to the initial survey, only 42 are listed here. Exercises 7–9 refer to these data.

7. Factor-analyze the Shetland and Orkney Youth Survey data:
   a. Use principal factoring *without iteration*. Construct a scree graph. How many factors appear worth retaining?

b. Use principal factoring *with iteration*, specifying two factors. Why is it important now to specify the number of factors?

c. Apply varimax (orthogonal) rotation to the two factors obtained from 7b. Circle the highest loading in each row (variable) of the factor-loading matrix. Interpret and suggest names for the factors.

d. Obliquely rotate the two factors from 7b. Again, circle the highest loading in each row. How do the results compare with your orthogonal-rotation results of 7c?

e. Calculate factor scores for the two orthogonal factors of 7c and the two oblique factors of 7d. Obtain a correlation matrix for these four variables, and discuss what it shows.

8. Draw a schematic diagram (like Figure 8.15) of the oblique factors from Exercise 7.

9. Apply maximum-likelihood factor analysis to the Shetland and Orkney Youth Survey data:

   a. How many factors should we retain? Obtain χ^2 tests of one versus more factors; of two versus more factors. Do these results agree with your informal scree-graph conclusion from Exercise 7a?

   b. Extract two factors, orthogonally rotate, and interpret the loadings. Calculate factor scores.

   c. Apply oblique rotation to the same two factors, and interpret the loadings. Calculate factor scores.

   d. Obtain an eight-variable matrix of correlations between the four factor scores of 9b–9c and the four scores of 7e. Did the principal factor and maximum-likelihood methods lead to similar or quite different factor scores?

Data from 1988 Shetland and Orkney Islands Youth Survey (Seyfrit, 1989).

Student	Famconf	Differ	Critical	Home	Badname	Famsac
1	2	5	4	2	4	1
2	5	5	4	3	5	2
3	5	3	4	4	4	4
4	4	5	3	1	5	3
5	2	5	4	2	3	2
6	2	3	4	2	4	3
7	4	3	5	4	2	4
8	4	5	2	3	2	3
9	4	2	2	3	2	3
10	4	2	3	2	4	3
11	5	4	4	2	4	3
12	3	3	3	3	2	3
13	5	3	3	4	2	5
14	3	4	5	5	4	4
15	4	5	4	4	5	3
16	3	2	3	5	4	3
17	4	4	5	4	4	3
18	4	3	3	2	2	4
19	3	3	2	3	3	2

Student	Famconf	Differ	Critical	Home	Badname	Famsac
20	4	3	5	4	4	4
21	4	1	2	4	1	4
22	3	2	2	3	2	3
23	3	3	4	3	3	4
24	4	2	3	5	3	4
25	3	3	3	4	4	2
26	4	2	3	2	3	3
27	4	2	3	5	1	2
28	5	1	2	5	1	5
29	4	3	3	3	3	3
30	3	3	3	3	3	1
31	4	3	3	3	3	3
32	4	2	3	3	2	3
33	4	3	3	3	3	3
34	1	3	4	2	2	3
35	4	2	2	3	3	2
36	4	2	2	3	2	3
37	3	5	4	2	5	3
38	3	3	4	2	4	1
39	4	3	4	1	5	3
40	3	2	3	4	2	3
41	5	3	4	4	3	3
42	4	3	3	2	3	3

Notes

1. As noted in Chapter 3, any variable X with mean $\bar{X}$ and standard deviation s_X can be transformed to a **standard score** Z, with mean $\bar{Z} = 0$ and standard deviation $s_Z = 1$, by the transformation

   $$Z = (X - \bar{X})/s_X$$

   For example, the mean log runoff is $\bar{X}_1 = 3.406$, with standard deviation $s_{X_1} = .496$. Log runoff in the Hokitika basin equals 3.948; in standardized form this becomes $Z_1 = (3.948 - 3.406)/.496 = 1.094$, indicating that Hokitika is 1.094 standard deviations above the mean. Arrow basin's log runoff equals 2.613, which is 1.599 standard deviations *below* the mean: $(2.613 - 3.406)/.496 = -1.599$. This linear transformation does not change correlations or graphical appearance, except by rescaling the axes.
2. The cosine of the angle between axes equals their correlation. Orthogonal axes, which meet at 90°, have correlation $r = \cos(90°) = 0$.
3. Alternatively, we could extract principal components of a modified covariance matrix **C***, which has the variables' shared variance (rather than total variance) along its major diagonal.

4. Coefficients on paths from factors (F_j) to variables (Z_k) are rotated factor loadings from Table 8.13. Correlations between factor scores appear along double-headed curved arrows. Coefficients on paths from each variable's error term (u_k) equal square roots of uniqueness.
5. Such distributions are common with environmental-concern measures. Environmentalism has become a "motherhood" issue, in the sense that few people will admit to being antienvironment, or in favor of pollution. Consequently, survey researchers often see a piling up of responses toward the high (proenvironment) end of the scale, as in Figure 8.16. Furthermore, because almost everybody claims similar views, such environmental-concern measures correlate weakly with demographic or political variables. However, environmentally conscious *behavior* (consumption habits, voting, activism, and so on) is less universal than environmentally conscious *attitudes*.
6. In keeping with Meffe and Sheldon's analysis, all variables in this table have been transformed. Velocity, depth, and width are $\log_e(X)$; other variables are arcsine $\sqrt{X}$. The arcsine $\sin^{-1}\sqrt{X}$ transformation stabilizes variances of binomially distributed variables.
7. The original survey used the opposite coding scheme (1 = strongly agree, 5 = strongly disagree), which is awkward to interpret.

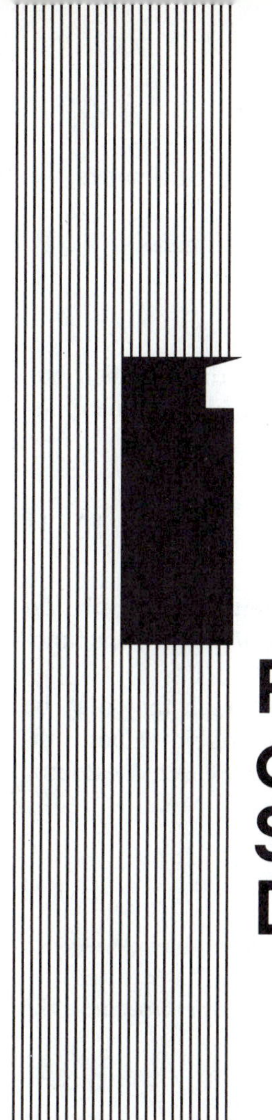

Appendix 1

Population and Sampling Distributions

Statisticians view sample data as representative of some real or imaginary population. A randomly selected sample may represent a real population, but even if the sample is nonrandom or the population is fictitious, the concept of representative samples remains analytically useful.

This appendix introduces expectation algebra, which offers tools for describing theoretical distributions, like the population distribution of a variable or the distribution of a statistic across all possible random samples (*sampling distribution*). Expectation algebra defines basic statistical properties of estimators, such as ordinary least squares (OLS). We conclude with a look at some important theoretical sampling distributions.

Expected Values

A sample mean equals the sum of values for each case, divided by the number of cases. This definition does not work well for populations, which may have an infinite number of cases. To define population means and related parameters, we turn to the algebra of expectations.

The *expected value* of variable Y, written $E[Y]$, is its population mean:

$$E[Y] = \mu_Y$$

A *discrete* variable can take on only certain distinct values. For discrete distributions, the expected value may be defined as the sum of each possible Y value times its probability:

$$E[Y] = \sum_{i=1}^{I} Y_i f(Y_i) = \mu_Y \qquad [A1.1]$$

where I is the number of discrete Y values. The probability (or population proportion) of Y value Y_i is denoted by $f(Y_i)$. $f(Y)$ is the *probability density* function of Y.

Continuous variables are not limited to discrete values (although they may be restricted to a certain range). Between any two continuous-variable values, infinitely many other values exist. We need calculus to define the expected value of a continuous variable:

$$E[Y] = \int_{-\infty}^{\infty} Yf(Y)\,dY = \mu_Y \qquad [A1.2]$$

Again, $f(Y)$ refers to the probability density function of Y.

We can define population means, but we often cannot calculate them because the probability density function is unknown. The sample mean

$$\bar{Y} = \frac{\Sigma Y_i}{n} \qquad [A1.3]$$

is an *unbiased estimator* of the population mean, $E[Y]$. (A later section defines "unbiased estimator.")

Expectation has the following properties. For any constants a and b, and any variables X and W:

$$E[a] = a \qquad [A1.4]$$

$$E[bX] = bE[X] \qquad [A1.5]$$

$$E[X + W] = E[X] + E[W] \qquad [A1.6]$$

Sample means have similar properties. The mean of a constant is the constant ([A1.4]); the mean of a constant times a variable equals the constant times the variable's mean ([A1.5]); and the mean of the sum of two variables equals the sum of the two variables' means ([A1.6]).

The basic properties of expectation lead to further deductions. For example, the expectation of any linear function of X equals the same linear function of $E[X]$:

$$E[a + bX] = E[a] + E[bX]$$
$$= a + bE[X]$$

The expectation of a linear function of X and W is

$$E[bX + cW] = E[bX] + E[cW]$$
$$= bE[X] + cE[W]$$

The expectation operator, $E[\ ]$, can be applied in this manner to any linear equation.

Covariance

Covariance, a basic measure of the relationship between X and Y, equals the expected value of the cross product [X minus the mean of X, times Y minus the mean of Y]:

$$\text{Cov}[X, Y] = E[(X - E[X])(Y - E[Y])]$$
$$= E[XY] - E[X]E[Y] \quad \text{[A1.7]}$$

An unbiased estimator of the population covariance is the sample covariance, s_{XY}:

$$s_{XY} = \frac{\Sigma\{(X_i - \bar{X})(Y_i - \bar{Y})\}}{n - 1} \quad \text{[A1.8]}$$

Algebraic properties of covariance follow from those of expectation. For any variables X, Y, and W, and any constants a and b:

$$\text{Cov}[a, Y] = 0 \quad \text{[A1.9]}$$
$$\text{Cov}[bX, Y] = b\,\text{Cov}[X, Y] \quad \text{[A1.10]}$$
$$\text{Cov}[X + W, Y] = \text{Cov}[X, Y] + \text{Cov}[W, Y] \quad \text{[A1.11]}$$

Sample covariances have similar properties.

The covariance of a variable with itself equals the *variance* (discussed in the next section):

$$\text{Cov}[X, X] = \text{Var}[X] \quad \text{[A1.12]}$$

Covariance is unaffected by the addition of a constant to either variable:

$$\begin{aligned}\text{Cov}[a + X, Y] &= \text{Cov}[a, Y] + \text{Cov}[X, Y]\\ &= \text{Cov}[X, Y]\end{aligned}$$

Covariances between sums of variables reduce to sums of covariances between their components:

$$\text{Cov}[X + W, Y] = \text{Cov}[X, Y] + \text{Cov}[W, Y]$$

or

$$\begin{aligned}\text{Cov}[X, Y - X] &= \text{Cov}[X, Y] - \text{Cov}[X, X]\\ &= \text{Cov}[X, Y] - \text{Var}[X]\end{aligned}$$

Variance

Expectation locates a distribution's center. Variance measures variation around that center. The variance of X is the expected value of squared deviations from the mean:

$$\begin{aligned}\text{Var}[X] &= \text{Cov}[X, X]\\ &= E[X - E[X]]^2 \qquad \text{[A1.13]}\end{aligned}$$

The sample variance s_X^2 provides an unbiased estimator of Var$[X]$:

$$s_X^2 = \frac{\Sigma(X_i - \bar{X})^2}{n - 1} \qquad \text{[A1.14]}$$

Basic algebraic properties include the following:

$$\text{Var}[a] = 0 \qquad \text{[A1.15]}$$

$$\text{Var}[bX] = b^2\,\text{Var}[X] \qquad \text{[A1.16]}$$

$$\text{Var}[X + W] = \text{Var}[X] + \text{Var}[W] + 2\,\text{Cov}[X, W] \qquad \text{[A1.17]}$$

These properties follow from the definitions of expectation and covariance. We can combine them for further applications, such as the variance of a linear function of X:

$$\begin{aligned}\text{Var}[a + bX] &= \text{Var}[a] + \text{Var}[bX]\\ &= b^2\,\text{Var}[X]\end{aligned}$$

or the variance of a linear function of X and W:

$$\mathrm{Var}[bX + cW] = b^2\,\mathrm{Var}[X] + c^2\,\mathrm{Var}[W] + 2bc\,\mathrm{Cov}[X, W]$$

Sample variances obey similar rules.

Further Definitions

A *covariance matrix* (also called a *variance-covariance matrix*) displays the variances and covariances of many variables at once. With three variables, a covariance matrix has the form

$$\begin{matrix} \mathrm{Var}[Y] & \mathrm{Cov}[Y, X] & \mathrm{Cov}[Y, W] \\ \mathrm{Cov}[X, Y] & \mathrm{Var}[X] & \mathrm{Cov}[X, W] \\ \mathrm{Cov}[W, Y] & \mathrm{Cov}[W, X] & \mathrm{Var}[W] \end{matrix}$$

Covariance is symmetrical: $\mathrm{Cov}[X, Y] = \mathrm{Cov}[Y, X]$. Consequently, the upper triangular part of a covariance matrix repeats the information of the lower triangle. Variances make up the *major diagonal* (upper left to lower right).

Many univariate, bivariate, and multivariate analyses require only the information in the covariance matrix. The raw data are not needed to find standard deviations, correlations, or multiple regression coefficients, for instance. Expectation, covariance, and variance are statistical building blocks from which other statistics can be defined. For example:

1. The *standard deviation* (σ) equals the square root of the variance:

$$\sigma_X = \sqrt{\mathrm{Var}[X]} \qquad [A1.18]$$

2. *Correlation* (ρ_{XY}) is a standardized covariance:[1]

$$\rho_{XY} = \frac{\mathrm{Cov}[X, Y]}{\sqrt{\mathrm{Var}[X]\,\mathrm{Var}[Y]}} \qquad [A1.19]$$

Correlation equals covariance if both variables are standard scores, with variances of 1 ([3.16]). A correlation matrix is laid out like a covariance matrix:

$$\begin{matrix} 1 & \rho_{YX} & \rho_{YW} \\ \rho_{XY} & 1 & \rho_{XW} \\ \rho_{WY} & \rho_{WX} & 1 \end{matrix}$$

Like covariance, correlation is symmetrical, so cells above the diagonal are redundant. A correlation matrix contains less information than a covariance matrix. It gives no indication of the variables' original scales.

3. In a two-variable regression the OLS *slope* or *regression coefficient* is

$$\beta_1 = \frac{\text{Cov}[X, Y]}{\text{Var}[X]} \qquad \text{[A1.20]}$$

The *Y-intercept* is

$$\beta_0 = E[Y] - \beta_1 E[X] \qquad \text{[A1.21]}$$

Equations [A1.20]–[A1.21] are for the *regression of* Y *on* X. A different slope and intercept result if we regress X on Y; unlike covariance and correlation, regression is not symmetrical.[2]

4. Multiple regression slopes (*partial regression coefficients*) can likewise be defined from variances and covariances. In the regression of Y on X and W, the OLS coefficient on X is

$$\beta_{YX.W} = \frac{\text{Var}[W]\,\text{Cov}[X, Y] - \text{Cov}[X, W]\,\text{Cov}[W, Y]}{\text{Var}[X]\,\text{Var}[W] - (\text{Cov}[X, W])^2} \qquad \text{[A1.22]}$$

Here $\beta_{YX.W}$ denotes "the regression of Y on X, controlling for W." (Simpler subscripting is employed elsewhere in this book.) The coefficient on W in the same regression will be

$$\beta_{YW.X} = \frac{\text{Var}[X]\,\text{Cov}[W, Y] - \text{Cov}[X, W]\,\text{Cov}[X, Y]}{\text{Var}[X]\,\text{Var}[W] - (\text{Cov}[X, W])^2} \qquad \text{[A1.23]}$$

We obtain unbiased estimates of these population parameters by substituting sample covariance and variance ([A1.8] and [A1.14]) into [A1.18]–[A1.23].

Continuing in this manner, using variance and covariance to define higher-order regression leads to unwieldy equations. Matrix algebra (Appendix 3) provides a better way to work with the large amounts of information needed for multivariate calculations.

Properties of Sampling Distributions

The *sampling distribution* of a statistic is its theoretical distribution over all possible random samples of a given size. Researchers typically have just one sample, but theories about sampling distributions guide what inferences they draw from their sample.

Sample statistics provide estimates of population parameters. A statistic b is an *unbiased* estimator of parameter β if

$$E[b] = \beta \qquad \text{[A1.24]}$$

This means that, over all possible random samples of size n, the average value of b equals the parameter we are trying to estimate. *Bias* equals the expected difference between statistic and parameter:

$$\text{bias} = E[b - \beta] \qquad \text{[A1.25]}$$

A biased estimator ($E[b] \neq \beta$, or $E[b - \beta] \neq 0$) tends on average to be too high or too low, relative to the population parameters. Other things being equal, we prefer unbiased estimators.

The *variance* of an estimator equals average squared deviation around its mean:

$$\text{Var}[b] = E[b - E[b]]^2 \qquad \text{[A1.26]}$$

The theoretical *standard error* equals the square root of the variance:

$$\sigma_b = \sqrt{\text{Var}[b]} \qquad \text{[A1.27]}$$

Other things being equal, we prefer an estimator with a low variance or standard error.

The *mean squared error* is the expected value of squared deviations from the parameter:

$$\begin{aligned} \text{MSE} &= E[b - \beta]^2 \\ &= \text{Var}[b] + \text{bias}^2 \end{aligned} \qquad \text{[A1.28]}$$

Note that, if b is an unbiased estimator,

$$\text{MSE} = \text{Var}[b]$$

Among unbiased estimators, the one with the lowest MSE or variance is the most efficient.

These definitions of bias, variance, and mean squared error apply to sampling distributions based on samples of any fixed, finite size. They are called *small-sample* properties. Sometimes the small-sample properties of an estimator are unknown, but we do know its *large-sample* or *asymptotic* properties: theoretical behavior as sample size approaches infinity. For example, b is an *asymptotically unbiased* estimator of β if

$$\lim_{n \to \infty} E[b] = \beta \qquad \text{[A1.29]}$$

This means that, as n (sample size) approaches infinity, the expected value of b approaches β. β is the mean of the *limiting distribution* of b.

For many estimators the sampling distribution would collapse to a single point, with zero variance, if sample size *reached* infinity. *Asymptotic variance* is the variance as sample size *approaches* infinity. Among asymptotically unbiased estimators of any parameter, that with the lowest asymptotic variance is called *asymptotically efficient*. *Asymptotic normality* means that, as $n \to \infty$, the distribution of b approaches normality.

We call b a *consistent* estimator of β if the probability that b is very close to β approaches 1 as n approaches infinity. Formally, for any small constant c:

$$\lim_{n \to \infty} P[|b - \beta| < c] = 1 \qquad \text{[A1.30]}$$

Consistency implies that, as sample size increases, we can be increasingly confident that b is close to β.

Ordinary Least Squares

We begin with a linear model for expected Y_i:

$$E[Y_i] = \beta_0 + \beta_1 X_{i1} + \beta_2 X_{i2} + \cdots + \beta_{K-1} X_{i,K-1} \qquad \text{[A1.31]}$$

and view X values as (hypothetically) fixed in repeated sampling. ε_i represents errors, which cause individual Y_i to differ from $E[Y_i]$:

$$Y_i = E[Y_i] + \varepsilon_i$$

If errors have zero mean:

$$E[\varepsilon_i] = 0 \qquad \text{for all } i \qquad \text{[A1.32]}$$

then ordinary least squares (OLS) estimates of the β parameters in [A1.31] should be unbiased:

$$E[b_k] = \beta_k \qquad \text{for } k = 0, 1, 2, \ldots, K - 1$$

To assess the likelihood of being close to the parameter, we want to know the estimator's variance. OLS variance and standard error estimates rest on two further assumptions:

Homoscedasticity: Errors have the same variance for every case:

$$\text{Var}[\varepsilon_i] = \text{Var}[\varepsilon] \qquad \text{for all } i \qquad \text{[A1.33]}$$

No autocorrelation: Errors for case i are independent of errors for case j:

$$\text{Cov}[\varepsilon_i, \varepsilon_j] = 0 \qquad \text{for } i \neq j \qquad \text{[A1.34]}$$

If [A1.33] is false (*heteroscedasiticity*) or if [A1.34] is false (*autocorrelation*), estimated standard errors may be biased, invalidating the usual hypothesis tests and confidence intervals.

Assuming fixed X and [A1.32]–[A1.34], the *Gauss-Markov Theorem* demonstrates that OLS is the most efficient linear unbiased estimator. Note the restriction to *linear unbiased estimators*; nonlinear or biased estimators sometimes perform better, even assuming [A1.31]–[A1.34].

Some Theoretical Distributions

Confidence intervals and hypothesis tests build upon theoretical sampling distributions. The normal (Gaussian) and three related distributions (χ^2, t, and F) are widely used.

The normal or Gaussian distribution is a theoretical probability distribution for variables that represent sums of many independent, identically distributed random variables. Certain sample statistics, such as means or regression coefficients, can be viewed in this way and hence should have approximately normal sampling distributions—given large enough samples.

Normal distributions are defined by the probability density function $f(Y)$:

$$f(Y) = \frac{1}{\sigma\sqrt{2\pi}} e^{-(Y-\mu)^2/2\sigma^2} \quad \text{[A1.35]}$$

Graphing $f(Y)$ against Y produces the familiar bell shape of the normal curve. There exists a unique normal distribution for every possible combination of mean (μ) and standard deviation (σ). The mean is the distribution's center, and standard deviation measures spread around this center.

A *standard normal distribution* has a mean of 0 and a standard deviation of 1. If Y is distributed normally with mean μ and standard deviation σ, then

$$Z = \frac{Y - \mu}{\sigma} \quad \text{[A1.36]}$$

follows a standard normal distribution (Figure A1.1). All normal distributions range from negative to positive infinity; only the range $\mu \pm 5\sigma$ is shown in Figure A1.1.

Sums of independent squared standard normal variables follow a *chi-square* (χ^2) distribution. If each Z_d is an independent standard normal variable ($d = 1, 2, \ldots \text{df}$), then

$$\chi^2_{\text{df}} = \sum_{d=1}^{\text{df}} Z_d^2 \quad \text{[A1.37]}$$

follows a χ^2 distribution with df degrees of freedom. There is a different χ^2 distribution for every possible value of df. Unlike normal variables, χ^2 variables can

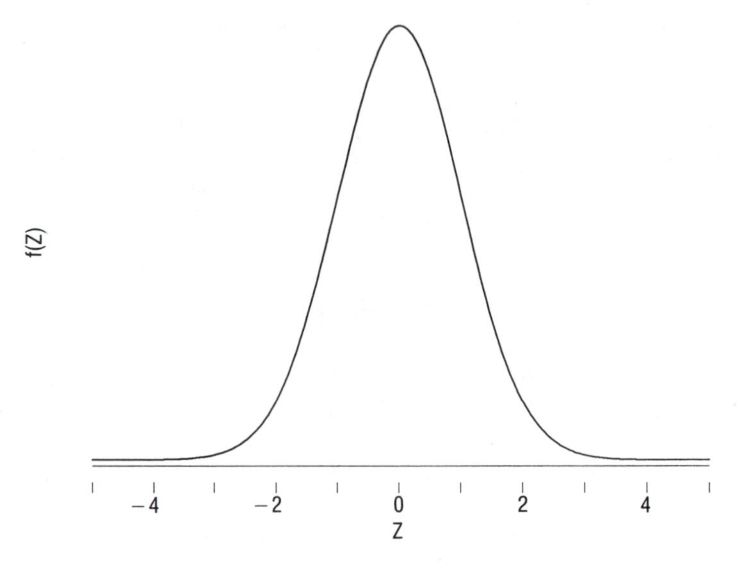

Figure A1.1 Standard normal distribution.

only be positive, and χ^2 distributions are positively skewed (Figure A1.2]. Their skew is most pronounced at low degrees of freedom.

If Z is a standard normal variable and s^2 is a chi-square variable, independent of Z and with df degrees of freedom, then the ratio

$$t = \frac{Z}{s/\sqrt{\text{df}}} \qquad \text{[A1.38]}$$

follows a t-distribution with df degrees of freedom. We construct t-statistics for hypothesis tests by expressing the distance between statistic and hypothesized parameter in estimated standard errors. t-distributions resemble the standard normal distribution but have heavier tails, especially with few degrees of freedom (Figure A1.3). With many degrees of freedom, t-distributions become approximately standard normal.

The ratio of two independent chi-square variables, each divided by its degrees of freedom, follows an *F-distribution.* If $s_1^2 \sim \chi^2_{\text{df}1}$ and $s_2^2 \sim \chi^2_{\text{df}2}$:

$$F^{\text{df}1}_{\text{df}2} = \frac{s_1^2/\text{df}1}{s_2^2/\text{df}2} \qquad \text{[A1.39]}$$

Two parameters, the numerator and denominator degrees of freedom, characterize an F-distribution. Like χ^2, an F-variable cannot be negative. F-tests can compare two estimated variances, such as the explained and residual variance in OLS.

Appendix 4 includes tables of critical values for the theoretical t- (Table A4.1), F- (Table A4.2), and χ^2- (Table A4.3) distributions. Standard normal distribution (Z) critical values equal the t_∞ critical values of Table A4.1.

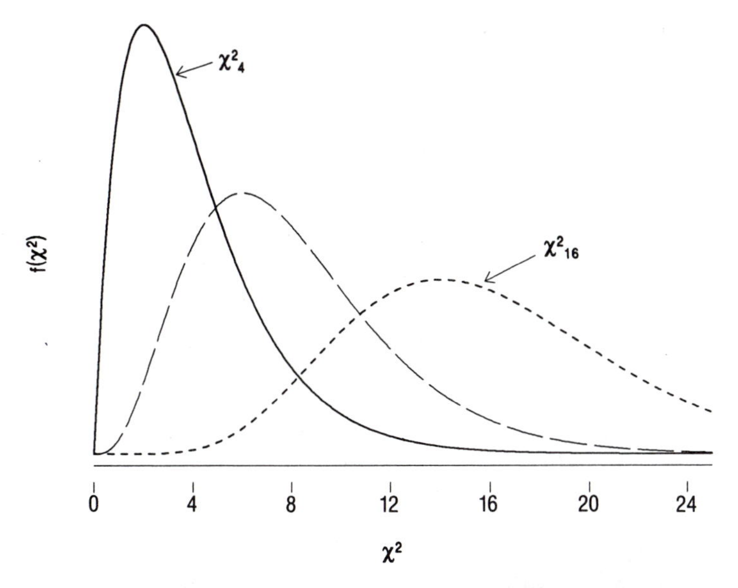

Figure A1.2 Chi-square (χ^2) distributions with 4, 8, and 16 df.

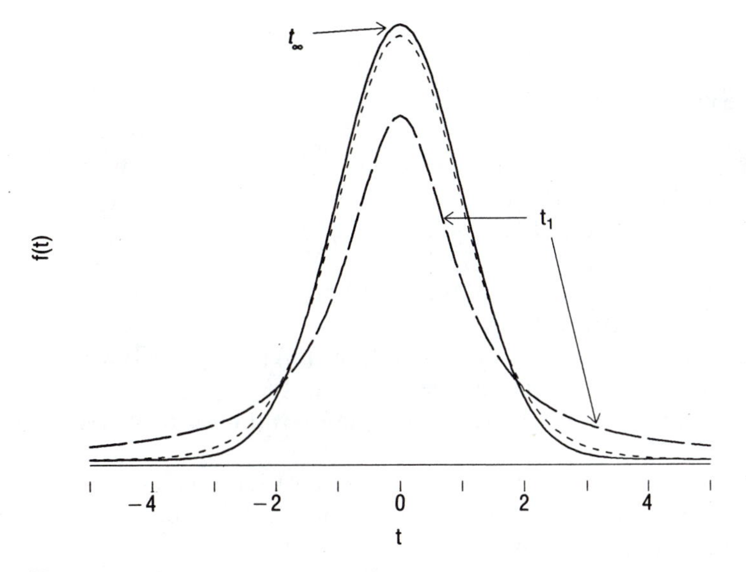

Figure A1.3 t-distributions with 1, 10, and infinite df.

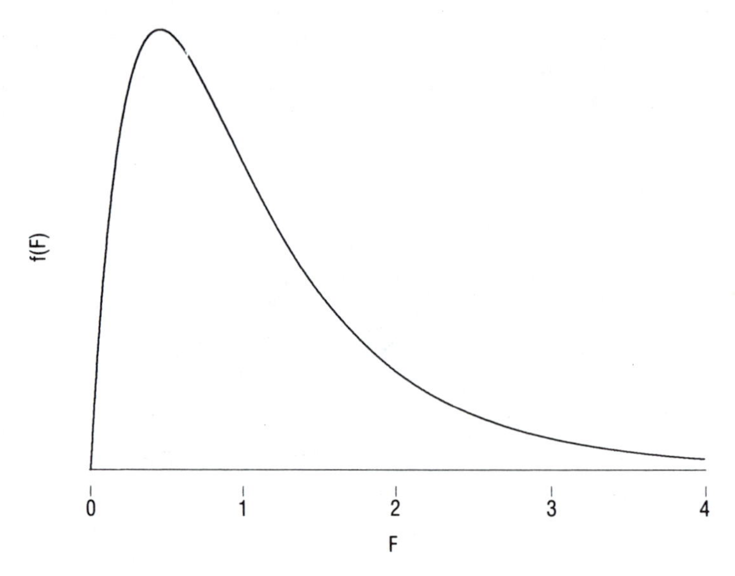

Figure A1.4 F-distribution with $df_1 = 4$ and $df_2 = 16$.

Exercises

1. Apply rules of expectation and covariance to simplify:
   a. $E[1201 + 47.5X]$
   b. $E[-11.4 + .176X + 3.9W]$
   c. $\text{Cov}[Y, 1201 + 47.5X]$
   d. $\text{Cov}[Y, -11.4 + .176X + 3.9W]$
   e. $\text{Var}[1201 + 47.5X]$
2. Use the properties of expectations ([A1.4]–[A1.6]) and the definition of covariance ([A1.7]) to derive Equations [A1.9]–[A1.11].
3. Use properties of expectation and covariance to derive Equations [A1.15]–[A1.17].
4. We wish to know the true values of a variable X, but our measurements contain some error (as all measurements do). Let X represent the true values, $\tilde{X}$ our *measured* values, and ε the errors. (For example, X might be individuals' true incomes, $\tilde{X}$ the incomes they claim in a survey, and ε whatever difference there is between reported ($\tilde{X}$) and true (X) incomes.) Measured values equal the true values plus error:

$$\tilde{X} = X + \varepsilon$$

   where $\tilde{X}$, X, and ε are all variables.
   a. What is the expected value of $\tilde{X}$? What therefore is required for $\tilde{X}$ to be an "unbiased estimator" of X, in the sense that both have the same expectation?

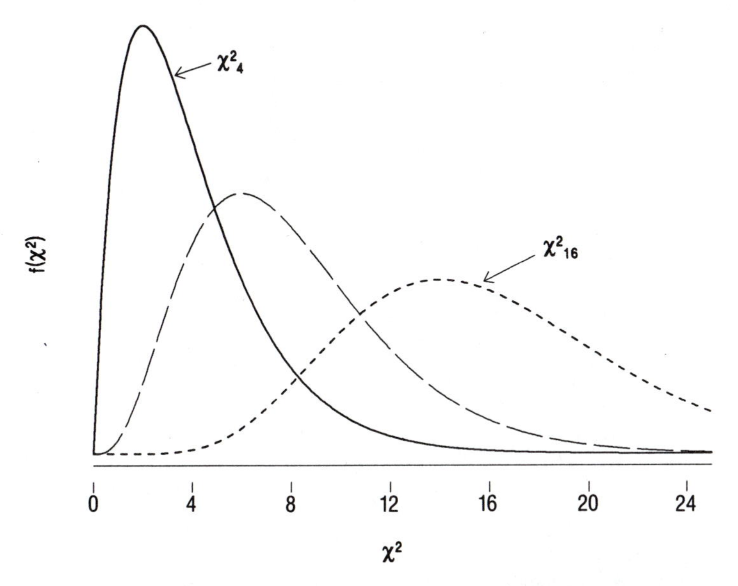

Figure A1.2 Chi-square (χ^2) distributions with 4, 8, and 16 df.

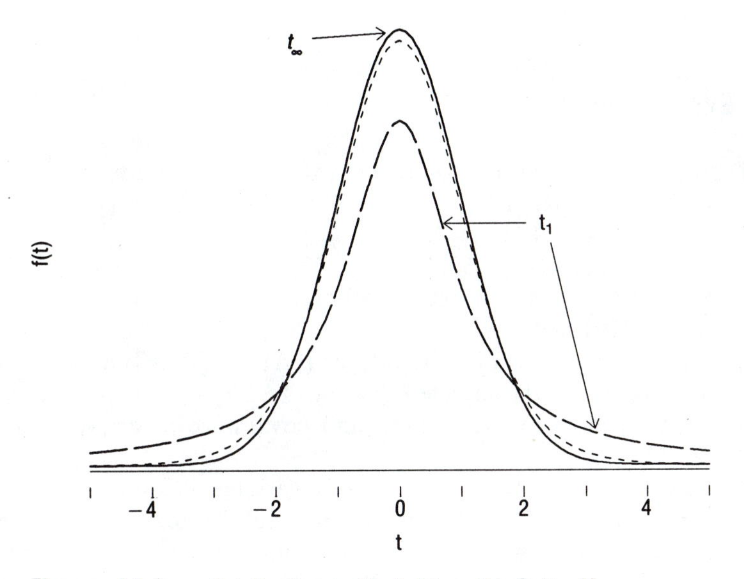

Figure A1.3 *t*-distributions with 1, 10, and infinite df.

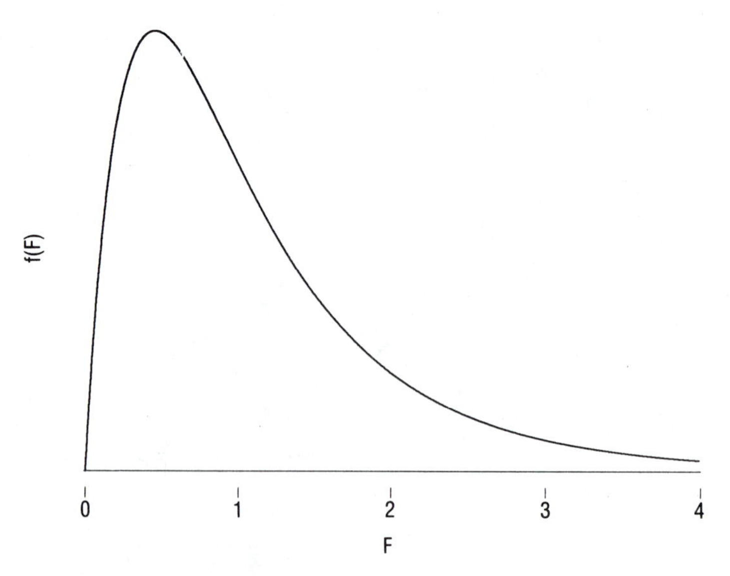

Figure A1.4 F-distribution with $df_1 = 4$ and $df_2 = 16$.

Exercises

1. Apply rules of expectation and covariance to simplify:
   a. $E[1201 + 47.5X]$
   b. $E[-11.4 + .176X + 3.9W]$
   c. $\text{Cov}[Y, 1201 + 47.5X]$
   d. $\text{Cov}[Y, -11.4 + .176X + 3.9W]$
   e. $\text{Var}[1201 + 47.5X]$
2. Use the properties of expectations ([A1.4]–[A1.6]) and the definition of covariance ([A1.7]) to derive Equations [A1.9]–[A1.11].
3. Use properties of expectation and covariance to derive Equations [A1.15]–[A1.17].
4. We wish to know the true values of a variable X, but our measurements contain some error (as all measurements do). Let X represent the true values, $\tilde{X}$ our *measured* values, and ε the errors. (For example, X might be individuals' true incomes, $\tilde{X}$ the incomes they claim in a survey, and ε whatever difference there is between reported ($\tilde{X}$) and true (X) incomes.) Measured values equal the true values plus error:

$$\tilde{X} = X + \varepsilon$$

   where $\tilde{X}$, X, and ε are all variables.
   a. What is the expected value of $\tilde{X}$? What therefore is required for $\tilde{X}$ to be an "unbiased estimator" of X, in the sense that both have the same expectation?

b. What is the variance $\tilde{X}$?
c. Assume that errors are *random*, meaning that $\text{Cov}[X, \varepsilon] = 0$. What does this imply about the relative variation in measured ($\tilde{X}$) and true (X) values?
d. If errors are *not* random and $\text{Cov}[X, \varepsilon] \neq 0$, does your generalization of part c still hold?

Notes

1. The Greek letter ρ (rho) traditionally denotes population correlations (sample correlations are denoted by r). In robustness literature (Chapter 6), ρ has a different, unrelated meaning.
2. If X, rather than Y, is the left-hand-side variable (*regression of X on Y*), the slope becomes

$$\beta_1 = \text{Cov}[X, Y]/\text{Var}[Y]$$

instead of [A1.20]. Similarly, the intercept becomes

$$\beta_0 = E[X] - \beta_1 E[Y]$$

instead of [A1.21].

Appendix 2

Computer-Intensive Methods

Mathematical reasoning from initial assumptions obtains theoretical sampling distributions for many statistics. For other statistics, however, or under less restrictive assumptions, mathematical reasoning may not be feasible. Computers play an increasingly important role in extending our knowledge about sampling distributions. The computer's speed in performing millions of calculations leads by "brute force" into realms that defeat theoretical understanding.

This appendix does not attempt an overview of computer-intensive methods. Instead, it gives examples, illustrating methods for investigating issues too complex for mathematical analysis. *Monte Carlo simulation* begins with artificial data and a model known to be true (because the analyst invented it) and then examines how estimators perform in trying to discover this model. *Bootstrapping* starts from real data and does not necessarily assume that we know the true model. Both methods make heavy use of computers' pseudorandom number generators, equations that can produce huge amounts of random data from any theoretical distribution.[1]

Monte Carlo Simulation

Monte Carlo simulations, named after the European gambling resort, use computers to create many samples of artificial data. In generating these data, the computer follows a model defined by the researcher. Many samples furnish many estimates of model parameters, forming an empirical view of the sampling distributions.

Consider a simple linear model:

$$E[Y_i] = \beta_0 + \beta_1 X_i \qquad \text{[A2.1]}$$

We wish to evaluate estimators of the β parameters. To do so by Monte Carlo simulation:

1. Choose values for the β parameters, and specify the distributions of X and errors. For example:
   Model 1—fixed normal X, normal errors:

   $$Y_i = 5 + 1X_i + \varepsilon_i \qquad \text{[A2.2]}$$

   where X_i values, fixed in repeated sampling, follow an approximately standard normal ($N(0, 1)$) distribution. Errors ($\varepsilon_i = Y_i - E[Y_i]$) are random, following independent $N(0, 1)$ distributions. The true parameter values are $\beta_0 = 5$ and $\beta_1 = 1$.
2. Program a computer to repeatedly generate samples of artificial data according to the model chosen, analyze each sample, and record the results. To obtain an $n = 100$ random sample from a population in which Model 1 is true: generate 100 random ε values from a standard normal distribution, use these to calculate 100 Y values according to [A2.2], and regress Y on X to estimate β_0 and β_1.

By repeating this operation many times, we learn how our regression method performs in estimating the true values, $\beta_0 = 5$ and $\beta_1 = 1$.

Model 1 (fixed normal X, normal ε) represents a "best case" regression problem. OLS should perform well in estimating its parameters. Figure A2.1 illustrates the general model using thousands of data points. Figure A2.2 shows four Monte Carlo $n = 100$ samples, each yielding a different OLS line. The OLS lines vary randomly around the true relation $E[Y_i] = 5 + 1X_i$. Although not obvious from the graph, the four samples all have the same 100 X values and differ only in their 100 random ε values.

Table A2.1 summarizes results from 1000 random samples like those in Figure A2.2. Each sample was generated according to Model 1 and analyzed by three different regression methods: OLS, robust M-estimation, and robust bounded-influence regression (see Chapter 6). All three methods performed well. Over 1000 samples, OLS estimates of β_1 ranged from about .666 to 1.353, with a mean close to the true parameter ($\beta_1 = 1$). Robust M-estimates ranged a little more widely (.646 to 1.354), with a slightly larger standard deviation (.106 versus .104), but also averaged close to $\beta_1 = 1$. The means of the Monte Carlo distributions provide a way to assess bias; their standard deviations assess efficiency.

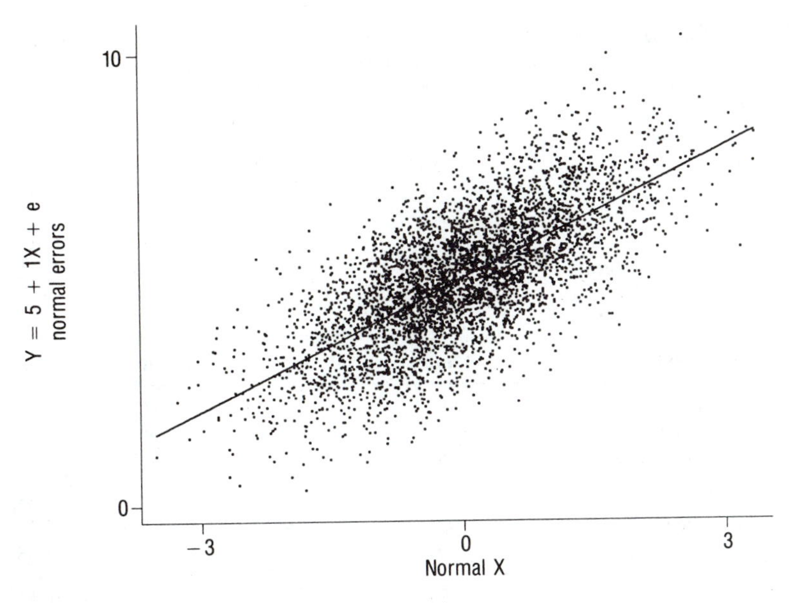

Figure A2.1 5000 data points illustrations.

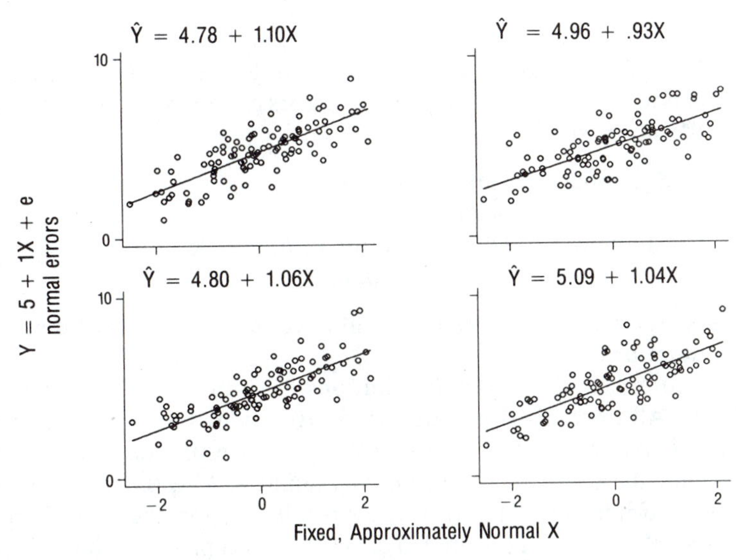

Figure A2.2 Four random $n = 100$ samples from Model 1.

Table A2.1 Monte Carlo simulation results: 1000 artificial samples of $n = 100$ cases each, based on Model 1 (fixed and approximately normal X; normal errors)

Model 1: $Y_i = 5 + 1X_i + \varepsilon_i$

where X fixed, approximately $N(0, 1)$, and

$\varepsilon_i \sim N(0, 1)$

1000-Sample Results:

Coefficient	Mean	Standard Deviation	Min	Max	Relative Efficiency[1]
OLS b_0	4.995	.097	4.673	5.306	100%
M-estimate b_0	4.995	.099	4.685	5.308	96%
Bounded-influence b_0	4.995	.101	4.667	5.332	92%
OLS b_1	.996	.104	.666	1.353	100%
M-estimate b_1	.996	.106	.646	1.354	95%
Bounded-influence b_1	.997	.114	.634	1.368	83%
OLS SE_{b_1}	.105	.010	.076	.134	
M-estimate SE_{b_1}	.108	.011	.079	.141	

[1] OLS variance as a percentage of estimator's variance; values below 100% indicate that the estimator performed worse than OLS.

Table A2.1 suggests that both OLS and M-estimate standard errors (SE_{b_1}) are unbiased: mean estimated standard errors closely resemble the standard deviations of the slope estimates:

$$\begin{aligned} \text{OLS: standard deviation of } b_1 &= .104 \\ \text{mean of } SE_{b_1} &= .105 \\ M\text{-estimation: standard deviation of } b_1 &= .106 \\ \text{mean of } SE_{b_1} &= .108 \end{aligned}$$

These standard errors *should* be unbiased, because Model 1 meets all the necessary assumptions.

Figure A2.3 shows the distributions of OLS and robust slope estimates graphically. All three distributions appear approximately normal and centered around $\beta_1 = 1$. As expected from Table A2.1, OLS estimates cluster a bit more closely around the true value than do M-estimates. M-estimates in turn vary less than bounded-influence estimates. These results support theoretical arguments that, under ideal conditions (normal i.i.d. errors), OLS provides the best estimates. Robust estimation protects against large errors, but with normal errors it needlessly downweights valid outliers and hence loses efficiency. In estimating β_1, the M-estimator proved to be only 95% as efficient as OLS, and bounded influence only 83% as efficient.[2]

We do not need a Monte Carlo simulation to learn all this; theoretical arguments reach the same conclusion on more general grounds. With normal i.i.d. errors,

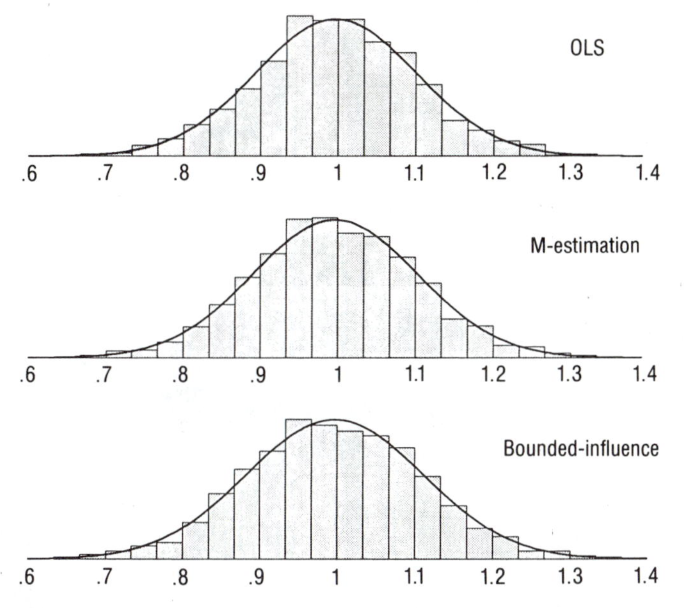

Figure A2.3 Model 1 b_1 estimates from 1000 Monte Carlo samples of $n = 100$ cases each.

theoretical properties of these estimators are well known.[3] Given nonnormal errors, however, their theoretical properties become less clear—particularly in small samples. There Monte Carlo research makes greater contributions.

To illustrate, we perform a second Monte Carlo experiment, identical to the first except with nonnormal errors. Instead of drawing 100% of the errors from an $N(0, 1)$ distribution, we draw only 90% from this distribution. The remaining 10% come from a skewed distribution with wider spread:

Model 2—fixed normal X, nonnormal errors:

$$Y_i = 5 + 1X_i + \varepsilon_i$$

where X_i values, fixed in repeated sampling, follow an approximately standard normal ($N(0, 1)$) distribution. Errors ($\varepsilon_i = Y_i - E[Y_i]$) are random and independent:

$$\varepsilon_i \sim \begin{cases} N(0, 1) & \text{with probability .9} \\ 3 \times (\chi_1^2 - 1) & \text{with probability .1} \end{cases}$$

where χ_1^2 is a chi-square distribution with one degree of freedom. This produces a *contaminated* distribution that may contain a small proportion of large positive errors (obvious in Figure A2.4).

Subtracting 1 from a χ_1^2 variable centers it on zero, so $E[\varepsilon_i] = 0$ for all i, and OLS estimates of β_0 and β_1 should still be unbiased.

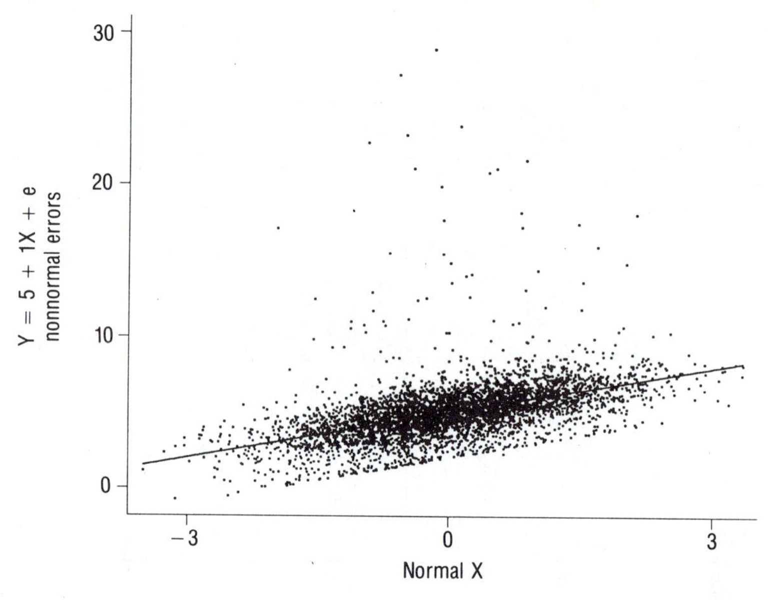

Figure A2.4 5000 data points illustrating Model 2.

Table A2.2 Monte Carlo simulation results: 1000 artificial samples of $n = 100$ cases each, based on Model 2 (fixed and approximately normal X; nonnormal errors)

Model 2: $Y_i = 5 + 1X_i + \varepsilon_i$

where X fixed, approximately $N(0, 1)$, and

$\varepsilon_i \sim N(0, 1)$ with probability .9

$\varepsilon_i \sim 3 \times (\chi_1^2 - 1)$ with probability .1

1000-Sample Results:

Coefficient	Mean	Standard Deviation	Min	Max	Relative Efficiency[1]
OLS b_0	4.990	.160	4.529	5.617	100%
M-estimate b_0	4.910	.115	4.501	5.279	193%
Bounded-influence b_0	4.911	.116	4.502	5.265	189%
OLS b_1	.998	.168	.313	1.757	100%
M-estimate b_1	.996	.121	.610	1.402	191%
Bounded-influence b_1	.997	.129	.595	1.413	170%
OLS SE_{b_1}	.161	.052	.092	.539	
M-estimate SE_{b_1}	.123	.013	.088	.167	

[1] OLS variance as a percentage of estimator's variance; values above 100% indicate that the estimator performed better than OLS.

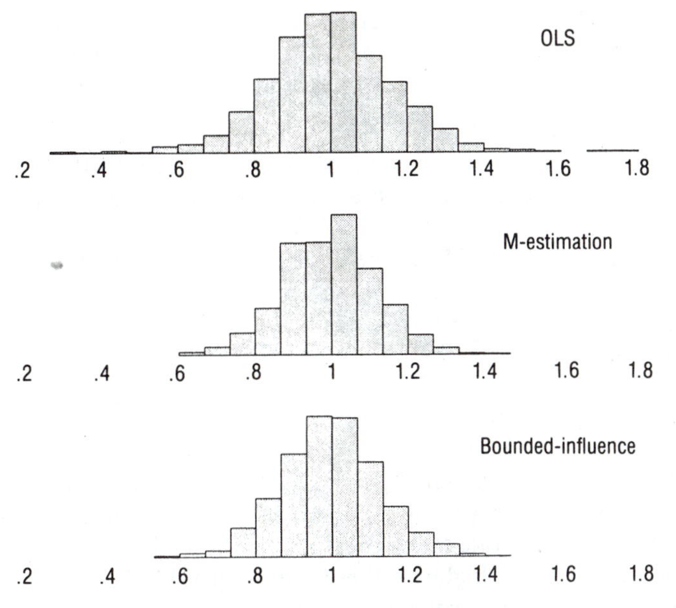

Figure A2.5 Model 2 b_1 estimates from 1000 Monte Carlo samples of $n = 100$ cases each.

Table A2.2 summarizes results from 1000 Model-2 samples. As expected, the slope estimators still appear unbiased (all $\bar{b}_1 \approx 1$), but their relative efficiency differs sharply from that observed with Model 1. Given nonnormal errors but fixed X values, M-estimation is the method of choice. Because it tracks outliers, OLS more often obtains estimates far from the true parameter values and, overall, exhibits almost twice the sampling variance of M-estimation. By unnecessarily down-weighting X-outliers, bounded-influence regression loses efficiency (although it still works much better than OLS).

Figure A2.5 graphs distributions of slope estimates from the experiment with Model 2, visualizing results from Table A2.2. Both robust methods exhibit much less sample-to-sample variation than does OLS. M-estimates cluster somewhat closer around $\beta_1 = 1$ than do bounded-influence estimates.

Model 2 violates one OLS assumption (normal errors) but leaves others intact—errors remain independent and identically distributed, and X values remain fixed. For an acid test of regression methods, we can specify a model that violates all of the classical assumptions:

Model 3—random nonnormal X, nonnormal errors:

$$Y_i = 5 + 1X_i + \varepsilon_i$$

where the distributions of both X and ε are mixtures of normal and χ_1^2:

$X_i, \varepsilon_i \sim N(0, 1)$ with probability .9

$X_i, \varepsilon_i \sim 3 \times (\chi_1^2 - 1)$ with probability .1

About 90% of the time, X_i and ε_i are drawn randomly from standard normal distributions. About 10% of the time, X_i *and* ε_i come from $3 \times (\chi_1^2 - 1)$ distributions. Thus large errors should occur more often with cases that are X-outliers (leverage points) as well.

With thousands of data points (Figure A2.6), Model 3's pathologies are obvious, but in smaller samples (including three of the four in Figure A2.7) they could easily go unnoticed.

Applied to Model 3, OLS yields skewed, heavy-tailed sampling distributions. Sometimes, extreme X values coincide with large positive errors, substantially steepening an OLS line (as at top right in Figure A2.7). Due to the nature of Model 3's X and ε distributions, other types of influence problems should rarely occur.

Table A2.3 and Figure A2.8 summarize results from 1000 Monte Carlo samples based on Model 3. Both robust estimators outperform OLS by wide margins, with bounded-influence regression gaining a clear advantage over M-estimation. Only bounded-influence regression downweights X-outliers and thereby lessens sample-to-sample variation due to leverage points.

Models 1, 2, and 3 highlight the strengths of OLS, M-estimation, and bounded-influence methods, respectively:

1. Under ideal conditions (normal i.i.d. errors), OLS works better than other methods. Heavy-tailed error distributions break OLS down, making it much worse than robust methods.
2. Robust M-estimation protects against nonnormal error distributions, *if* large errors do not coincide with leverage points.
3. Because it downweights leverage points, bounded-influence regression copes with occasional large errors in X and/or Y.

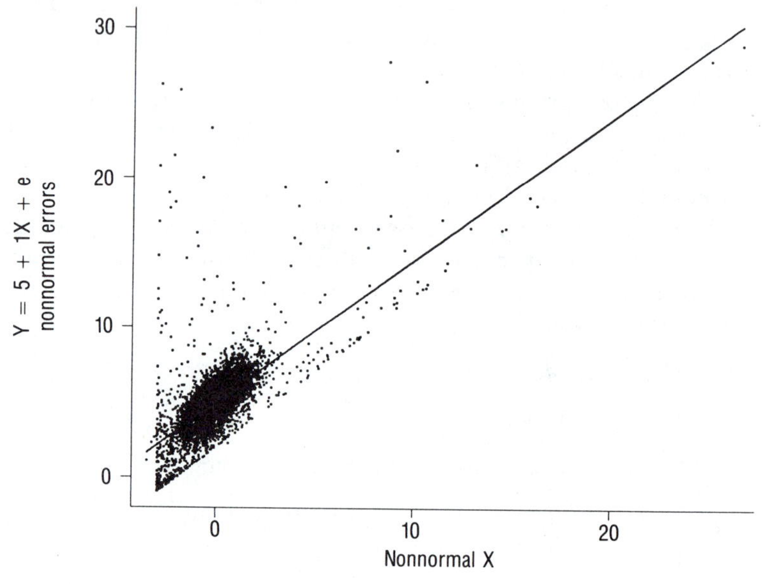

Figure A2.6 5000 data points illustrating Model 3.

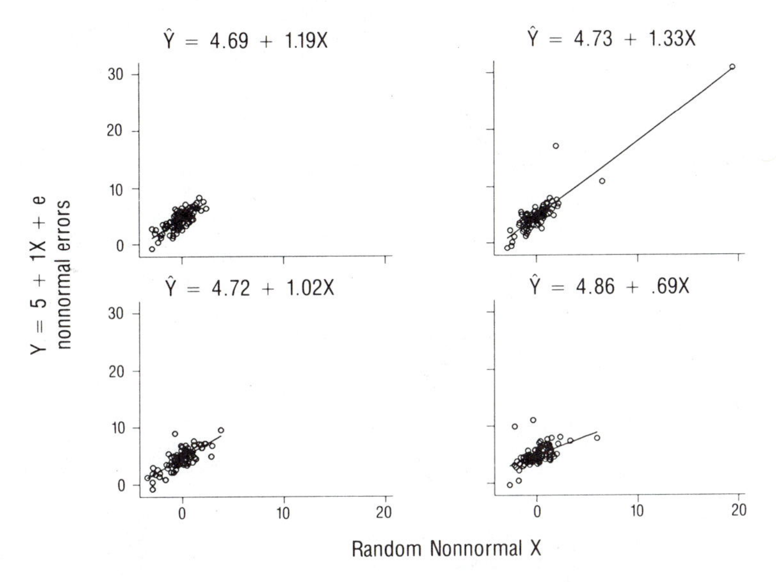

Figure A2.7 Four random $n = 100$ samples from Model 3.

Table A2.3 Monte Carlo simulation results: 1000 artificial samples of $n = 100$ cases each, based on Model 3 (random and nonnormal normal X; nonnormal errors)

Model 3: $Y_i = 5 + 1X_i + \varepsilon_i$

where $X_i, \varepsilon_i \sim N(0, 1)$ with probability .9

$X_i, \varepsilon_i \sim 3 \times (\chi_1^2 - 1)$ with probability .1

1000-Sample Results:

Coefficient	Mean	Standard Deviation	Min	Max	Relative Efficiency[1]
OLS b_0	5.001	.163	4.576	5.718	100%
M-estimate b_0	4.920	.119	4.587	5.296	189%
Bounded-influence b_0	4.945	.114	4.614	5.293	204%
OLS b_1	1.004	.225	.320	2.627	100%
M-estimate b_1	1.033	.144	.672	1.501	243%
Bounded-influence b_1	1.054	.125	.639	1.476	322%
OLS SE_{b_1}	.107	.041	.029	.369	
M-estimate SE_{b_1}	.082	.020	.019	.142	

[1] OLS variance as a percentage of estimator's variance; values above 100% indicate that the estimator performed better than OLS.

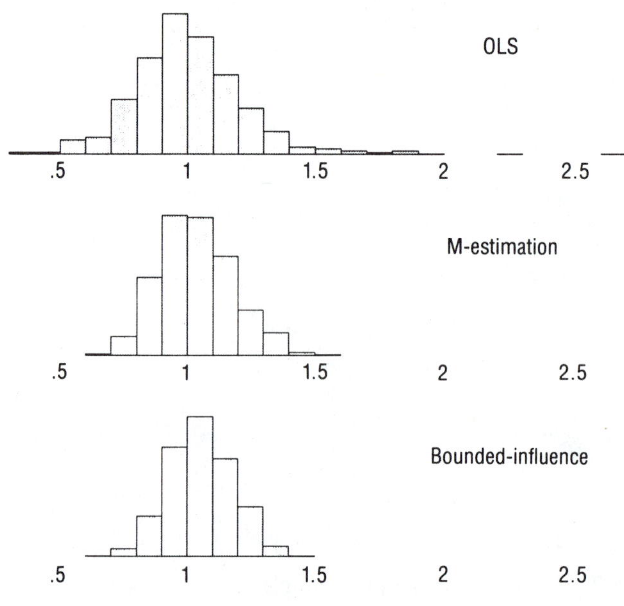

Figure A2.8 Model 3 b_1 estimates from 1000 Monte Carlo samples of $n = 100$ cases each.

Relatively minor inefficiency (on the order of 5–10%) results when we choose a method more robust than we need—for example, *M*-estimation with normal errors. Greater problems can arise from choosing a method that is not robust enough (for example, OLS with nonnormal errors). Table A2.1–A2.3 thus illustrate not only Monte Carlo work but also the goal of robustness:

> A good robust method works almost as well as the best classical method under ideal conditions, and much better than classical methods when conditions are not ideal.

Errors in Model 3 are neither independent of X nor identically distributed. These problems bias the standard errors calculated by both OLS and *M*-estimation:

OLS: standard deviation of $b_1 = .255$
mean of $\text{SE}_{b_1} = .107$
M-estimation: standard deviation of $b_1 = .144$
mean of $\text{SE}_{b_1} = .082$

Both standard error estimates understate actual sampling variability—the mean OLS SE_{b_1} value equals less than half the observed standard deviation of b_1. This demonstrates how, under false assumptions, estimated standard errors (and the usual hypothesis tests or confidence intervals) could lead research astray. With unrealistically low standard error estimates, we might often declare chance results "statistically significant."

Theoretically derived standard errors (even robust standard errors) depend heavily on underlying assumptions. The next section describes computer-intensive methods that seek to estimate standard errors with less reliance on assumptions.

Bootstrap Methods

In a chapter of *Some Recent Advances in Statistics*, Bradley Efron writes:

> The next 20 years should be exciting ones for statisticians. My prediction is for a partial replacement of parametric models, and the accompanying mathematical calculations we have become used to, by computer-intensive methods. These methods will replace "theory from a book," typified by t-tables and F-tables, by "theory from scratch," generated anew by the computer for each new data analysis problem [Oliveira and Epstein, 1982: 180–181].

That is, statisticians will rely less on theoretical sampling distributions like the normal, χ^2, t, and F, whose appropriateness for any given problem always rests on untestable assumptions. Instead, they will construct appropriate sampling distributions empirically, using the data at hand. Efron developed *bootstrap* methods to accomplish this goal (Efron, 1982; Efron and Tibshirani, 1986; for an introduction see Stine, 1990).

Like Monte Carlo simulation, bootstrapping requires hundreds of times more calculations than classical methods, so its rise follows the availability of fast computers. Efron's prediction derives partly from confidence that tomorrow's computers will be faster still. Basically we bootstrap a sample of size n by drawing many random samples, each of them also size n, choosing cases from the original sample *with replacement*. Thus a typical bootstrap sample leaves out some cases from the original data but could include other cases two, three, four, or more times. Analyzing B bootstrap samples generates a new data set containing B different estimates of the parameters of interest, each based on a bootstrap sample of size n.

Let β represent the parameter of interest (for example, a population regression coefficient). Applying an estimator such as OLS to the original sample yields b, an estimate of β. Applying the same estimator to any one bootstrap sample yields an estimate denoted b^*. $F(b)$ represents the true (unknown) sampling distribution of b, and $\hat{F}(b^*)$ the observed distribution of b^*, over B bootstrap samples. If certain assumptions hold:

1. The observed bootstrap distribution $\hat{F}(b^*)$ approximates a maximum-likelihood estimate of $F(b)$.
2. The bootstrap bias, $\bar{b}^* - b$, estimates the true bias $E(b) - \beta$.
3. The bootstrap standard deviation, s_{b^*}, estimates the true standard error of b, σ_b.

We study $\hat{F}(b^*)$ for evidence about the shape, bias, and variability of the true sampling distribution $F(b)$. The assumptions required depend on the bootstrap method used but are often less restrictive than those needed to derive theoretical standard errors.

A simple regression bootstrap called *residual resampling* proceeds as follows:

1. Perform regression with the original sample; calculate predicted values ($\hat{Y}$) and residuals (e).[4]
2. Randomly resample the residuals, but leave X and $\hat{Y}$ values unchanged.

3. Construct new Y^* values by adding the original predicted values to the bootstrap residuals (after fattening those residuals as described in Note 4): $Y^* = \hat{Y} + e^*$.
4. Regress Y^* on the original X variable(s).
5. Repeat steps 2–4 $B = 1000$ or more times (for confidence intervals, $B \geq 2000$ is recommended).

We then study the distributions of bootstrap regression estimates b_0^*, b_1^*, and so on across B bootstrap samples.

Residual resampling assumes fixed X values and i.i.d. (but not necessarily normal) errors. That is, it assumes that the residual found for the ith case could equally well have occurred with the jth case instead; residual resampling randomly reassigns the original-sample residuals to new cases. The n sets of X values from the original sample remain unchanged in each bootstrap sample. OLS makes similar assumptions. When these assumptions hold, bootstrapping OLS should yield approximately the same standard errors and sampling distributions expected from classical theory.

When we use methods with unknown small-sample properties, bootstrapping provides methods for approximating sampling distributions empirically. Bootstrapping also steps into the theoretical gap when the usual assumptions seem dubious. For example, *data resampling* bootstraps regression without assuming fixed X or identically distributed errors:

1. Randomly choose samples of size n, sampling complete cases $(Y, X_1, X_2, \ldots)$ from the original data with replacement.
2. Within each bootstrap sample, regress Y on the X variable(s) as usual.
3. Repeat steps 1 and 2 1000 or more times (2000 or more times for confidence intervals).

As before, the bootstrap distributions of b_0^*, b_1^*, ... values contain clues about the true sampling distributions.

Residual and data resampling often lead to divergent results, which underlines the key role of assumptions. Residual resampling makes most of the usual OLS assumptions. Data resampling, which makes fewer assumptions, can produce bootstrap distributions strikingly unlike those of residual resampling or theoretical ideals.

Bootstrap Distributions

Monte Carlo simulations reveal how different estimators perform with artificial data, where the true parameters are known. Bootstrapping helps explore estimator performance with real data, where parameters are unknown. For example, Table A2.4 summarizes results from bootstrapping six regression models introduced in Chapter 6. Residual resampling and reestimation, repeated $B = 1000$ times

Table A2.4 Bootstrap comparisons of OLS and two robust estimators (residual resampling, $B = 1000$).[1]

		OLS				Robust M-Estimator				Bounded Influence		
Y	X	$\bar{b}^*$	SE_b	s_{b^*}	PSD*	$\bar{b}^*$	SE_b	s_{b^*}	PSD*	$\bar{b}^*$	s_{b^*}	PSD*
Mortality (Table 6.1) $n = 60$	log HC	15.49	6.87	7.27	7.26	19.34	6.55	6.79	6.50	32.01	7.46	7.52
Mortality (Table 6.2) $n = 60$	log HC	17.61	4.64	4.45	4.45	17.96	4.62	4.44	4.62	19.56	4.65	4.79
	precip	2.35	.64	.59	.59	2.32	.64	.59	.60	2.64	.63	.64
	jan	−2.13	.50	.48	.47	−2.11	.50	.48	.48	−1.86	.52	.52
	educat	−18.01	6.20	5.87	5.84	−19.06	6.19	5.91	5.67	−18.42	5.97	5.82
	$\sqrt{\text{nonw}}$	27.33	4.40	4.43	4.56	26.22	4.31	4.35	4.45	24.32	4.48	4.39
Log(PCB) (Table 6.4) $n = 37$	log PCB	.850	.054	.054	.049	.987	.015	.016	.014	.987	.018	.016
LT ratio (Exercise 3.12) $n = 15$	age	.072	.045	.046	.046	.089	.016	.042	.022	.088	.041	.021
	employ	.194	.141	.145	.141	.349	.050	.141	.067	.349	.141	.068
	expose	.138	.556	.560	.550	.462	.196	.544	.262	.455	.545	.259
Distance (Exercise 6.5) $n = 9$	dosewt	−26654	51292	51525	33444	−40834	22367	45916	23242	—	—	—[2]
Nitrogen (Exercise 6.7) $n = 20$	agric	.0085	.0158	.0160	.0143	−.0072	.0102	.0126	.0107	−.0136	.0085	.0072
	forest	−.0085	.0145	.0149	.0134	−.0239	.0093	.0118	.0099	−.0290	.0079	.0068
	urban	.0293	.0276	.0281	.0268	−.0025	.0179	.0233	.0203	.1179	.0255	.0148

[1] $\bar{b}^*$ denotes mean coefficient estimate over 1000 bootstrap samples. SE_b is the original-sample standard error, estimated using [3.19] (OLS) or [6.21]–[6.23] (M-estimation). s_{b^*} is the standard deviation, and PSD* the pseudo-standard deviation (IQR/1.35), of the 1000 bootstrap coefficients.
[2] When $n < 10$, the bounded-influence estimator described in Chapter 6 is no different from M-estimation, unless we set c^H below the 90th percentile.

for each model, produced $6 \times 3 \times 1000 = 18{,}000$ regressions. Since each robust regression in turn required many iterations of weighted least squares, Table A2.4 actually represents more than 150,000 regressions. The term "computer intensive" aptly describes such work!

Bootstrap results provide improved standard error estimates in situations in which original-sample standard errors are either untrustworthy due to false assumptions or unavailable due to theoretical or computational complexity. For most of the OLS estimates and the larger-sample ($n > 30$) M-estimates in Table A2.4, original-sample standard errors (SE_b) agree reasonably well with bootstrap standard deviations (s_{b^*}). But the usual M-estimate standard errors derive from asymptotic theory; their small-sample validity is doubtful. Bootstrap standard deviations in Table A2.4 suggest that, with smaller ($n \leq 20$) samples, the M-estimate SE_b substantially underestimates sampling variability. The bootstrap s_{b*} then provides a better standard error estimate, which is needed for credible comparisons, confidence intervals, or tests.

Bootstrapping thus suggests when theoretically derived standard errors become suspect and provides a superior alternative. When we do not have theoretically derived standard errors, as with the bounded-influence estimates in Table A2.4, bootstrapping may provide the only standard error estimates.

Robust estimators' small-sample sampling distributions frequently exhibit heavier-than-normal tails. Figure A2.9 illustrates with bootstrap coefficients on beryllium exposure, from the LT-ratio study of mine workers introduced in Chapter 3 (Exercise 3.12). The three boxplots depict distributions summarized in Table A2.4, making it plain that (1) the middles of the robust estimate distributions are much more compact (less variation among most of the samples), but (2) a few robust estimates vary more widely.

Pseudostandard deviation (PSD = IQR/1.35) provides a resistant measure of spread for such heavy-tailed distributions. PSD reflects spread of the middle 50% of a distribution, ignoring what goes on in the tails. M-estimate standard errors in Table A2.4 are often closer to bootstrap PSD* than to bootstrap standard deviation s_{b*}. Comparing standard deviation with PSD furnishes clues about distribution shape. Assuming rough symmetry, the more $s_{b*} >$ PSD*, the more heavy tailed the bootstrap distribution is relative to a normal distribution. $s_{b*} \approx$ PSD* indicates approximate normality (if also symmetrical).

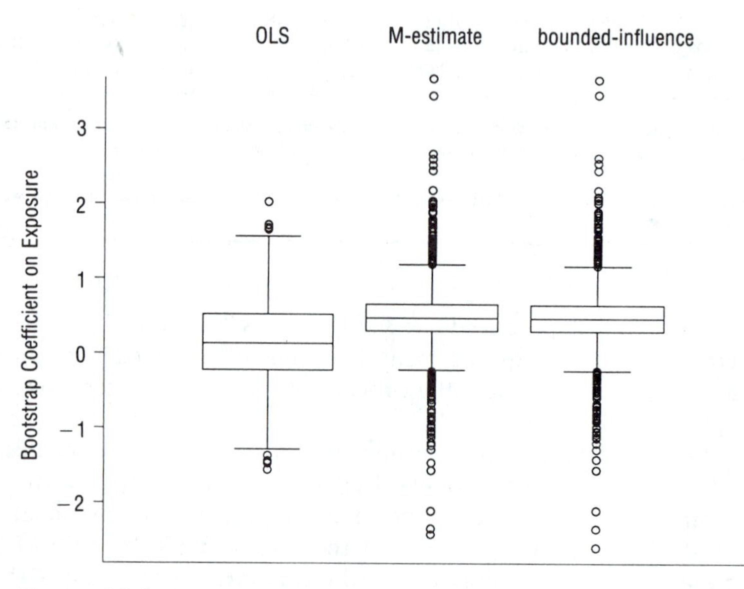

Figure A2.9 Coefficients on mine worker beryllium exposure from B = 1000 bootstrap samples.

This discussion implies a second use for bootstrap results: to compare the performance of different estimators *applied to the sample at hand*. SE_b, s_{b*}, and PSD* all assess sampling variability. For the five-predictor mortality example (second from top in Table A2.4), OLS and M-estimation produce similar estimates, both with slightly less sampling variability than bounded influence. Applied to these relatively well-behaved data, the three estimators' relative performance resembles that with Model 1 in Table A2.1.

When the sample includes severe X-outliers (leverage points), as does the river/nitrogen sample (bottom in Table A2.4), bounded-influence estimates vary least and OLS most. This ordering corresponds to that observed with Model 3 in Table A2.3. With other outlier problems (PCBs, LT ratio in Table A2.4) the two robust methods perform similarly—both generally better than OLS (as happened with Model 2 in Table A2.2). Thus bootstrap and Monte Carlo results show general agreement about the behavior of these three regression methods. Bootstrapping provides evidence about how each method performs with the researcher's own data, not just with idealized models.

Substantial differences between original-sample results and bootstrap means warn that the original-sample results are unstable, very sensitive to small changes in the data. For example, in the one-predictor mortality model, the mean of the OLS bootstrap distribution ($\bar{b}^* = 15.49$) is almost twice the original-sample estimate ($b = 7.97$). Several leverage cases (Californian cities) with unusually low mortality pull down the original-sample slope (see Figure 6.1). The combination of unusual Y values and high leverage makes these cases influential. Residual resampling randomly reassigns residuals. When the large Californian residuals attach to lower-leverage cases, and the high-leverage cases get more ordinary residuals, these cases exert much less influence—allowing the mortality/pollution line to steepen. The mean bootstrap OLS slope in this instance more closely resembles the original-sample robust slope and provides a better summary of the relationship than does the original-sample OLS slope.

Among unbiased estimators of the same parameter, the estimator with the lowest variance is best. If the estimators estimate different parameters, however, it is harder to compare or choose between them. In the Monte Carlo simulations of Tables A2.1–A2.3, OLS and robust regression estimate the same (known) parameters, so sampling variance allowed straightforward comparisons. With the natural data of Table A2.4, the different regression methods often appear to estimate different parameters.

Figure A2.10 illustrates this problem using the mortality/log hydrocarbon pollution example (at top in Table A2.4). The three estimators produce similarly spread-out, but differently centered, bootstrap distributions. Bounded-influence regression finds the strongest effect of air pollution on mortality. M-estimation yields the most stable solution, in the sense that it changes least when we include four other predictors. But, since these three methods evidently estimate different and unknown parameters, we cannot statistically determine which method is best. The analyst must make this decision on substantive grounds, perhaps aided by bootstrap evidence regarding the stability of results.

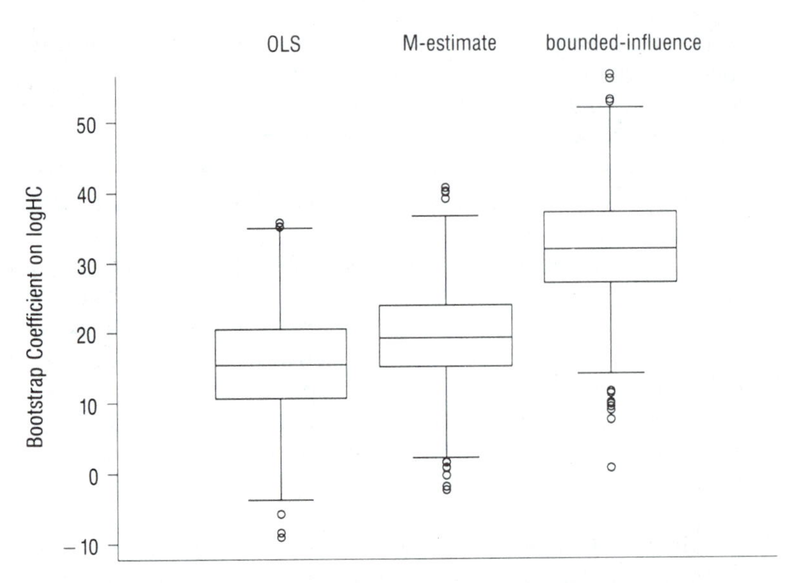

Figure A2.10 Coefficients on log hydrocarbon pollution, from $B = 1000$ bootstrap samples.

Residual Versus Data Resampling

Unlike residual resampling, data resampling does not assume i.i.d. errors. Since it allows for other possibilities, and also admits random X values as a new source of sample-to-sample variation, data resampling often yields results quite different from those expected under the usual regression assumptions. Applied to data with outliers, for example, data resampling tends to find multimodal sampling distributions.

Stine (1990) recommends basing the choice of residual versus data resampling on how the data were collected. If the fixed-X assumption is realistic, and we could readily collect a new sample with all the same X values but different Y's, then residual resampling makes sense. For example, a data set at the end of this appendix lists precipitation measured at 102 rain-gauge sites; we could revisit all sites in other years and collect a new set of Y values (precipitation) to go with unchanged X values (elevation). On the other hand, sometimes X varies as randomly as Y—for example, with the 1984 and 1985 PCB contamination levels described in Chapter 6. Then data resampling makes better substantive sense. In either case, we want the mechanism of bootstrap resampling to mimic the way in which our sample was originally chosen from the population.

Figure A2.11 contrasts residual and data resampling. Four histograms show distributions of regression slopes obtained by bootstrapping the PCBs data of Figure 6.17. Residual resampling produced the two histograms at left in Figure A2.11. Of these two, the distribution of robust (M-estimate) coefficients appears much less spread out; by downweighting outliers, robust estimation limits sample-to-

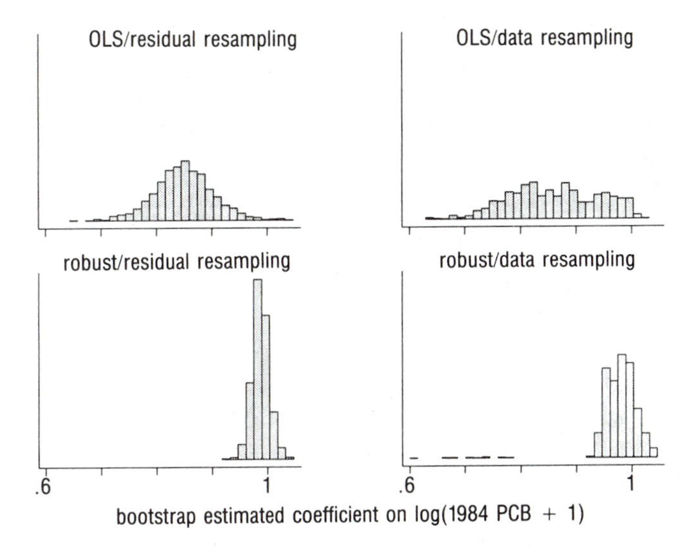

Figure A2.11 Bootstrap distributions comparing residual and data resampling.

sample variation. Both residual and resampling distributions resemble normal curves, and both have standard deviations close to the original-sample estimated standard errors (third example in Table A2.4).

At right in Figure A2.11 are corresponding distributions obtained by data resampling. The general point that robust estimates vary less remains true, but everything else has changed. Neither distribution looks remotely normal; both show evidence of negative skew with multiple peaks and gaps. Data resampling produces some bootstrap samples without the outliers seen in Figure 6.17. These samples yield a steep OLS or robust slope. Other bootstrap samples include the outliers, which make the OLS slope less steep—accounting for the several peaks seen in the OLS distribution at top right. *M*-estimation usually downweighted the outliers; even when they appeared in a bootstrap sample, *M*-estimation ignored them and found a steep slope. But some bootstrap samples included Boston Harbor (a high-pollution outlier) four or more times. Then these outliers "captured" the robust line and caused it to downweight other cases instead—producing the few very-low-slope estimates in the robust/data resampling distribution.[5]

Data resampling tends to produce distributions with greater spread and complex nonnormality, contrasting with approaches that assume i.i.d. errors (residual resampling and standard theory). If errors are not really i.i.d., then data resampling may provide the most realistic guidance. The following section considers confidence intervals from data resampling.

Bootstrap Confidence Intervals

When theory-based confidence intervals and tests seem untrustworthy, bootstrapping provides an alternative, data-based approach. The bootstrap distribution represent our best guess about the shape and spread of the true sampling distribution. However, it usually centers around b, not necessarily around the true parameter β. This complicates confidence-interval construction; we cannot simply use (for example) the bootstrap 5th and 95th percentiles as endpoints for a 90% confidence interval. The intuitively appealing and widely used bootstrap-percentile approach encounters two problems:

1. it is theoretically backward; and
2. even after correcting for backwardness, bootstrap-percentile confidence intervals do not seem to work.[6]

Figure A2.12 displays a data-resampling bootstrap distribution of $B = 2000$ OLS slope estimates for the log PCB regression, noting the positions of the original-sample slope ($b = .8508$) and also of the 5th ($b^*_{.05} = .7327$) and 95th ($b^*_{.95} = .9888$) bootstrap percentiles.[7] Using this information to find an approximate 90% confidence interval:

$$b - (b^*_{.95} - b) < \beta < b - (b^*_{.05} - b) \qquad \text{[A2.3]}$$

$$.8508 - (.9888 - .8508) < \beta < .8508 - (.7327 - .8508)$$

$$.7128 < \beta < .9689$$

This interval appears at bottom in Figure A2.12. The distance from b to the interval's *left* endpoint equals the distance from b to the bootstrap distribution's 95th percentile (*right* tail). The distance from b to the interval's right endpoint likewise equals the distance from b to the bootstrap distribution's 5th percentile (left tail). The logic of confidence intervals requires this counterintuitive reversal.[8]

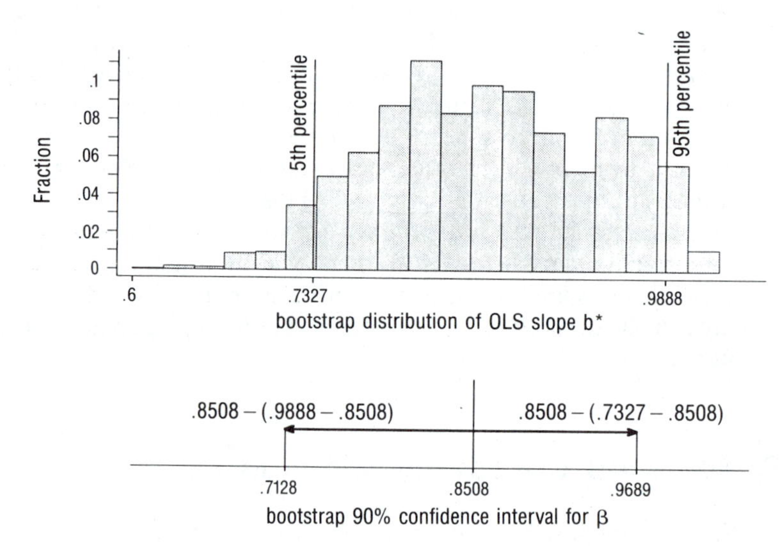

Figure A2.12 Using bootstrap percentiles to construct a 90% confidence interval.

Intervals could be formed in this fashion from any bootstrap percentiles. *Monte Carlo simulations suggest, however, that bootstrap-percentile confidence intervals often have less than the nominal coverage probability.*

Two more elaborate methods attempt to work around this failure: *percentile-t* (also called *studentized percentiles*) and *accelerated bias correction* (BC_a). BC_a presents formidable computational challenges and is not described here.[9] Percentile-t methods are simpler and have been recommended on theoretical and empirical grounds (Hall, 1988; Owen, 1988). Instead of simply obtaining a slope b^* from each bootstrap sample, we subtract the original-sample slope b and divide by the bootstrap sample's standard error (SE_{b^*}) to obtain a *studentized* slope:

$$t^* = \frac{b^* - b}{SE_{b^*}} \qquad \text{[A2.4]}$$

We then use percentiles of the t^*-distribution to construct confidence intervals.

For example, a percentile-t 90% confidence interval is

$$b - t^*_{.95}s < \beta < b - t^*_{.05}s \qquad \text{[A2.5]}$$

where $t^*_{.05}$ denotes the 5th percentile of the bootstrap distribution of t^* and $t^*_{.95}$ is its 95th percentile; s represents an estimated standard error of b. Possible estimators for s include:

1. the original-sample estimated standard error (SE_b);
2. the standard deviation (s^*_b) of the bootstrap distribution of b^*; or
3. the pseudostandard deviation ($PSD^* = IQR^*/1.35$) of the bootstrap distribution of b^*.

s^*_b provides a more robust estimate than SE_b, without ignoring the tails of the distribution, as does PSD^*.

From the same $B = 2000$ bootstrap samples displayed in Figure A2.12, we obtain:

5th percentile of t^*: $t^*_{.05} = -2.1535$

95th percentile of t^*: $t^*_{.95} = 8.3369$

standard deviation of b^*: $s_{b^*} = .0804$

These lead to the 90% confidence interval:

$$b - t^*_{.95}s^*_b < \beta < b - t^*_{.05}s^*_b$$

$$.8508 - 8.3369 \times .0804 < \beta < .8508 - (-2.1535 \times .0804)$$

$$.1805 < \beta < 1.0239$$

This percentile-t interval is much wider than that obtained directly from percentiles of b^*—and indeed is wider than the whole range of the bootstrap distribution

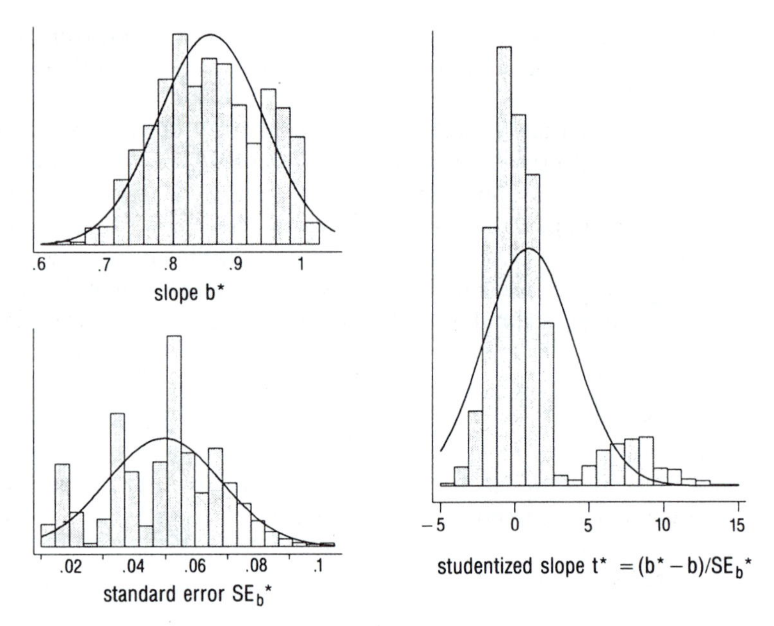

Figure A2.13 Bootstrap slope, standard error, and studentized slope (for percentile-t) distributions.

$(.6049 \leq b^* \leq 1.0224)$. The interval's excessive width indicates a problem with this method.

Details appear in Figure A2.13. The bootstrap distribution of OLS standard errors (lower left) exhibits skew, sharp peaks and gaps, and erratic variation—the largest SE_{b*} values are ten times the smallest. This pattern results from outliers, which have even greater effects on estimated OLS standard errors than on slopes. Combining slope and standard error estimates according to [A2.3] produces the bimodal t^*-distribution at right in Figure A2.13. Its main peak includes bootstrap slopes centered around the original-sample slope. The distinct minor peak comes from bootstrap samples that did not happen to include the original sample's outliers (consult Figure 6.17). Without these outliers we obtain (1) steeper slopes and (2) smaller standard errors, leading to (3) much higher studentized slopes.

The percentile-t method thus depends on the quality of the bootstrap samples' estimated standard errors (SE_{b*}). Where these estimates are unstable, as with OLS in the presence of outliers, the percentile-t method tends to form wide intervals. Robust standard errors should reduce the problem; unfortunately, one motivation for bootstrapping in the first place is to obtain better standard errors. Theoretically derived standard errors generally require more assumptions, and possess less robustness, than the statistics they describe.

To get more stable estimates of standard errors, we might consider bootstrapping the bootstrap: within each bootstrap sample, we bootstrap again to estimate the standard error. With outlier-filled data, a resistant estimate such as PSD* might perform better than s_{b*}. Nested bootstrapping requires at least 100 times more computation than an ordinary bootstrap.

Tukey's *jackknife* method (see Mosteller and Tukey, 1977) may provide decent standard error estimates with somewhat less effort. One way to obtain jackknife estimates for standard errors of regression coefficients is as follows:

1. Starting with a sample of size n, repeat the regression n times, each time omitting one case. Thus the first regression employs $n - 1$ cases: cases 2 through n (omitting case 1). The second regression employs a second sample of size $n - 1$: cases 1 and 3 through n (omitting case 2), and so on.
2. Let $b_{(i)}$ represent the coefficient from the regression omitting case i, and $b_{(.)}$ the mean of all n jackknife coefficient estimates:

$$b_{(.)} = \frac{\sum_{i=1}^{n} b_{(i)}}{n} \qquad \text{[A2.6]}$$

The jackknife estimate of the standard error of b equals

$$\sqrt{\frac{n-1}{n} \sum_{i=1}^{n} [b_{(i)} - b_{(.)}]^2} \qquad \text{[A2.7]}$$

Repeating steps 1–2 to estimate a standard error for each bootstrap sample requires nB iterations, compared with at least $100B$ iterations for a nested bootstrap.[10]

The name "jackknife" connotes a rugged, all-purpose tool. Like bootstrapping, jackknifing seeks improved, data-based estimates of standard errors and other statistics. Bootstrapping often performs better; the jackknife's chief advantage is computational simplicity. The jackknife is really a special case of the bootstrap, but one that provides no picture of the sampling distribution itself. See Efron (1982) for a more detailed comparison of jackknife and bootstrap.

Evaluating Confidence Intervals

The bootstrap appears to be a powerful tool, liberating researchers from the restrictions of unrealistic assumptions. But how well does it actually work? One way to explore this question is to embed bootstrapping within a Monte Carlo simulation. With Monte Carlo simulation, unlike with most real data, we know the true parameters and therefore can objectively judge a statistical method's success. For example, we can check whether a "90% confidence" interval procedure really does produce intervals that contain the parameter about 90% of the time.

Table A2.5 gives results from such an experiment. One thousand Monte Carlo samples (each with $n = 100$ cases) were generated according to our ill-behaved Model 3 (see Table A2.3). For each of these 1000 samples, "90% confidence"

Table A2.5 Monte Carlo evaluation of 90% confidence intervals from OLS estimates of β_1, Model 3 (nonnormal X, nonnormal and non-i.i.d. errors)[1]

Confidence Interval Method	Percent of "90%" Intervals Actually Containing $\beta_1 = 1$	Median Interval Width
Standard t-table, Equation [2.27]	64%	.34
Bootstrap: 5th and 95th percentiles, incorrect	84%	.48
Bootstrap: hybrid, Equation [A2.3]	86%	.48
Bootstrap: percentile-t, Equations [A2.4], [A2.5]	90%	.70

[1] Based on 1000 Monte Carlo samples of $n = 100$ cases each. Within each Monte Carlo sample, $B = 1000$ bootstrap iterations occurred, so this table summarizes 1,001,000 regressions.

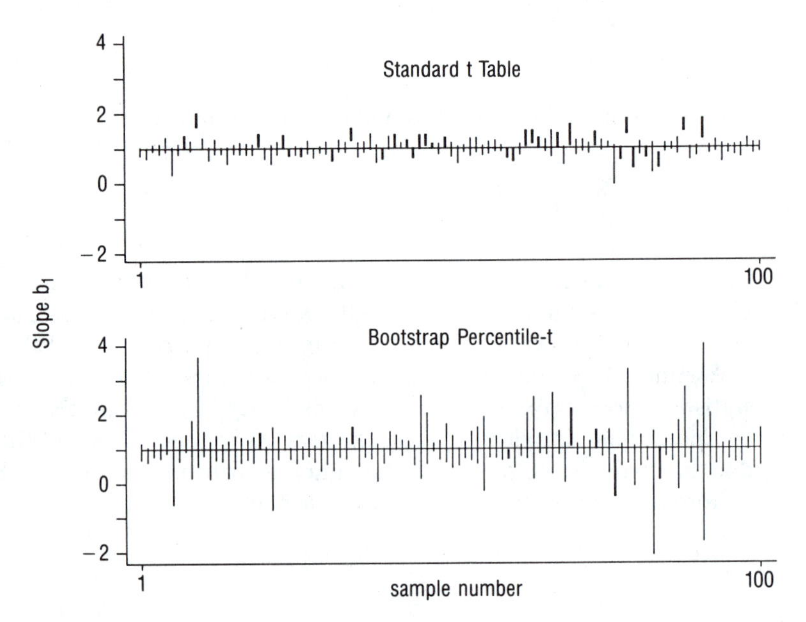

Figure A2.14 Monte Carlo evaluation of standard and bootstrap confidence intervals; "wrong" (dark) intervals do not include $\beta_1 = 1$.

intervals were found from OLS slopes by four different methods: the usual t-tables and three bootstrap procedures. Bootstrapping required resampling the data 1000 times for each Monte Carlo sample, so Table A2.5 summarizes more than a million regressions.[11]

The standard t-table method fails badly: only 64% of these "90% confidence" intervals include $\beta_1 = 1$. Its failure is not surprising, since Model 3 violates basic assumptions upon which this method rests. OLS slope estimates for Model 3 exhibit much greater sampling variation than would be predicted by the usual standard error, SE_b.

Intervals based on the 5th and 95th bootstrap percentiles are wider but still fall short of 90% coverage. The correct "hybrid" procedure (Equation [A2.3]) performs slightly better than the incorrect (but often recommended) method of defining the interval directly from the 5th and 95th percentiles. These two procedures yield identical-width intervals, but [A2.3] should give better coverage with asymmetrical sampling distributions like that produced by Model 3.

We earlier saw that percentile-t methods sometimes yield very wide intervals. Table A2.5 suggests that these wide intervals more accurately reflect the true sampling variability, however. In about 90% of the 1000 Monte Carlo samples, 90% confidence intervals based on the bootstrap percentile-t actually contained β_1. The almost perfect success of the percentile-t procedure here may be coincidental, and in any event this represents only one experiment with one model—far from a demonstration that the procedure always works (it does not). The results do encourage further research along these lines, however.

Figure A2.14 illustrates standard t-table and bootstrap percentile-t confidence intervals from the first 100 Monte Carlo samples of Table A2.5. Percentile-t intervals more often include $\beta_1 = 1$ (horizontal line); "wrong" intervals are shown darkly. The percentile-t method buys greater accuracy by constructing wider intervals—sometimes very much wider.

Computer-Intensive Methods in Research

The examples in this appendix illustrate a few of the capabilities that computer-intensive methods add to research. Monte Carlo simulation can investigate how any procedure works with any definable model. The bootstrap investigates estimator performance by resampling real data. Theoretically derived standard errors attempt to describe sampling behavior too, but they may offer less realistic assumptions.

Bootstrapping provides alternative standard error estimates and confidence intervals. Such confidence intervals provide hypothesis tests as well. For example, does the null-hypothesis parameter value lie outside of a $100(1 - \alpha)$ confidence interval? If so, we can reject H_0 in favor of a two-tailed H_1, at level α. Apart from formal tests and intervals, bootstrapping forces us to think qualitatively about the shape of our sampling distribution—which may be far from bell shaped, for substantively or statistically interesting reasons.

Outliers have complex effects in bootstrap analysis. They sometimes dictate the shape of the sampling distribution. Seeing this effect encourages closer thinking about what outliers represent. Do they reflect random but unexpectedly large errors, or something else—cases from another population, omitted variables, nonlinear relationships, nonconstant error variance, errors correlated with X values, or what? Each possible answer implies a different solution. Cases from other populations should be excluded; omitted variables or nonlinearity requires changes in the model. Heteroscedasticity or correlated errors imply that residual resampling (and conventional standard errors) will be misleading. With bootstrapping we also have more information about *how much difference* it makes that outliers or other problems exist.

Bootstrapping investigates the interaction between data and method: we learn more about how an estimator (or several estimators) performs applied to the data at hand. This adds a new dimension to exploratory data analysis, furnishing insights into both method and data. An important application of Monte Carlo simulation involves evaluating new ideas like the bootstrap, and delineating conditions under which these ideas succeed.

Exercises

1. Perform a Monte Carlo simulation to compare sampling behavior of the mean and median of a variable from a normal (Gaussian) population distribution. Generate 200 artificial samples, consisting of $n = 50$ cases each, by writing a program that repeats the following operation 200 times:
   a. Generate 50 random values of Y, drawn from a normal distribution with $\mu = 25$ and $\sigma = 2$.
   b. Calculate the mean and median of this $n = 50$ sample, and record them in a data file.

   The end result should be a new data set containing 200 means and 200 medians. Comment on the distributions of these sample statistics, including their means, standard deviations, and approximate normality.
2. Theoretically, both sample mean and sample median are unbiased estimators of μ when data come from a normally distributed population. Based on your results from Exercise 1, which statistic (mean or median) appears to be more efficient given a normal distribution?
3. Repeat Exercise 1, this time drawing 200 $n = 50$ samples from a contaminated normal population, 95% $N(25, 2)$ and 5% $N(25, 20)$.
4. Theoretically, both sample mean and sample median are unbiased estimators of μ when data come from any symmetrical population distribution. Based on your results from Exercise 3, which statistic (mean or median) appears to be more efficient given this heavy-tailed but symmetrical distribution?
5. Design your own Monte Carlo simulation to compare the performance of two or more estimators applied to a model of your choosing. Discuss your findings.

Following are data on elevation (X) and precipitation (Y) recorded by rain gauges at 102 sites in the northeastern United States (from Dingman, Seely-Reynolds, and Reynolds, 1988). These data form the basis for bootstrap analysis in Exercises 6–8. Calculations for these three exercises could be done together to save computing time.

Data from 102 northeastern rain-gauge sites

Site		State	Elevation (meters)	Mean Annual Precipitation (cm)
1	Berlin	NH	284	97.5
2	Beth	NH	421	86.8
3	Bradford	NH	296	112.4
4	Dixville Notch	NH	482	117.0
5	Durham	NH	23	109.8
6	Errol	NH	390	96.8
7	First C.L.	NH	506	111.1
8	Fitzwater	NH	354	112.8
9	Franklin Falls	NH	131	100.2
10	Hanover	NH	184	93.1
11	Keene	NH	146	102.4
12	Lakeport	NH	171	105.4
13	Lakeport 2	NH	153	94.5
14	Lebanon	NH	171	88.4
15	McDowell D.	NH	294	117.9
16	Marlow	NH	357	96.2
17	Massabesic	NH	76	101.3
18	Milan	NH	360	96.8
19	Milford	NH	92	114.3
20	Mount Washington	NH	1910	228.4
21	Nashua	NH	57	109.9
22	Newport	NH	239	97.5
23	Peterboro	NH	311	112.0
24	Pinkham Notch	NH	613	147.1
25	Plymouth	NH	171	109.2
26	S. Danbury	NH	284	107.2
27	S. Lyndeboro	NH	198	114.4
28	W. Rumney	NH	171	111.4
29	Woodstock	NH	220	114.7
30	Bell. Falls	VT	92	102.6
31	Chelsea	VT	244	93.1
32	Chittenden	VT	329	106.9
33	Cornwall	VT	104	86.4
34	Dorset	VT	299	118.9
35	Enos. Falls	VT	129	104.9
36	Gilman	VT	259	85.0
37	Montpelier	VT	343	86.2
38	Newfane	VT	128	110.9
39	Newport	VT	233	101.5
40	Peru	VT	509	129.1
41	Rochester	VT	253	110.2
42	Rutland	VT	189	88.7
43	St. Johnsbury	VT	213	92.5
44	Salisbury	VT	113	94.5
45	Sears Station	VT	475	135.4
46	S. Londonderry	VT	500	108.9
47	S. Newbury	VT	143	87.7
48	Union V.	VT	141	89.8
49	Vernon	VT	69	110.6

Site		State	Elevation (meters)	Mean Annual Precipitation (cm)
50	W. Burke	VT	274	103.5
51	W'ham	VT	442	130.4
52	Heath	MA	485	128.1
53	Tully Lake	MA	209	108.6
54	Birch Hill	MA	256	99.8
55	Ash'ham	MA	362	120.6
56	Lawrence	MA	17	108.1
57	Haverhill	MA	18	108.4
58	Newburyport	MA	6	114.2
59	Middle Dam	ME	445	92.4
60	Albany	NY	84	90.8
61	Grafton	NY	475	111.7
62	Salem	NY	150	99.6
63	Whitehall	NY	36	93.3
64	Elizabethtown	NY	180	87.3
65	Peru	NY	156	73.5
66	Plainfield	MA	495	121.3
67	Petersham	MA	332	107.9
68	Gardner	MA	338	110.0
69	Fitchburg	MA	122	116.3
70	Middleton	MA	32	108.9
71	Ipswich	MA	24	119.7
72	Rockport	MA	24	115.4
73	Dublin	NH	466	110.2
74	E. Deering	NH	241	111.9
75	Epping	NH	49	113.1
76	Mount Sunapee	NH	384	104.5
77	Otter Brook	NH	207	102.7
78	Walpole	NH	281	96.6
79	Wolfeboro	NH	219	98.5
80	Bethel	VT	201	98.0
81	Danville	VT	424	106.6
82	H'ton Center	VT	214	106.9
83	Manchester	VT	244	113.9
84	Mays Mills	VT	251	131.6
85	Morrisville	VT	189	106.8
86	Mount Mansfield	VT	1205	173.8
87	St. A. Bay	VT	34	87.9
88	Waitsfield	VT	250	110.3
89	Waterbury	VT	247	104.6
90	Wardsboro	VT	299	114.9
91	Adams	MA	229	104.8
92	Shel. Falls	MA	143	116.6
93	Turners Falls	MA	59	110.7
94	Win'don	MA	311	110.8
95	Sanford	ME	86	119.6
96	Bridgton	ME	183	117.6
97	S. Andover	ME	202	108.9
98	Upper Dam	ME	453	88.2
99	Berlin	NY	293	108.6
100	Battenville	NY	119	94.5
101	Ticonderoga	NY	50	86.4
102	Plats	NY	53	79.7

6. Write a program to find the median and mean elevation for 200 bootstrap samples (use data resampling, each $n = 102$ bootstrap sample chosen by random sampling *with replacement*). Describe the resulting bootstrap distributions of these two statistics.
7. Repeat Exercise 6 with the median and mean precipitation.
8. Bootstrap ($B = 200$) the regression of precipitation on elevation, using:
   a. residual resampling.
   b. data resampling.
   Graph and prepare a table presenting the results. Discuss your findings.
9. Reliable confidence intervals generally require $B = 2000$ or more bootstrap samples, but for practice we can use the $B = 200$ samples of Exercise 8. Find studentized slopes (Equation [A2.4]), and use [A2.5] to construct 90% confidence intervals for β_1, based on the bootstrap distributions of Exercise 8a and 8b. Calculate a theoretical ($b \pm tSE_b$) 90% confidence interval, and describe how it differs from your bootstrap-based confidence interval.

Notes

1. For a description of the pseudorandom generator used with examples in this appendix, see Hamilton (1990b).
2. Given normal errors, OLS is theoretically the most efficient of all unbiased estimators. As described in Chapter 6, the M-estimator was tuned to achieve about 95% of OLS's normal-errors efficiency; Table A2.1 suggests that this tuning worked as intended. Bounded-influence regression gives up further efficiency because it downweights the information from X-outliers.
3. For further Monte Carlo demonstrations of the theoretical properties of OLS, see Hanushek and Jackson (1977).
4. To improve estimation of the error variance, divide sample residuals by $\sqrt{1 - K/n}$. Then use

   $$e^*/\sqrt{1 - K/n}$$

   in place of e^* in steps 2–5.
5. This breakdown reveals a weakness in the robust estimator used. Bootstrapping can supplement Monte Carlo simulations as a tool for evaluating the statistical properties of estimators.
6. Peter Hall (1988:928) pointed out that "using the percentile method critical point amounts to looking up the wrong tables backward." Note 8 explains why the percentile-critical method is "backward." Percentile reversal (what Hall terms the hybrid method) overcomes backwardness but remains "looking up the wrong tables," like using a standard normal table instead of a t-table. In theory, the percentile-t method described later looks up the "right tables."
7. This distribution resembles that at upper right in Figure A2.7 but is based on 2000 instead of 1000 bootstrap samples. Doubling the number of samples brings little change in distribution shape but improves estimates of tail percentiles.

8. In the unknown sampling distribution, 90% of the b values fall between the 5th ($b_{.05}$) and 95th ($b_{.95}$) percentiles:

$$b_{.05} < b < b_{.95}$$

We can express this interval as a certain distance above and below the true parameter β:

$$\beta + (b_{.05} - \beta) < b < \beta + (b_{.95} - \beta) \quad \text{[A2.8a]}$$

To simplify notation, let $d_{.05} = b_{.05} - \beta$ represent the distance from β to low endpoint and $d_{.95} = b_{.95} - \beta$ the distance from β to high endpoint. Then [A2.8a] can be written:

$$\beta + d_{.05} < b < \beta + d_{.95} \quad \text{[A2.8b]}$$

Confidence intervals rearrange this inequality to isolate β:

$$\beta + d_{.05} < b \rightarrow \beta < b - d_{.05}$$

$$b < \beta + d_{.95} \rightarrow b - d_{.95} < \beta$$

so

$$b - d_{.95} < \beta < b - d_{.05} \quad \text{[A2.9]}$$

The apparent reversal of endpoints from [A2.8b] to [A2.9] results from putting β in the middle, rather than b. *If* the sampling distribution is symmetrical, so that $|d_{.05}| = d_{.95}$, this reversal makes no difference.

Confidence-interval endpoints such as $b - d_{.95}$ and $b - d_{.05}$ can be estimated in several ways, including the following:

a. from percentiles of the bootstrap distribution of b^* ([A2.3]);
b. from percentiles of the bootstrap-t distribution of t^* ([A2.5]), together with an estimated standard error; or
c. on the basis of theory.

For example, under the Central Limit Theorem, a $1 - \alpha$ confidence interval for the mean is

$$\bar{X} - Z_{1-\alpha}\sigma/\sqrt{n} < \mu < \bar{X} - Z_{\alpha}\sigma/\sqrt{n}$$

where Z_α denotes the 100α percentile of the standard normal distribution. Normal-distribution symmetry ($|Z_\alpha| = Z_{1-\alpha}$) makes the reversal unnoticeable in this example.

9. See Efron and Tibshirani (1986); Hall (1988).

10. Mosteller and Tukey (1977) caution, however, that the jackknife may not work well with outliers. The performance of various standard error estimators for percentile-t confidence intervals with ill-behaved data remains an interesting avenue for future research.
11. For reliable bootstrap confidence intervals, $B = 2000$ or 4000 bootstrap repetitions are generally recommended. This two-level experiment performed only $B = 1000$ repetitions per sample, to keep the computing time within reasonable bounds.

Appendix 3

Matrix Algebra

Matrix algebra permits a compact, general formulation of regression analysis. For example, a regression model with any number of X variables is described by the simple matrix equation $\mathbf{Y} = \mathbf{XB} + \mathbf{e}$ (explained below), and coefficient estimates are obtained by solving $\mathbf{B} = (\mathbf{X}'\mathbf{X})^{-1}\mathbf{X}'\mathbf{Y}$. Without matrix algebra, the estimating equations change with each added variable (for example, compare [A1.20] and [A1.22]).

Matrix calculations are difficult by hand but easily done by computers. This appendix introduces some basic ideas and applications.

Basic Ideas

A *matrix* is any rectangular array of numbers. For example:

$$\mathbf{A} = \begin{bmatrix} 3 & 11 \\ 19 & 11 \\ 77 & 80 \end{bmatrix}$$

The *dimensions* of matrix **A** are 3×2, meaning that it has three rows and two columns. The *element* in the first row, first column of **A** is $a_{11} = 3$. The element in the second row, first column is $a_{21} = 19$. Subscripts denote the row and column of each element, with row listed first (a_{rc}):

$$\mathbf{A} = \begin{bmatrix} a_{11} & a_{12} \\ a_{21} & a_{22} \\ a_{31} & a_{32} \end{bmatrix}$$

Two matrices are *equal* if they have the same dimensions, with the same elements at each row and column position.

The *transpose* of matrix **A** is written **A**′. Each row of **A** becomes a column of **A**′:

$$\mathbf{A} = \begin{bmatrix} 3 & 11 \\ 19 & 11 \\ 77 & 80 \end{bmatrix} \qquad \mathbf{A}' = \begin{bmatrix} 3 & 19 & 77 \\ 11 & 11 & 80 \end{bmatrix}$$

Since **A** has dimensions 3×2, **A**′ has dimensions 2×3.

A *square matrix* has the same number of rows and columns. **C** is a square 3×3 matrix:

$$\mathbf{C} = \begin{bmatrix} c_{11} & c_{12} & c_{13} \\ c_{21} & c_{22} & c_{23} \\ c_{31} & c_{32} & c_{33} \end{bmatrix}$$

The elements c_{11}, c_{22}, and c_{33} lie on the *major diagonal* of **C**.

A matrix with only one row, or with only one column, is called a *vector*. If **Y** is a *column vector*, then **Y**′ is a *row vector*:

$$\mathbf{Y} = \begin{bmatrix} 4 \\ 2 \\ 6 \\ 5 \\ 0 \end{bmatrix} \qquad \mathbf{Y}' = [4 \quad 2 \quad 6 \quad 5 \quad 0]$$

A *scalar* is an individual number, or a matrix with just one row and one column. Examples of scalars are:

$$a = 42 \qquad b = 5.11$$

Scalar algebra consists of the familiar methods for working with (adding, multiplying, rearranging, and so on) individual numbers rather than entire matrices of numbers.

Matrix Addition and Multiplication

We can add two matrices if they have the same dimensions. Corresponding elements are added together:

$$\mathbf{A}+\mathbf{E}=\begin{bmatrix}3 & 11\\ 19 & 11\\ 77 & 80\end{bmatrix}+\begin{bmatrix}1 & 2\\ 3 & 4\\ 5 & 6\end{bmatrix}$$

$$=\begin{bmatrix}3+1 & 11+2\\ 19+3 & 11+4\\ 77+5 & 80+6\end{bmatrix}$$

$$=\begin{bmatrix}4 & 13\\ 22 & 15\\ 82 & 86\end{bmatrix}$$

The sum has the same dimensions as the added matrices. As in scalar algebra, $\mathbf{A}+\mathbf{E}=\mathbf{E}+\mathbf{A}$. Subtraction is just negative addition.

Any matrix can be multiplied by a scalar. Multiplication of matrix **A** by scalar $s=2$ is performed as follows:

$$s\mathbf{A}=2\begin{bmatrix}3 & 11\\ 19 & 11\\ 77 & 80\end{bmatrix}$$

$$=\begin{bmatrix}2\times 3 & 2\times 11\\ 2\times 19 & 2\times 11\\ 2\times 77 & 2\times 80\end{bmatrix}$$

$$=\begin{bmatrix}6 & 22\\ 38 & 22\\ 154 & 160\end{bmatrix}$$

The product has the dimensions of the original matrix. As in scalar algebra, $s\mathbf{A}=\mathbf{A}s$. Division by scalar s is the same as multiplication by $1/s$.

Matrix multiplication less closely resembles its scalar counterpart. To multiply two matrices, **E** and **F**:

$$\mathbf{EF}=\begin{bmatrix}1 & 2\\ 3 & 4\\ 5 & 6\end{bmatrix}\begin{bmatrix}11 & 12 & 13\\ 14 & 15 & 16\end{bmatrix}$$

$$=\begin{bmatrix}(1\times 11)+(2\times 14) & (1\times 12)+(2\times 15) & (1\times 13)+(2\times 16)\\ (3\times 11)+(4\times 14) & (3\times 12)+(4\times 15) & (3\times 13)+(4\times 16)\\ (5\times 11)+(6\times 14) & (5\times 12)+(6\times 15) & (5\times 13)+(6\times 16)\end{bmatrix}$$

$$=\begin{bmatrix}39 & 42 & 45\\ 89 & 96 & 103\\ 139 & 150 & 161\end{bmatrix}$$

The first-row, first-column element of the product matrix equals the sum of each first-row element of **E** times the corresponding first-column element of **F**. More generally, each element of the product matrix equals the sum of row elements from the first matrix times column elements of the second matrix.

Note the dimensions involved:

$$\underset{3 \times 2}{\mathbf{E}} \quad \underset{2 \times 3}{\mathbf{F}} = \underset{3 \times 3}{\mathbf{G}}$$

Matrix multiplication is possible only if the number of columns in the first matrix equals the number of rows in the second (both are 2 above). We then say that their "inner dimensions" are equal. The resulting product matrix will have their "outer dimensions": the number of rows in the first matrix and the number of columns in the second (above, 3×3).

In general terms, the dimensions for matrix multiplication are

$$\underset{r \times q}{\mathbf{A}} \quad \underset{q \times c}{\mathbf{B}} = \underset{r \times c}{\mathbf{C}}$$

Unlike in scalar algebra, with matrices it is usually *not* true that $\mathbf{AB} = \mathbf{BA}$. Unless $c = r$, **BA** is not even possible, since it would involve unequal inner dimensions:

$$\underset{q \times c}{\mathbf{B}} \quad \underset{r \times q}{\mathbf{A}} = \{\text{impossible if } c \neq r\}$$

It may be possible to multiply a vector and a matrix; for example,

$$\underset{r \times c}{\mathbf{A}} \quad \underset{c \times 1}{\mathbf{B}} = \underset{r \times 1}{\mathbf{C}}$$

A row and a column vector can also be multiplied:

$$\underset{r \times 1}{\mathbf{A}} \quad \underset{1 \times c}{\mathbf{B}} = \underset{r \times c}{\mathbf{C}}$$

or

$$\underset{1 \times c}{\mathbf{A}} \quad \underset{c \times 1}{\mathbf{B}} = \underset{1 \times 1}{\mathbf{C}}$$

Multiplying a row vector times a column vector (last example) results in a 1×1 matrix or scalar.

An *identity matrix* is a square matrix with 1's on the major diagonal and 0's elsewhere. For example, a 3 × 3 identity matrix, denoted $\mathbf{I}_3$, is

$$\mathbf{I}_3 = \begin{bmatrix} 1 & 0 & 0 \\ 0 & 1 & 0 \\ 0 & 0 & 1 \end{bmatrix}$$

In matrix multiplication, identity matrices behave somewhat as 1 does in scalar multiplication. That is, any matrix **A** will be unchanged if multiplied by the appropriate identity matrix. If **A** is $r \times c$, it will be unchanged when either:

1. premultiplied by an $\mathbf{I}_r$ matrix: $\underset{r \times r}{\mathbf{I}} \quad \underset{r \times c}{\mathbf{A}} = \underset{r \times c}{\mathbf{A}}$
2. postmultiplied by an $\mathbf{I}_c$ matrix: $\underset{r \times c}{\mathbf{A}} \quad \underset{c \times c}{\mathbf{I}} = \underset{r \times c}{\mathbf{A}}$

Matrix Inversion

There is no such operation as "matrix division." Multiplication by an inverse matrix takes its place. The inverse of square $r \times r$ matrix **C**, written $\mathbf{C}^{-1}$, is an $r \times r$ matrix such that

$$\mathbf{CC}^{-1} = \mathbf{C}^{-1}\mathbf{C} = \mathbf{I}_r$$

For example, the inverse of

$$\mathbf{C} = \begin{bmatrix} 10 & 4 \\ 3 & 11 \end{bmatrix}$$

is

$$\mathbf{C}^{-1} = \begin{bmatrix} .112245 & -.04082 \\ -.03061 & .102041 \end{bmatrix}$$

because

$$\mathbf{CC}^{-1} = \mathbf{I}_2$$

$$\begin{bmatrix} 10 & 4 \\ 3 & 11 \end{bmatrix} \begin{bmatrix} .112245 & -.04082 \\ -.03061 & .102041 \end{bmatrix} = \begin{bmatrix} 1 & 0 \\ 0 & 1 \end{bmatrix}$$

Inversion is defined only for square matrices. Computers readily perform the necessary calculations.[1]

Not all square matrices can be inverted. If any row or column in a matrix is a linear function of other rows or columns, the matrix is termed *singular*, and no inverse exists. A matrix for which an inverse does exist is called *nonsingular*.

Regression in Matrix Form

A regression with n cases and $K - 1$ X variables involves the following matrices:

$\mathbf{Y}$ ($n \times 1$), a column vector of observed Y values:

$$\mathbf{Y} = \begin{bmatrix} Y_1 \\ Y_2 \\ Y_3 \\ \vdots \\ Y_n \end{bmatrix}$$

$\hat{\mathbf{Y}}$ ($n \times 1$), a column vector of predicted Y values:

$$\hat{\mathbf{Y}} = \begin{bmatrix} \hat{Y}_1 \\ \hat{Y}_2 \\ \hat{Y}_3 \\ \vdots \\ \hat{Y}_n \end{bmatrix}$$

$\mathbf{X}$ ($n \times K$), a matrix with observed X values in columns 2–K. Generally, the first column of $\mathbf{X}$ consists of 1's:

$$\mathbf{X} = \begin{bmatrix} 1 & X_{11} & X_{12} & X_{13} & \cdots & X_{1K-1} \\ 1 & X_{21} & X_{22} & X_{23} & \cdots & X_{2K-1} \\ 1 & X_{31} & X_{32} & X_{33} & \cdots & X_{3K-1} \\ \vdots & \vdots & \vdots & \vdots & \vdots & \vdots \\ 1 & X_{n1} & X_{n2} & X_{n3} & \cdots & X_{nK-1} \end{bmatrix}$$

$\mathbf{B}$ ($K \times 1$), a column vector of sample regression coefficients; the first element is the Y-intercept b_0:

$$\mathbf{B} = \begin{bmatrix} b_0 \\ b_1 \\ b_2 \\ b_3 \\ \vdots \\ b_{K-1} \end{bmatrix}$$

$\mathbf{e}$ ($n \times 1$), a column vector of sample residuals for each of the n cases:

$$\mathbf{e} = \begin{bmatrix} e_1 \\ e_2 \\ e_3 \\ \vdots \\ e_n \end{bmatrix}$$

The regression model is

$$\mathbf{Y} = \mathbf{XB} + \mathbf{e} \quad \text{[A3.1]}$$

so

$$\hat{\mathbf{Y}} = \mathbf{XB} \quad \text{[A3.2]}$$

We estimate ordinary least squares coefficients by solving

$$\mathbf{B} = (\mathbf{X}'\mathbf{X})^{-1}\mathbf{X}'\mathbf{Y} \quad \text{[A3.3]}$$

An inverse of the square matrix $(\mathbf{X}'\mathbf{X})$ exists only if none of the columns in $\mathbf{X}$ is a linear function of other columns—that is, only if we do not have perfect multicollinearity. If we did have perfect multicollinearity, there would be no $(\mathbf{X}'\mathbf{X})^{-1}$ and we could not uniquely define $\mathbf{B}$, the vector of sample regression coefficients.

Combining [A3.2] and [A3.3], we have

$$\begin{aligned}\hat{\mathbf{Y}} &= \mathbf{X}(\mathbf{X}'\mathbf{X})^{-1}\mathbf{X}'\mathbf{Y} \\ &= \mathbf{HY} \quad \text{[A3.4]}\end{aligned}$$

where $\mathbf{H} = \mathbf{X}(\mathbf{X}'\mathbf{X})^{-1}\mathbf{X}'$ is the *hat matrix* $(n \times n)$, which "puts the hat" on Y. The *leverage* of case i, h_i, equals the ith diagonal element of $\mathbf{H}$.

The estimated variance-covariance matrix of coefficients is

$$\mathbf{S} = s_e^2(\mathbf{X}'\mathbf{X})^{-1} \quad \text{[A3.5]}$$

where s_e^2 is the residual variance, $s_e^2 = \text{RSS}/(n - K)$. Estimated standard errors of the regression cofficients (SE_{b_k}) equal square roots of elements on the major diagonal of $\mathbf{S}$. Off-diagonal elements are estimated covariances between estimated coefficients.

Correlations between estimated coefficients, also found from $\mathbf{S}$, provide a check for multicollinearity. The correlation between any two coefficients, b_j and b_k, is

$$r_{b_j b_k} = \frac{s_{jk}}{\sqrt{s_{jj}}\sqrt{s_{kk}}} \quad \text{[A3.6]}$$

where s_{jk} denotes the (j, k) element of variance-covariance matrix $\mathbf{S}$. That is, s_{jj} equals the estimated variance of coefficient b_j, $s_{jj} = \text{SE}_{b_j}^2$; s_{kk} equals the estimated variance of coefficient b_k; and s_{jk} equals the covariance between these two coefficients.

An Example

Following are data on estimated decommissioning costs of five nuclear power plants (from Brown, Chandler, Flavin, Pollack, Postel, Starke, and Wolf, 1986):

Site	Y: Estimated Decommissioning Costs (million $)	X_1 Capacity (megawatts)	X_2 Years in Operation
1 Elk River	14	24	6
2 Windscale	64	33	18
3 Humboldt Bay 3	55	65	13
4 Shippingport	98	72	25
5 Dresden 1	95	210	18

Regressing decommissioning costs on capacity and years in operation yields

$$Y_i = -11.4 + 0.176X_{i1} + 3.9X_{i2} + e_i$$

In matrix form, $\mathbf{Y} = \mathbf{XB} + \mathbf{e}$:

$$\underset{\mathbf{Y}}{\begin{bmatrix} 14 \\ 64 \\ 55 \\ 98 \\ 95 \end{bmatrix}} = \underset{\mathbf{X}}{\begin{bmatrix} 1 & 24 & 6 \\ 1 & 33 & 18 \\ 1 & 65 & 13 \\ 1 & 72 & 25 \\ 1 & 210 & 18 \end{bmatrix}} \underset{\mathbf{B}}{\begin{bmatrix} -11.4 \\ 0.176 \\ 3.9 \end{bmatrix}} + \underset{\mathbf{e}}{\begin{bmatrix} -2.22 \\ -0.61 \\ 4.26 \\ -0.77 \\ -0.76 \end{bmatrix}}$$

Note the column of 1's in **X**, so, with $K-1$ X variables, its dimensions are $n \times K$. The equation for predicted Y is $\hat{\mathbf{Y}} = \mathbf{XB}$:

$$\underset{\hat{\mathbf{Y}}}{\begin{bmatrix} 16.224 \\ 64.608 \\ 50.74 \\ 98.772 \\ 95.76 \end{bmatrix}} = \underset{\mathbf{X}}{\begin{bmatrix} 1 & 24 & 6 \\ 1 & 33 & 18 \\ 1 & 65 & 13 \\ 1 & 72 & 25 \\ 1 & 210 & 18 \end{bmatrix}} \underset{\mathbf{B}}{\begin{bmatrix} -11.4 \\ 0.176 \\ 3.9 \end{bmatrix}}$$

These matrices are simple enough that we could confirm the arithmetic by hand. For example, by the definition of matrix multiplication, the first element of $\hat{\mathbf{Y}}$ is

$$(1 \times -11.4) + (24 \times 0.176) + (6 \times 3.9) = 16.224$$

The $K \times K$ matrix $\mathbf{X'X}$ can also be found by hand:

$$\underset{\mathbf{X'}}{\begin{bmatrix} 1 & 1 & 1 & 1 & 1 \\ 24 & 33 & 65 & 72 & 210 \\ 6 & 18 & 13 & 25 & 18 \end{bmatrix}} \underset{\mathbf{X}}{\begin{bmatrix} 1 & 24 & 6 \\ 1 & 33 & 18 \\ 1 & 65 & 13 \\ 1 & 72 & 25 \\ 1 & 210 & 18 \end{bmatrix}} = \underset{\mathbf{X'X}}{\begin{bmatrix} 5 & 404 & 80 \\ 404 & 55174 & 7163 \\ 80 & 7163 & 1478 \end{bmatrix}}$$

Elements of $\mathbf{X'X}$ are

$$\begin{bmatrix} n & \Sigma X_1 & \Sigma X_2 \\ \Sigma X_1 & \Sigma X_1^2 & \Sigma X_1 X_2 \\ \Sigma X_2 & \Sigma X_1 X_2 & \Sigma X_2^2 \end{bmatrix}$$

Means, variances, and covariances of the X variables are easily obtained from $\mathbf{X'X}$, called a *cross-product matrix*. We multiply the first row or first column by $1/n$ to find means. To get a matrix of *deviation scores*, $\mathbf{D}$:

$$\mathbf{D} = \mathbf{X} - \mathbf{NX'X} \qquad \text{[A3.7]}$$

where $\mathbf{N}$ is an $n \times K$ matrix with $1/n$ in the first column and zeros elsewhere; for this example

$$\mathbf{N} = \begin{bmatrix} 1/5 & 0 & 0 \\ 1/5 & 0 & 0 \\ 1/5 & 0 & 0 \\ 1/5 & 0 & 0 \\ 1/5 & 0 & 0 \end{bmatrix}$$

and

$$\mathbf{D} = \begin{bmatrix} 0 & X_1 - \bar{X}_1 & X_2 - \bar{X}_2 \\ 0 & X_1 - \bar{X}_1 & X_2 - \bar{X}_2 \\ 0 & X_1 - \bar{X}_1 & X_2 - \bar{X}_2 \\ 0 & X_1 - \bar{X}_1 & X_2 - \bar{X}_2 \\ 0 & X_1 - \bar{X}_1 & X_2 - \bar{X}_2 \end{bmatrix}$$

$$= \begin{bmatrix} 0 & -56.8 & -10 \\ 0 & -47.8 & 2 \\ 0 & -15.8 & -3 \\ 0 & -8.8 & 9 \\ 0 & 129.2 & 2 \end{bmatrix}$$

Elements of **D** are deviations of each X value from its mean. From **D** we calculate a variance-covariance matrix, **C**:

$$\mathbf{C} = \left[\frac{1}{n-1}\right]\mathbf{D}'\mathbf{D} \qquad \text{[A3.8]}$$

C contains variances and covariances of the *variables*, in contrast to **S** ([A3.5]), which contains variances and covariances of *coefficient estimates*. With two variables, X_1 and X_2, elements of **C** are

$$\mathbf{C} = \begin{bmatrix} 0 & 0 & 0 \\ 0 & s_1^2 & s_{12} \\ 0 & s_{12} & s_2^2 \end{bmatrix}$$

The correlation between X_1 and X_2, r_{12}, is

$$r_{12} = \frac{s_{12}}{\sqrt{s_1^2}\sqrt{s_2^2}} \qquad \text{[A3.9]}$$

Further steps require matrix inversion, which is difficult to perform by hand. Equation [A3.3] finds the vector of estimated coefficients, and [A3.5] estimates their standard errors.

Regression from Correlation Matrices

A correlation matrix plus a table of means and standard deviations contain enough information for basic regression calculations.[2] These calculations permit reanalysis of some published research.

A vector of standardized regression coefficients ("beta weights"), **B***, follows from the matrix of correlations among X variables, $\mathbf{R}_X$, and the vector of correlations between Y and the X's, $\mathbf{R}_Y$:

$$\mathbf{B}^* = \mathbf{R}_X^{-1}\mathbf{R}_Y \qquad \text{[A3.10]}$$

$$\begin{bmatrix} b_1^* \\ b_2^* \\ b_2^* \\ \cdot \\ \cdot \\ \cdot \\ b_{K-1}^* \end{bmatrix} = \begin{bmatrix} 1 & r_{12} & r_{13} & \cdots & r_{1,K-1} \\ r_{21} & 1 & r_{23} & \cdots & \cdot \\ r_{31} & r_{32} & 1 & \cdots & \cdot \\ \cdot & \cdot & \cdot & \cdots & \cdot \\ \cdot & \cdot & \cdot & \cdots & \cdot \\ \cdot & \cdot & \cdot & \cdots & \cdot \\ r_{K-1,1} & \cdot & \cdot & \cdots & 1 \end{bmatrix}^{-1} \begin{bmatrix} r_{1Y} \\ r_{2Y} \\ r_{3Y} \\ \cdot \\ \cdot \\ \cdot \\ r_{K-1,Y} \end{bmatrix}$$

where r_{12} is the correlation betwen X_1 and X_2, and so on.

We calculate the unstandardized regression coefficient on X_k, b_k, from the standardized coefficient b_k^*:

$$b_k = b_k^*\left(\frac{s_Y}{s_k}\right) \qquad [A3.11]$$

where s_Y is the standard deviation of Y and s_k is the standard deviation of X_k. The Y-intercept b_0 is

$$b_0 = \bar{Y} - \sum_{k=1}^{K-1} b_k \bar{X}_k \qquad [A3.12]$$

R is the $K \times K$ matrix of correlations among all variables, Y and X's:

$$\mathbf{R} = \begin{bmatrix} 1 & r_{Y1} & r_{Y2} & r_{Y3} & \cdots & r_{Y,K-1} \\ r_{1Y} & 1 & r_{12} & r_{13} & \cdots & r_{1,K-1} \\ r_{2Y} & r_{21} & 1 & r_{23} & \cdots & r_{2,K-1} \\ \cdot & \cdot & \cdot & \cdot & \cdots & \cdot \\ \cdot & \cdot & \cdot & \cdot & \cdots & \cdot \\ \cdot & \cdot & \cdot & \cdot & \cdots & \cdot \\ \cdot & \cdot & \cdot & \cdot & \cdots & \cdot \\ r_{K-1,Y} & \cdot & \cdot & \cdot & \cdots & 1 \end{bmatrix}$$

The inverse of **R**, $\mathbf{R}^{-1}$, is likewise $K \times K$. Let v_Y represent the diagonal element of $\mathbf{R}^{-1}$ corresponding to Y. The coefficient of determination R^2 (Y regressed on all X's) is

$$R^2 = 1 - \left(\frac{1}{v_Y}\right) \qquad [A3.13]$$

The total sum of squares for Y can be found from Y's standard deviation:

$$\text{TSS}_Y = (n-1)s_Y^2 \qquad [A3.14]$$

The explained sum of squares (ESS) and residual sum of squares (RSS) are as follows:

$$\text{ESS} = \text{TSS}_Y(R^2) \qquad [A3.15]$$

$$\text{RSS} = \text{TSS}_Y(1 - R^2) = \text{TSS}_Y - \text{ESS} \qquad [A3.16]$$

These sums of squares lead to an F-statistic (Equation [3.29]) and the standard deviation of the residuals, e_e (Equation [2.11]).

Let w_k represent the kth diagonal element of $\mathbf{R}_X^{-1}$. The coefficient of determination R_k^2 (X_k regressed on all other X variables) is

$$R_k^2 = 1 - \left(\frac{1}{w_k}\right) \qquad \text{[A3.17]}$$

x_k's standard deviation s_k implies the total sum of squares for X_k:

$$\text{TSS}_k = (n - 1)s_k^2 \qquad \text{[A3.18]}$$

From these quantities we estimate the standard error of b_k:

$$\text{SE}_{b_k} = \frac{s_e}{\sqrt{\text{TSS}_k(1 - R_k^2)}} \qquad \text{[A3.19]}$$

Alternatively, $\text{SE}_{b_k} = s_e\sqrt{c_k}$, where c_k is the kth diagonal element of the inverse of the variance-covariance matrix of X variables.

Thorough diagnostic work, as described in Chapter 4, is impossible without the raw data. Nonetheless, we may be able to carry an analysis further than the original authors, based only on their published correlations and summary statistics.[3]

Some statistical programs can calculate regression directly from correlation or covariance matrices, which saves much work. Otherwise, the calculations above can be accomplished with a programming language or spreadsheet that performs matrix multiplication and inversion.

Further Definitions

This section offers brief definitions of some additional matrix concepts often encountered in statistics, and particularly in Chapter 8 of this book. For more detail consult a matrix-oriented statistics text such as Everitt and Dunn (1983).

The *rank* of a matrix is its maximum number of linearly independent (not multicollinear) columns. If all columns are dependent, the matrix is of *full rank.*

A *diagonal matrix* is square, with nonzero elements only on the major diagonal. For example:

$$\begin{bmatrix} 4 & 0 & 0 \\ 0 & 26 & 0 \\ 0 & 0 & 50 \end{bmatrix}$$

An *orthogonal matrix* yields an identity matrix when multiplied by its own transpose. $\mathbf{A}$ is orthogonal if

$$\mathbf{AA}' = \mathbf{I}$$

implying that:

$$\mathbf{A}' = \mathbf{A}^{-1}$$

The *trace* of a square matrix is the sum of elements on the major diagonal. If

$$\mathbf{G} = \begin{bmatrix} g_{11} & g_{12} & g_{13} \\ g_{21} & g_{22} & g_{23} \\ g_{31} & g_{32} & g_{33} \end{bmatrix}$$

then the trace of **G** is

$$\text{trace}(\mathbf{G}) = g_{11} + g_{22} + g_{33}$$

The scalar λ_j is the jth *eigenvalue*, and the vector a_j the jth *eigenvector*, of a square matrix **G** if

$$\underset{J \times J}{\mathbf{G}}\underset{J \times 1}{\mathbf{a}_j} = \underset{1 \times 1}{\lambda_j}\underset{J \times 1}{\mathbf{a}_j}$$

A $J \times J$ matrix has J eigenvectors and J associated eigenvalues (although some of these eigenvalues may be duplicates or zero).

Exercises

1. Transpose

$$\begin{bmatrix} 1 & 2 & 3 \\ 4 & 5 & 6 \\ 7 & 8 & 9 \\ 10 & 11 & 12 \end{bmatrix}$$

2. Multiply

$$\begin{bmatrix} 2 & 3 & 4 \\ 5 & 6 & 7 \end{bmatrix} \begin{bmatrix} 10 & 11 \\ 20 & 21 \\ 30 & 31 \end{bmatrix}$$

3. Multiply

$$\begin{bmatrix} 2 & 3 & 4 \\ 5 & 6 & 7 \end{bmatrix} \begin{bmatrix} 1 & 0 & 0 \\ 0 & 1 & 0 \\ 0 & 0 & 1 \end{bmatrix}$$

4. If **A** and **B** are both 8×3 matrices, which of the following operations are possible? (Assume that singularity is not a problem.) If possible, what are the dimensions of the result?
   a. **AB**
   b. **BA**
   c. **AB′**
   d. $\mathbf{A}^{-1}$
   e. $(\mathbf{A}'\mathbf{A})^{-1}$

5. Use a spreadsheet or programming language capable of matrix algebra to find the following for the reactor decommissioning data:
   a. The matrix of regression coefficients, **B** (use [A3.3]).
   b. The hat matrix, **H**, and leverage for each case, h_i (use [A3.4]).
   c. The variance-covariance matrix of coefficients, **S** (use [A3.5]).
   d. Carry out *t*-tests on each coefficient. Do capacity and years in operation have significant effects on reactor decommissioning costs?
6. A study (Zupan, 1973) collected data on air pollution and demographic characteristics in 21 New York counties. Variables included:

   nox—nitrous-oxide emissions, in metric tons per square mile
   high—percentage of households with high incomes
   density—people per square mile

   The following is a correlation matrix, with means and standard deviations, for these three variables.

	Nox	*High*	*Density*
nox	1.0000		
high	−0.6792	1.0000	
density	0.8769	−0.7064	1.000
mean	0.1249	27.2857	9660.838
s.d.	0.1030	6.8348	15921.45

   Notice the negative correlation between *high* and *nox*: wealthy counties had less air pollution. Is this correlation spurious, a consequence of both variables' relation with *density*? Find out by regressing *nox* on *high* and *density*.
   a. Obtain the vector of standardized regression coefficients, **B*** (Equation [A3.10]).
   b. Write out the *un*standardized regression equation (coefficients from [A3.11] and [A3.12]).
   c. Find the coefficient of determination (Equation [A3.13]).
   d. Use a *t*-test to decide whether, after adjusting for *density*, *high* remains a significant ($\alpha < .05$) predictor of *nox*. Is *density*'s effect significant? (Standard errors from [A3.19]; *t*-statistics from [3.23].)

Notes

1. Simple algebra obtains the inverse of a 2 × 2 matrix. For example, the inverse of **C**, $\mathbf{C}^{-1}$, must have elements x, y, z, and w such that

$$\begin{bmatrix} 10 & 4 \\ 3 & 11 \end{bmatrix} \begin{bmatrix} x & y \\ z & w \end{bmatrix} = \begin{bmatrix} 1 & 0 \\ 0 & 1 \end{bmatrix}$$

By the definition of matrix multiplication this implies:

$$10z + 4x = 1 \qquad 10y + 4w = 0$$

$$3x + 11z = 0 \qquad 3y + 11w = 1$$

We find the two unknowns x and z by solving the two left-hand equations:

$$3x + 11z = 0$$
$$11z = -3x$$
$$z = (-3/11)x$$

$$10x + 4z = 1$$
$$10x + 4(-3/11)x = 1$$
$$(10 - 12/11)x = 1$$
$$x = 1/(10 - 12/11)$$
$$x = .112244898$$

$$z = (-3/11)x$$
$$z = (-3/11).112244898$$
$$= -.03061224$$

Similar work obtains y and w. Rounding off, we arrive at

$$\mathbf{C}^{-1} = \begin{bmatrix} .112245 & -.04082 \\ -.03061 & .10204 \end{bmatrix}$$

With enough effort, we can also invert larger matrices by this method. Matrix-oriented texts describe more efficient, though less intuitive, methods of matrix inversion.

2. Calculations are more direct if we have the cross-product matrix **(X′X)** or variance-covariance matrix (Equation [A3.8]), rather than just the correlation matrix.
3. The accuracy of this approach depends heavily on the accuracy of the published statistics. Since these are usually rounded off, our results will be just approximations. Surprisingly often, calculations turn out to be impossible due to errors in the published tables.

4 Appendix

Statistical Tables

Table A4.1 Critical values for student's t-distribution

Probability $P(|t| \leq c)$ for confidence intervals

Probability $P(|t| \geq c)$ for two-sided tests

Probability $P(t \geq c)$ for one-sided tests

Probability

	.50	.80	.90	.95	.98	.99	.995	.998	.999	Confidence Intervals
	.50	.20	.10	.05	.02	.01	.005	.002	.001	Two-Sided Tests
df	.25	.10	.05	.025	.01	.005	.0025	.001	.0005	One-Sided Tests
1	1.000	3.078	6.314	12.706	31.821	63.637	127.32	318.31	636.62	
2	.816	1.886	2.920	4.303	6.965	9.925	14.089	22.326	31.598	
3	.765	1.638	2.353	3.182	4.541	5.841	7.453	10.213	12.924	
4	.741	1.533	2.132	2.776	3.747	4.604	5.598	7.173	8.610	
5	.727	1.476	2.015	2.571	3.365	4.032	4.773	5.893	6.869	
6	.718	1.440	1.943	2.447	3.143	3.707	4.317	5.208	5.959	
7	.711	1.415	1.895	2.365	2.998	3.499	4.020	4.785	5.408	
8	.706	1.397	1.860	2.306	2.896	3.355	3.833	4.501	5.041	
9	.703	1.383	1.833	2.262	2.821	3.250	3.690	4.297	4.781	
10	.700	1.372	1.812	2.228	2.764	3.169	3.581	4.144	4.537	
11	.697	1.363	1.796	2.201	2.718	3.106	3.497	4.025	4.437	
12	.695	1.356	1.782	2.179	2.681	3.055	3.428	3.930	4.318	
13	.694	1.350	1.771	2.160	2.650	3.012	3.372	3.852	4.221	
14	.692	1.345	1.761	2.145	2.624	2.977	3.326	3.787	4.140	
15	.691	1.341	1.753	2.131	2.602	2.947	3.286	3.733	4.073	
16	.690	1.337	1.746	2.120	2.583	2.921	3.252	3.686	4.015	
17	.689	1.333	1.740	2.110	2.567	2.898	3.222	3.646	3.965	
18	.688	1.330	1.734	2.101	2.552	2.878	3.197	3.610	3.922	
19	.688	1.328	1.729	2.093	2.539	2.861	3.174	3.579	3.883	
20	.687	1.325	1.725	2.086	2.528	2.845	3.153	3.552	3.850	
21	.686	1.323	1.721	2.080	2.518	2.831	3.135	3.527	3.819	
22	.686	1.321	1.717	2.074	2.508	2.819	3.119	3.505	3.792	
23	.685	1.319	1.714	2.069	2.500	2.807	3.104	3.485	3.767	
24	.685	1.318	1.711	2.064	2.492	2.797	3.091	3.467	3.745	
25	.684	1.316	1.708	2.060	2.485	2.787	3.078	3.450	3.725	
26	.684	1.315	1.706	2.056	2.479	2.779	3.067	3.435	3.707	
27	.684	1.314	1.703	2.052	2.473	2.771	3.057	3.421	3.690	
28	.683	1.313	1.701	2.048	2.467	2.763	3.047	3.408	3.674	
29	.683	1.311	1.699	2.045	2.462	2.756	3.038	3.396	3.659	
30	.683	1.310	1.697	2.042	2.457	2.750	3.030	3.385	3.646	
40	.681	1.303	1.684	2.021	2.423	2.704	2.971	3.307	3.551	
60	.679	1.296	1.671	2.000	2.390	2.660	2.915	3.232	3.460	
120	.677	1.289	1.658	1.980	2.358	2.617	2.860	3.160	3.373	
∞	.674	1.282	1.645	1.960	2.326	2.576	2.807	3.090	3.291	

Source: Abridged from Table 12 of *Biometrika Tables for Statisticians*, Vol. 1, edited by E. S. Pearson and H. O. Hartley (London: Cambridge University Press, 1962).

Table A4.2 Critical values for the F-distribution

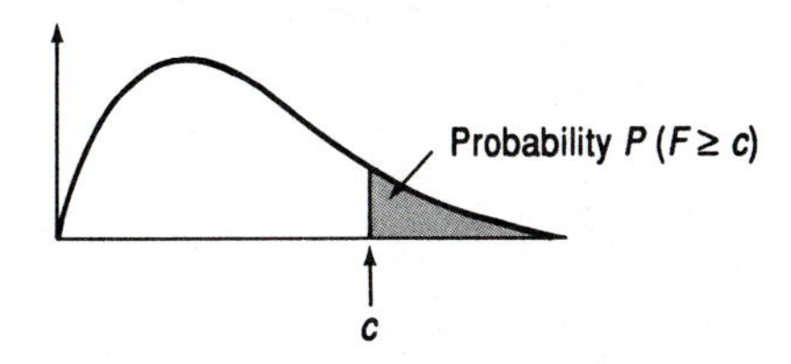

Denominator Degrees of Freedom, df_2	P	Numerator Degrees of Freedom, df_1 1	2	3	4	5	6	8	10	20	40	∞
1	.25	5.83	7.50	8.21	8.58	8.82	8.98	9.19	9.32	9.58	9.71	9.85
	.10	39.86	49.50	53.59	55.83	57.24	58.20	59.44	60.19	61.74	62.53	63.33
	.05	161.4	199.5	215.7	224.6	230.2	234.0	238.9	241.9	248.0	251.1	254.3
2	.25	2.57	3.00	3.15	3.23	3.28	3.31	3.35	3.38	3.43	3.45	3.48
	.10	8.53	9.00	9.16	9.24	9.29	9.33	9.37	9.39	9.44	9.47	9.49
	.05	18.51	19.00	19.16	19.25	19.30	19.33	19.37	19.40	19.45	19.47	19.50
	.01	98.50	99.00	99.17	99.25	99.30	99.33	99.37	99.40	99.45	99.47	99.50
	.001	998.5	999.0	999.2	999.2	999.3	999.3	999.4	999.4	999.4	999.5	999.5
3	.25	2.02	2.28	2.36	2.39	2.41	2.42	2.44	2.44	2.46	2.47	2.47
	.10	5.54	5.46	5.39	5.34	5.31	5.28	5.25	5.23	5.18	5.16	5.13
	.05	10.13	9.55	9.28	9.12	9.01	8.94	8.85	8.79	8.66	8.59	8.53
	.01	34.12	30.82	29.46	28.71	28.24	27.91	27.49	27.23	26.69	26.41	26.13
	.001	167.0	148.5	141.1	137.1	134.6	132.8	130.6	129.2	126.4	125.0	123.5
4	.25	1.81	2.00	2.05	2.06	2.07	2.08	2.08	2.08	2.08	2.08	2.08
	.10	4.54	4.32	4.19	4.11	4.05	4.01	3.95	3.92	3.84	3.80	3.76
	.05	7.71	6.94	6.59	6.39	6.26	6.16	6.04	5.96	5.80	5.72	5.63
	.01	21.20	18.00	16.69	15.98	15.52	15.21	14.80	14.55	14.02	13.75	13.46
	.001	74.14	61.25	56.18	53.44	51.71	50.53	49.00	48.05	46.10	45.09	44.05
5	.25	1.69	1.85	1.88	1.89	1.89	1.89	1.89	1.89	1.88	1.88	1.87
	.10	4.06	3.78	3.62	3.52	3.45	3.40	3.34	3.30	3.21	3.16	3.10
	.05	6.61	5.79	5.41	5.19	5.05	4.95	4.82	4.74	4.56	4.46	4.36
	.01	16.26	13.27	12.06	11.39	10.97	10.67	10.29	10.05	9.55	9.29	9.02
	.001	47.18	37.12	33.20	31.09	29.75	28.84	27.64	26.92	25.39	24.60	23.79
6	.25	1.62	1.76	1.78	1.79	1.79	1.78	1.78	1.77	1.76	1.75	1.74
	.10	3.78	3.46	3.29	3.18	3.11	3.05	2.98	2.94	2.84	2.78	2.72
	.05	5.99	5.14	4.76	4.53	4.39	4.28	4.15	4.06	3.87	3.77	3.67
	.01	13.75	10.92	9.78	9.15	8.75	8.47	8.10	7.87	7.40	7.14	6.88
	.001	35.51	27.00	23.70	21.92	20.81	20.03	19.03	18.41	17.12	16.44	15.75
7	.25	1.57	1.70	1.72	1.72	1.71	1.71	1.70	1.69	1.67	1.66	1.65
	.10	3.59	3.26	3.07	2.96	2.88	2.83	2.75	2.70	2.59	2.54	2.47
	.05	5.59	4.74	4.35	4.12	3.97	3.87	3.73	3.64	3.44	3.34	3.23
	.01	12.25	9.55	8.45	7.85	7.46	7.19	6.84	6.62	6.16	5.91	5.65
	.001	29.25	21.69	18.77	17.19	16.21	15.52	14.63	14.08	12.93	12.33	11.70

(continued)

Table A4.2 continued

Denominator Degrees of Freedom, df_2	*P*	Numerator Degrees of Freedom, df_1										
		1	**2**	**3**	**4**	**5**	**6**	**8**	**10**	**20**	**40**	**∞**
8	*.25*	1.54	1.66	1.67	1.66	1.66	1.65	1.64	1.63	1.61	1.59	1.58
	.10	3.46	3.11	2.92	2.81	2.73	2.67	2.59	2.54	2.42	2.36	2.29
	.05	5.32	4.46	4.07	3.84	3.69	3.58	3.44	3.35	3.15	3.04	2.93
	.01	11.26	8.65	7.59	7.01	6.63	6.37	6.03	5.81	5.36	5.12	4.86
	.001	25.42	18.49	15.83	14.39	13.49	12.86	12.04	11.54	10.48	9.92	9.33
9	*.25*	1.51	1.62	1.63	1.63	1.62	1.61	1.60	1.59	1.56	1.54	1.53
	.10	3.36	3.01	2.81	2.69	2.61	2.55	2.47	2.42	2.30	2.23	2.16
	.05	5.12	4.26	3.86	3.63	3.48	3.37	3.23	3.14	2.94	2.83	2.71
	.01	10.56	8.02	6.99	6.42	6.06	5.80	5.47	5.26	4.81	4.57	4.31
	.001	22.86	16.39	13.90	12.56	11.71	11.13	10.37	9.89	8.90	8.37	7.81
10	*.25*	1.49	1.60	1.60	1.59	1.59	1.58	1.56	1.55	1.52	1.51	1.48
	.10	3.28	2.92	2.73	2.61	2.52	2.46	2.38	2.32	2.20	2.13	2.06
	.05	4.96	4.10	3.71	3.48	3.33	3.22	3.07	2.98	2.77	2.66	2.54
	.01	10.04	7.56	6.55	5.99	5.64	5.39	5.06	4.85	4.41	4.17	3.91
	.001	21.04	14.91	12.55	11.28	10.48	9.92	9.20	8.75	7.80	7.30	6.76
12	*.25*	1.46	1.56	1.56	1.55	1.54	1.53	1.51	1.50	1.47	1.45	1.42
	.10	3.18	2.81	2.61	2.48	2.39	2.33	2.24	2.19	2.06	1.99	1.90
	.05	4.75	3.89	3.49	3.26	3.11	3.00	2.85	2.75	2.54	2.43	2.30
	.01	9.33	6.93	5.95	5.41	5.06	4.82	4.50	4.30	3.86	3.62	3.36
	.001	18.64	12.97	10.80	9.63	8.89	8.38	7.71	7.29	6.40	5.93	5.42
14	*.25*	1.44	1.53	1.53	1.52	1.51	1.50	1.48	1.46	1.43	1.41	1.38
	.10	3.10	2.73	2.52	2.39	2.31	2.24	2.15	2.10	1.96	1.89	1.80
	.05	4.60	3.74	3.34	3.11	2.96	2.85	2.70	2.60	2.39	2.27	2.13
	.01	8.86	5.51	5.56	5.04	4.69	4.46	4.14	3.94	3.51	3.27	3.00
	.001	17.14	11.78	9.73	8.62	7.92	7.43	6.80	6.40	5.56	5.10	4.60
16	*.25*	1.42	1.51	1.51	1.50	1.48	1.48	1.46	1.45	1.40	1.37	1.34
	.10	3.05	2.67	2.46	2.33	2.24	2.18	2.09	2.03	1.89	1.81	1.72
	.05	4.49	3.63	3.24	3.01	2.85	2.74	2.59	2.49	2.28	2.15	2.01
	.01	8.53	6.23	5.29	4.77	4.44	4.20	3.89	3.69	3.26	3.02	2.75
	.001	16.12	10.97	9.00	7.94	7.27	6.81	6.19	5.81	4.99	4.54	4.06
18	*.25*	1.41	1.50	1.49	1.48	1.46	1.45	1.43	1.42	1.38	1.35	1.32
	.10	3.01	2.62	2.42	2.29	2.20	2.13	2.04	1.98	1.84	1.75	1.66
	.05	4.41	3.55	3.16	2.93	2.77	2.66	2.51	2.41	2.19	2.06	1.92
	.01	8.29	6.01	5.09	4.58	4.25	4.01	3.71	3.51	3.08	2.84	2.57
	.001	15.38	10.39	8.49	7.46	6.81	6.35	5.76	5.39	4.59	4.15	3.67

Table A4.2 continued

Denominator Degrees of Freedom, df_2	P	Numerator Degrees of Freedom, df_1 1	2	3	4	5	6	8	10	20	40	∞
20	.25	1.40	1.49	1.48	1.46	1.45	1.44	1.42	1.40	1.36	1.33	1.29
	.10	2.97	2.59	2.38	2.25	2.16	2.09	2.00	1.94	1.79	1.71	1.61
	.05	4.35	3.49	3.10	2.87	2.71	2.60	2.45	2.35	2.12	1.99	1.84
	.01	8.10	5.85	4.94	4.43	4.10	3.87	3.56	3.37	2.94	2.69	2.42
	.001	14.82	9.95	8.10	7.10	6.46	6.02	5.44	5.08	4.29	3.86	3.38
30	.25	1.38	1.45	1.44	1.42	1.41	1.39	1.37	1.35	1.30	1.27	1.23
	.10	2.88	2.49	2.28	2.14	2.05	1.98	1.88	1.82	1.67	1.57	1.46
	.05	4.17	3.32	2.92	2.69	2.53	2.42	2.27	2.16	1.93	1.79	1.62
	.01	7.56	5.39	4.51	4.02	3.70	3.47	3.17	2.98	2.55	2.30	2.01
	.001	13.29	8.77	7.05	6.12	5.53	5.12	4.58	4.24	3.49	3.07	2.59
40	.25	1.36	1.44	1.42	1.40	1.39	1.37	1.35	1.33	1.28	1.24	1.19
	.10	2.84	2.44	2.23	2.09	2.00	1.93	1.83	1.76	1.61	1.51	1.38
	.05	4.08	3.23	2.84	2.61	2.45	2.34	2.18	2.08	1.84	1.69	1.51
	.01	7.31	5.18	4.31	3.83	3.51	3.29	2.99	2.80	2.37	2.11	1.80
	.001	12.61	8.25	6.60	5.70	5.13	4.73	4.21	3.87	3.15	2.73	2.23
60	.25	1.35	1.42	1.41	1.38	1.37	1.35	1.32	1.30	1.25	1.21	1.15
	.10	2.79	2.39	2.18	2.04	1.95	1.87	1.77	1.71	1.54	1.44	1.29
	.05	4.00	3.15	2.76	2.53	2.37	2.25	2.10	1.99	1.75	1.59	1.39
	.01	7.08	4.98	4.13	3.65	3.34	3.12	2.82	2.63	2.20	1.94	1.60
	.001	11.97	7.76	6.17	5.31	4.76	4.37	3.87	3.54	2.83	2.41	1.89
120	.25	1.34	1.40	1.39	1.37	1.35	1.33	1.30	1.28	1.22	1.18	1.10
	.10	2.75	2.35	2.13	1.99	1.90	1.82	1.72	1.65	1.48	1.37	1.19
	.05	3.92	3.07	2.68	2.45	2.29	2.17	2.02	1.91	1.66	1.50	1.25
	.01	6.85	4.79	3.95	3.48	3.17	2.96	2.66	2.47	2.03	1.76	1.38
	.001	11.38	7.32	5.79	4.95	4.42	4.04	3.55	3.24	2.53	2.11	1.54
∞	.25	1.32	1.39	1.37	1.35	1.33	1.31	1.28	1.25	1.19	1.14	1.00
	.10	2.71	2.30	2.08	1.94	1.85	1.77	1.67	1.60	1.42	1.30	1.00
	.05	3.84	3.00	2.60	2.37	2.21	2.10	1.94	1.83	1.57	1.39	1.00
	.01	6.64	4.61	3.78	3.32	3.02	2.80	2.51	2.32	1.88	1.59	1.00
	.001	10.83	6.91	5.42	4.62	4.10	3.74	3.27	2.96	2.27	1.84	1.00

Source: Abridged from Table 18 of *Biometrika Tables for Statisticians*, Vol. 1, edited by E. S. Pearson and H. O. Hartley (London: Cambridge University Press, 1962).

Table A4.3 Critical values for the chi-square distribution

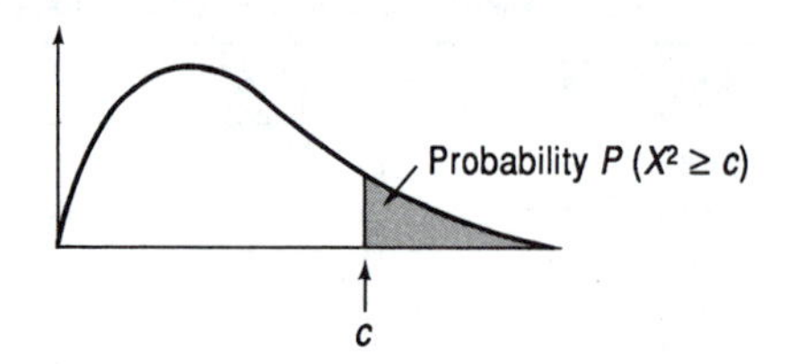

	Probability							
df	.500	.250	.100	.050	.025	.010	.005	.001
1	.455	1.323	2.706	3.841	5.024	6.635	7.879	10.83
2	1.386	2.773	4.605	5.991	7.378	9.210	10.60	13.82
3	2.366	4.108	6.251	7.815	9.348	11.34	12.84	16.27
4	3.357	5.385	7.779	9.488	11.14	13.28	14.86	18.47
5	4.351	6.626	9.236	11.07	12.83	15.09	16.75	20.52
6	5.348	7.841	10.64	12.59	14.45	16.81	18.55	22.46
7	6.346	9.037	12.02	14.07	16.01	18.48	20.28	24.32
8	7.344	10.22	13.36	15.51	17.53	20.09	21.96	26.12
9	8.343	11.39	14.68	16.92	19.02	21.67	23.59	27.88
10	9.342	12.55	15.99	18.31	20.48	23.21	25.19	29.59
11	10.34	13.70	17.28	19.68	21.92	24.72	26.76	31.26
12	11.34	14.85	18.55	21.03	23.34	26.22	28.30	32.91
13	12.34	15.98	19.81	22.36	24.74	27.79	29.82	34.53
14	13.34	17.12	21.06	23.68	26.12	29.14	31.32	36.12
15	14.34	18.25	22.31	25.00	27.49	30.58	32.80	37.70
16	15.34	19.37	23.54	26.30	28.85	32.00	34.27	39.25
17	16.34	20.49	24.77	27.59	30.19	33.41	35.72	40.79
18	17.34	21.60	25.99	28.87	31.53	34.81	37.16	42.31
19	18.34	22.72	27.20	30.14	32.85	36.19	38.58	43.82
20	19.34	23.83	28.41	31.41	34.17	37.57	40.00	45.32
21	20.34	24.93	29.62	33.67	35.48	38.93	41.40	46.80
22	21.34	26.04	30.81	33.92	36.78	40.29	42.80	48.27
23	22.34	27.14	32.01	35.17	38.08	41.64	44.18	49.73
24	23.34	28.24	33.20	36.42	39.36	42.98	45.56	51.18
25	24.34	29.34	34.38	37.65	40.65	44.31	46.93	52.62
26	25.34	30.43	35.56	38.89	41.92	45.64	48.29	54.05
27	26.34	31.53	36.74	40.11	43.19	46.96	49.64	55.48
28	27.34	32.62	37.92	41.34	44.46	48.28	50.99	56.89
29	28.34	33.71	39.09	42.56	45.72	49.59	52.34	58.30
30	29.34	34.80	40.26	43.77	46.98	50.89	53.67	59.70
40	39.34	45.62	51.81	55.76	59.34	63.69	66.77	73.40
50	49.33	56.33	63.17	67.50	71.42	76.15	79.49	86.66
60	59.33	66.98	74.40	79.08	83.30	88.38	91.95	99.61
70	69.33	77.58	85.53	90.53	95.02	100.4	104.2	112.3
80	79.33	88.13	96.58	101.9	106.6	112.3	116.3	124.8
90	89.33	98.65	107.6	113.1	118.1	124.1	128.3	137.2
100	99.33	109.1	118.5	124.3	129.6	135.8	140.2	149.4

Source: Abridged from Table 8 of *Biometrika Tables for Statisticians*, Vol. 1, edited by E. S. Pearson and H. O. Hartley (London: Cambridge University Press, 1962).

Table A4.4 Critical values for the Durbin-Watson Test for Autocorrelation $P = .05$[1]

Sample Size	Number of X Variables (K − 1)									
	1		2		3		4		5	
n	d_L	d_U	d_L	d_U	d_L	d_U	d_L	d_U	d_L	d_U
15	1.08	1.36	.95	1.54	.82	1.75	.69	1.97	.56	2.21
16	1.10	1.37	.98	1.54	.86	1.73	.74	1.93	.62	2.15
17	1.13	1.38	1.02	1.54	.90	1.71	.78	1.90	.67	2.10
18	1.16	1.39	1.05	1.53	.93	1.69	.82	1.87	.71	2.06
19	1.18	1.40	1.08	1.53	.97	1.68	.86	1.85	.75	2.02
20	1.20	1.41	1.10	1.54	1.00	1.68	.90	1.83	.79	1.99
21	1.22	1.42	1.13	1.54	1.03	1.67	.93	1.81	.83	1.96
22	1.24	1.43	1.15	1.54	1.05	1.66	.96	1.80	.86	1.94
23	1.26	1.44	1.17	1.54	1.08	1.66	.99	1.79	.90	1.92
24	1.27	1.45	1.19	1.55	1.10	1.66	1.01	1.78	.93	1.90
25	1.29	1.45	1.21	1.55	1.12	1.66	1.04	1.77	.95	1.89
26	1.30	1.46	1.22	1.55	1.14	1.65	1.06	1.76	.98	1.88
27	1.32	1.47	1.24	1.56	1.16	1.65	1.08	1.76	1.01	1.86
28	1.33	148	1.26	1.56	1.18	1.65	1.10	1.75	1.03	1.85
29	1.34	1.48	1.27	1.56	1.20	1.65	1.12	1.74	1.05	1.84
30	1.35	1.49	1.28	1.57	1.21	1.65	1.14	1.74	1.07	1.83
31	1.36	1.50	1.30	1.57	1.23	1.65	1.16	1.74	1.09	1.83
32	1.37	1.50	1.31	1.57	1.24	1.65	1.18	1.73	1.11	1.82
33	1.38	1.51	1.32	1.58	1.26	1.65	1.19	1.73	1.13	1.81
34	1.39	1.51	1.33	1.58	1.27	1.65	1.21	1.73	1.15	1.81
35	1.40	1.52	1.34	1.58	1.28	1.65	1.22	1.73	1.16	1.80
36	1.41	1.52	1.35	1.59	1.29	1.65	1.24	1.73	1.18	1.80
37	1.42	1.53	1.36	1.59	1.31	1.66	1.25	1.72	1.19	1.80
38	1.43	1.54	1.37	1.59	1.32	1.66	1.26	1.72	1.21	1.79
39	1.43	1.54	1.38	1.60	1.33	1.66	1.27	1.72	1.22	1.79
40	1.44	1.54	1.39	1.60	1.34	1.66	1.29	1.72	1.23	1.79
45	1.48	1.57	1.43	1.62	1.38	1.67	1.34	1.72	1.29	1.78
50	1.50	1.59	1.46	1.63	1.42	1.67	1.38	1.72	1.34	1.77
55	1.53	1.60	1.49	1.64	1.45	1.68	1.41	1.72	1.38	1.77
60	1.55	1.62	1.51	1.65	1.48	1.69	1.44	1.73	1.41	1.77
65	1.57	1.63	1.54	1.66	1.50	1.70	1.47	1.73	1.44	1.77
70	1.58	1.64	1.55	1.67	1.52	1.70	1.49	1.74	1.46	1.77
75	1.60	1.65	1.57	1.68	1.54	1.71	1.51	1.74	1.49	1.77
80	1.61	1.66	1.59	1.69	1.56	1.72	1.53	1.74	1.51	1.77
85	1.62	1.67	1.60	1.70	1.57	1.72	1.55	1.75	1.52	1.77
90	1.63	1.68	1.61	1.70	1.59	1.73	1.57	1.75	1.54	1.78
95	1.64	1.69	1.62	1.71	1.60	1.73	1.58	1.75	1.56	1.78
100	1.65	1.69	1.63	1.72	1.61	1.74	1.59	1.76	1.57	1.78

[1] Tests the null hypothesis of no positive first-order autocorrelation, $H_0: r_{t,t-1} \leq 0$. Reject H_0 (and accept $H_1: r_{t,t-1} > 0$) if $d < d_L$; do not reject H_0 if $d > d_U$. The test is inconclusive if $d_L < d < d_U$.

Table A4.4 continued : $P = .01$

Sample Size	Number of X Variables ($K - 1$)									
	1		2		3		4		5	
n	d_L	d_U	d_L	d_U	d_L	d_U	d_L	d_U	d_L	d_U
15	.81	1.07	.70	1.25	.59	1.46	.49	1.70	.30	1.96
16	.84	1.09	.74	1.25	.63	1.44	.53	1.66	.44	1.90
17	.87	1.10	.77	1.25	.67	1.43	.57	1.63	.48	1.85
18	.90	1.12	.80	1.26	.71	1.42	.61	1.60	.52	1.80
19	.93	1.13	.83	1.26	.74	1.41	.65	1.58	.56	1.77
20	.95	1.15	.86	1.27	.77	1.41	.68	1.57	.60	1.74
21	.97	1.16	.89	1.27	.80	1.41	.72	1.55	.63	1.71
22	1.00	1.17	.91	1.28	.83	1.40	.75	1.54	.66	1.69
23	1.02	1.19	.94	1.29	.86	1.40	.77	1.53	.70	1.67
24	1.04	1.20	.96	1.30	.88	1.41	.80	1.53	.72	1.66
25	1.05	1.21	.98	1.30	.90	1.41	.83	1.52	.75	1.65
26	1.07	1.22	1.00	1.31	.93	1.41	.85	1.52	.78	1.64
27	1.09	1.23	1.02	1.32	.95	1.41	.88	1.51	.81	1.63
28	1.10	1.24	1.04	1.32	.97	1.41	.90	1.51	.83	1.62
29	1.12	1.25	1.05	1.33	.99	1.42	.92	1.51	.85	1.61
30	1.13	1.26	1.07	1.34	1.01	1.42	.94	1.51	.88	1.61
31	1.15	1.27	1.08	1.34	1.02	1.42	.96	1.51	.90	1.60
32	1.16	1.28	1.10	1.35	1.04	1.43	.98	1.51	.92	1.60
33	1.17	1.29	1.11	1.36	1.05	1.43	1.00	1.51	.94	1.59
34	1.18	1.30	1.13	1.36	1.07	1.43	1.01	1.51	.95	1.59
35	1.19	1.31	1.14	1.37	1.08	1.44	1.03	1.51	.97	1.59
36	1.21	1.32	1.15	1.38	1.10	1.44	1.04	1.51	.99	1.59
37	1.22	1.32	1.16	1.38	1.11	1.45	1.06	1.51	1.00	1.59
38	1.23	1.33	1.18	1.39	1.12	1.45	1.07	1.52	1.02	1.58
39	1.24	1.34	1.19	1.39	1.14	1.45	1.09	1.52	1.03	1.58
40	1.25	1.34	1.20	1.40	1.15	1.46	1.10	1.52	1.05	1.58
45	1.29	1.38	1.24	1.42	1.20	1.48	1.16	1.53	1.11	1.58
50	1.32	1.40	1.28	1.45	1.24	1.49	1.20	1.54	1.16	1.59
55	1.36	1.43	1.32	1.47	1.28	1.51	1.25	1.55	1.21	1.59
60	1.38	1.45	1.35	1.48	1.32	1.52	1.28	1.56	1.25	1.60
65	1.41	1.47	1.38	1.50	1.35	1.53	1.31	1.57	1.28	1.61
70	1.43	1.49	1.40	1.52	1.37	1.55	1.34	1.58	1.31	1.61
75	1.45	1.50	1.42	1.53	1.39	1.56	1.37	1.59	1.34	1.62
80	1.47	1.52	1.44	1.54	1.42	1.57	1.39	1.60	1.36	1.62
85	1.48	1.53	1.46	1.55	1.43	1.58	1.41	1.60	1.39	1.63
90	1.50	1.54	1.47	1.56	1.45	1.59	1.43	1.61	1.41	1.64
95	1.51	1.55	1.49	1.57	1.47	1.60	1.45	1.62	1.42	1.64
100	1.52	1.56	1.50	1.58	1.48	1.60	1.46	1.63	1.44	1.65

Reproduced in part from Tables 4, 5, and 6 of J. Durbin and G. S. Watson, Testing for correlation in least squares regression. II. *Biometrika 38*, 159–178(1951) with permission of the Biometrika Trustees.

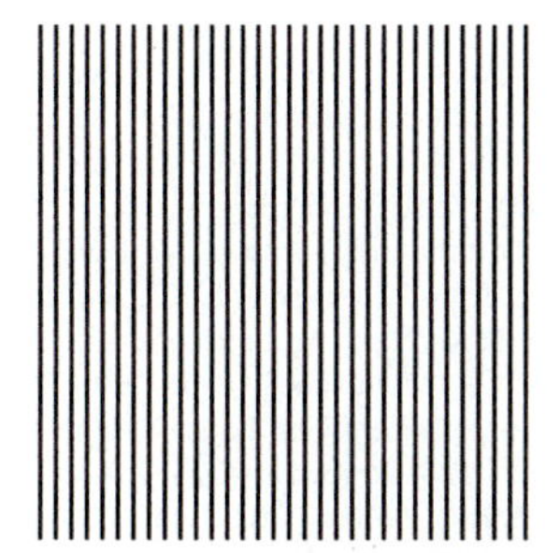

References

ACHEN, Christopher H. (1982). *Interpreting and Using Regression.* Beverly Hills: Sage.

AGRESTI, Alan (1990). *Categorical Data Analysis.* New York: Wiley.

ALDRICH, John H., and Forrest D. Nelson (1984). *Linear Probability, Logit, and Probit Models.* Beverly Hills: Sage.

ALLEN, David M., and Foster B. Cady (1982). *Analyzing Experimental Data by Regression.* Belmont, CA: Wadsworth.

ANDREWS, D. F., P. J. Bickel, F. R. Hampel, P. J. Huber, W. H. Rogers, and J. W. Tukey (1972). *Robust Estimates of Location.* Princeton, NJ: Princeton University Press.

ARCHER, Victor F. (1987). "Association of lung cancer mortality with Precambrian granite." *Archives of Environmental Health, 42*: 87–91.

ARCHER, Victor F. (1988). "Response to Cohen." *Archives of Environmental Health, 43*(4): 314–315.

BACHE, C. A., J. W. Serum, W. D. Youngs, and D. J. Lisk (1972). "Polychlorinated biphenyl residues: Accumulation in Cayuga Lake trout with age." *Science, 117*: 1192–1193.

BATES, Douglas M., and Donald G. Watts (1988). *Nonlinear Regression Analysis and Its Applications.* New York: Wiley.

BEATTY, J. Kelly, Brian O'Leary, and Andrew Chaikin, Eds, (1981). *The New Solar System.* Cambridge, MA: Sky.

BELSLEY, David A., Edwin Kuh, and Roy E. Welsch (1980). *Regression Diagnostics: Identifying Influential Data and Sources of Collinearity.* New York: Wiley.

BENJAMINI, Yoav (1988). "Opening the box of a boxplot." *The American Statistician*, November, *42*(4): 257–262.

BLOCKER, T. Jean, and Douglas Lee Eckberg (1989). "Environmental issues as women's issues: General concerns and local hazards." *Social Science Quarterly*, *70*(3): 586–593.

BOLLEN, Kenneth A. (1989). *Structural Equations with Latent Variables.* New York: Wiley.

BROWN, Lester R., William U. Chandler, Christopher Flavin, Cynthia Pollock, Sandra Postel, Linda Starke, and Edward C. Wolf (1986). *State of the World 1986.* New York: W. W. Norton.

CAIRNS, David K. (1988). "The regulation of seabird colony size: A hinterland model." *The American Naturalist*, *134*(1): 141–146.

CAPUZZO, Judith M., and Franz E. Anderson (1973). "The use of modern chromium accumulation to determine estuary sedimentation rates." *Marine Geology*, *14*: 225–235.

CARMINES, Edward G., and Richard A. Zeller (1979). *Reliability and Validity Assessment*. Beverly Hills: Sage.

CARROLL, Raymond J., and A. H. Welsh (1988). "A note on asymmetry and robustness in linear regression." *The American Statistician, 42*(4): 285–287.

CHAMBERS, John M., William S. Cleveland, Beat Kleiner, and Paul A. Tukey, Eds. (1983). *Graphical Methods for Data Analysis*. Belmont, CA: Wadsworth.

CHARPIN, D., J. P. Kleisbauer, J. Fondarai, B. Graland, A, Viala, and F. Gouezo (1988). "Respiratory symptoms and air pollution changes in children: The Gardanne coal-basin study." *Archives of Environmental Health, 43*(1): 22–27.

CLEVELAND, William S. (1985). *The Elements of Graphing Data*. Monterey, CA: Wadsworth.

CLEVELAND, William S., and Marylyn E. McGill (1988). *Dynamic Graphics for Statistics*. Pacific Grove, CA: Brooks/Cole.

COHEN, Bernard L. (1988). Letter to editor. *Archives of Environmental Health, 43*(4): 313–314.

COMMITTEE on Monitoring and Assessment of Trends in Acid Deposition (1986). *Acid Deposition: Long-Term Trends*. Washington, DC: National Academy Press.

COOK, R. Dennis, and Sanford Weisberg (1982). *Residuals and Influence in Regression*. New York: Chapman and Hall.

COUNCIL on Environmental Quality (1985). *Environmental Quality: 1985*. Washington, DC: Council on Evironmental Quality.

COUNCIL on Environmental Quality (1988). *Environmental Quality: 1987–1988*. Washington, DC: Council on Environmental Quality.

DEBLOIS, E.M., and W. C. Leggett (1991). "Functional response and potential impact of invertebrate predators on benthic fish eggs: Analysis of the *Calliopius laeviusculus*-capelin (*Mallotus villosus*) predator-prey system." *Marine Ecology Progress Series, 69*: 205–216.

DINGMAN, S. Lawrence, Diana M. Seely-Reynolds, and Robert C. Reynolds III (1988). "Application of kriging to estimating mean annual precipitation in a region of orographic influence." *Water Resources Bulletin, 24*(2): 329–339.

DRAPER, N. R. and H. Smith (1981). *Applied Regression Analysis*. New York: Wiley.

EFRON, Bradley (1982). *The Jackknife, the Bootstrap, and Other Resampling Plans*. Philadelphia: Society for Industrial and Applied Mathematics.

EFRON, Bradley, and R. Tibshirani (1986). "Bootstrap methods for standard errors, confidence intervals, and other measures of statistical accuracy." *Statistical Science, 1*(1): 54–77.

EVERITT, B.S., and G. Dunn (1983). *Advanced Methods of Data Exploration and Modelling*. Exeter, NH: Heinemann.

FRIGGE, Michael, David C. Hoaglin, and Boris Iglewicz (1989). "Some implementations of the boxplot." *The American Statistician*, February, *53*(1): 50–54.

FURNESS, R. W., and T. R. Birkhead (1984). "Seabird colony distributions suggest competition for food supplies during the breeding season." *Nature, 311*: 655–656.

GALLANT, A. Ronald (1987). *Nonlinear Statistical Models*. New York: Wiley.

GLAISTER, David H., and Nita L. Miller (1990). "Cerebral tissue oxygen status and psychomotor performance during lower body negative pressure (LBNP)." *Aviation, Space, and Environmental Medicine, 61*(2): 99–105.

GOODALL, Colin (1983). "Examining residuals," pp. 211–246 in David C. Hoaglin, Frederick Mosteller, and John W. Tukey (Eds.), *Understanding Robust and Exploratory Data Analysis*. New York: Wiley.

HAITH, Douglas A, (1976). "Land use and water quality in New York Rivers." *Journal of the Environmental Engineering Division, ASCE, 102*: 1–15.

HALL, Peter (1988). "Theoretical comparison of bootstrap confidence intervals." *The Annals of Statistics, 16*(3): 927–953.

HAMILTON, Lawrence C. (1990a). *Modern Data Analysis: A First Course in Applied Statistics*. Pacific Grove, CA: Brooks/Cole.

HAMILTON, Lawrence C. (1990b). *Statistics with Stata*. Pacific Grove, CA: Brooks/Cole.

HANSEN, Lars P. (1982). "Large sample properties of generalized method of moments estimators." *Econometrica, 50*: 1029–1054.

HANUSHEK, Eric A., and John E. Jackson (1977). *Statistical Methods for Social Scientists*. New York: Academic Press.

HICKS, D. Murray, M. J. McSaveney, and T. J. Chinn (1990). "Sedimentation in proglacial Ivory Lake, Southern Alps, New Zealand." *Arctic and Alpine Research, 22*(1): 26–42.

HOAGLIN, David C., Boris Iglewicz, and John W. Tukey (1986). "Performance of some resistant rules for outlier labeling." Journal of the American Statistical Association 81 (396): 991–999.

HOAGLIN, David C., Frederick Mosteller, and John W. Tukey, Eds. (1983). *Understanding Robust and Exploratory Data Analysis*. New York: Wiley.

HOAGLIN, David C., Frederick Mosteller, and John W. Tukey, Eds. (1985). *Exploring Data Tables, Trends, and Shapes*. New York: Wiley.

HOSMER, David W., and Stanley Lemeshow (1989). *Applied Logistic Regression*. New York: Wiley.

HUBER, Peter J. (1981). *Robust Statistics*. New York: Wiley.

HURD, Michael L. (1986). *The Geochemistry of Chromium in the Great Bay Estuary and the Gulf of New Hampshire* (Master's Thesis). Durham: University of New Hampshire.

IANNACCHIONE, Anthony T., and Donald G. Puglio (1979). *Methane Content and Geology of the Hartshorne Coalbed in Haskell and Le Flore Counties, Okla.*, RI 8407. Washington, DC: U. S. Bureau of Mines.

JOHNSTON, J. (1972). *Econometric Methods* (2nd ed.). New York: McGraw–Hill.

JÖRESKOG, K. G., and D. Sörbom (1979). *Advances in Factor Analysis and Structural Equation Modeling*. Cambridge, MA: Abt.

KLEINBAUM, David G., and Lawrence L. Kupper (1978). *Applied Regression Analysis and Other Multivariable Methods*. North Scituate, MA: Duxbury.

LAUNER, Robert L., and Andrew F. Siegel, Eds. (1982). *Modern Data Analysis*. New York: Academic Press.

LIOY, Paul L., Jed M. Waldman, Arthur Greenberg, Ronald Harkov, and Charles Pietarninen (1988). "The total human environmental exposure study (THEES) to Benzo (a) pyrene: Comparison of the inhalation and food pathways." *Archives of Environmental Health, 43*(4): 304–312.

MCCLEARY, Richard, Richard A. Hay, Jr., Errol E. Meidinger, and David McDowall (1980). *Applied Time Series Analysis for the Social Sciences*. Beverly Hills: Sage.

MCDONALD, G. C., and Ayers, J. A. (1978). "Some applications of the 'Chernoff faces': A technique for graphically representing multivariate data." In P. C. C. Wang (Ed.), *Graphical Representation of Multivariate Data*. New York: Academic Press, pp. 183–197.

MCDONALD, G. C., and Schwing, R. C. (1973). "Instabilities of regression estimates relating air pollution to mortality." *Technometrics, 15*: 463–481.

MCKELVEY, R. D., and W. Zavoina (1976). "A statistical model for the analysis of ordinal level dependent variables." *Journal of Mathematical Sociology, 4*: 103–120.

MEFFE, Gary K., and Andrew L. Sheldon (1988). "The influence of habitat structure on fish assemblage composition in Southeastern blackwater streams." *The American Midland Naturalist, 120*(2): 225–236.

MOORE, David S. (1990). "Uncertainty," pp. 95–137 in Lynn Arthur Steen (Ed.), *On the Shoulders of Giants: New Approaches to Numeracy*. Washington, DC: National Academy Press.

MOSTELLER, Frederick, and John W. Tukey (1977). *Data Analysis and Regression.* Reading MA: Addison-Wesley.

OFFICE of Population Censuses and Surveys (1987). *Period and Cohort Birth Order Statistics.* London: Her Majesty's Stationary Office.

OKAEME, A. N., E. A. Agbelusi, J. Mshelbwala, M. Wari, A. Ngulge, and M. Haliru (1988). "Effects of Immobilon and Revivon in the immobilization of western kob (*Kobus kob kob*)." *African Journal of Ecology, 26*: 63–67.

OLIVEIRA, J. Tiago de, and Benjamin Epstein (1982). *Some Recent Advances in Statistics.* New York: Academic.

OWEN, Art (1988). "Small sample central confidence intervals for the mean." Technical Report 302, Stanford University Department of Statistics.

RABINOWITZ, Michael, Allen Leviton, Herbert Needleman, David Bellinger, and Christine Waternaux (1985). "Evironmental correlates of infant blood levels in Boston." *Environmental Research, 38*: 96–107.

RABINOWITZ, Michael, Herbert Needleman, Michael Burley, Hollister Finch, and John Rees (1984). "Lead in umbilical blood, indoor air, tap water, and gasoline in Boston." *Archives of Environmental Health, 39*(4): 299–301.

RATHBUN, Galen B. (1988). "Fixed-wing airplane versus helicopter surveys of manatees (*trichechus manatus*)." *Marine Mammal Science, 4*(1): 71–75.

RAWLINGS, John O. (1988). *Applied Regression Analysis: A Research Tool.* Pacific Grove, CA: Wadsworth.

REY, William J. J. (1983). *Introduction to Robust and Quasi-Robust Statistical Methods.* New York: Springer-Verlag.

RISK, Michael J., and Paul W. Sammarco (1991). "Cross-shelf trends in skeletal density of the massive coral *Porites lobata* from the Great Barrier Reef." *Marine Ecology Progress Series, 69*: 195–200.

ROM, William N., James E. Lockey, Ki Moon Bang, Charles Dewitt, and Richard E. Johns (1983). "Reversible beryllium sensitization in a prospective study of beryllium workers." *Archives of Environmental Health, 38*(5): 303–307.

ROUSSEEUW, Peter J., and Annick M. Leroy (1987). *Robust Regression and Outlier Deletion.* New York: Wiley.

ROUSSEEUW, Peter J., and Bert C. Van Zomeren (1990). "Unmasking multivariate outliers and leverage points." *Journal of the American Statistical Association, 85*(411): 633–639.

RUBEN, John A., and Arthur J. Boucot (1989). "The origin of the lungless salamanders (Amphibia: plethodontidae)." *The American Naturalist, 134*(2): 161–169.

RUPPERT, David, and Douglas G. Simpson (1990). "Comment on Rousseeuw and Van Zomern." *Journal of the American Statistical Association, 85*(411): 644–646.

SEBER, G. A. F., and C. J. Wild (1989). *Nonlinear Regression.* New York: Wiley.

SEYFRIT, Carole L. (1989). "North Sea oil and Scotland's youth: High School students' attitudes toward community, family, and the oil industry." Paper presented at annual meeting of the American Sociological Association, San Franciso.

SHULTZ, Steven D. (1987). *Willingness to Pay for Groundwater Protection in Dover, NH: A Contingent Valuation Approach* (Master's Thesis). Durham: University of New Hampshire.

SIMPSON, D. G., D. Ruppert, and R. J. Carroll (1989). "One-step GM-estimates for regression with bounded influence and high breakdown-point." Technical Report 859, School of Operations Research and Industrial Engineering, Cornell University.

SOLER, Alfonso (1990). "Dependence on latitude of the relation between the diffuse fraction of solar radiation and the ratio of global-to-extraterrestrial radiation for monthly average daily values." *Solar Energy, 44*(5): 297–302.

STACEY, Conway Ivan, William Stanley Perriman, and Susan Whitney (1985). "Organochlorine pesticide residue levels in human milk: Western Australia, 1979–1980." *Archives of Environmental Health*, March/April, *40*(2): 102–108.

STINE, Robert (1990). "An introduction to bootstrap methods," pp. 325–373 in John Fox and J. Scott Long (Eds.), *Modern Methods of Data Analysis.* Beverly Hills: Sage.

STREET, James O., Raymond J. Carroll, and David Ruppert (1988). "A note on computing robust regression estimates via iteratively reweighted least squares." *The American Statistician*, May, *42*(2): 152–154.

TEMBO, A. (1987). "Population status of the hippopotamus on the Luangwa River." *African Journal of Ecology, 25*: 71–77.

THEIL, Henri (1971). *Principles of Econometrics.* New York: Wiley.

TRAYNOR, G. W., M. G. Apte, A. R. Carruthers, J. F. Dillworth, D. T. Grimsrud, and L. A. Gundel (1987). "Indoor air pollution due to emissions from wood-burning stoves." *Environmental Science and Technology, 21*(7): 691–696.

TUFTE, Edward R. (1983). *The Visual Display of Quantitative Information.* Cheshire, CT: Graphics Press.

TUKEY, John W. (1977). *Exploratory Data Analysis.* Reading, MA: Addison-Wesley.

UYAR, Bulent, and Orhan Erdem (1990). "Regression procedures in SAS: Problems?" *The American Statistician, 44*(4): 296–301.

VELLEMAN, Paul F. (1982). "Applied nonlinear smoothing," pp. 141–177 in Samuel Leinhardt (Ed.), *Sociological Methodology 1982.* San Francisco: Jossey-Bass.

VELLEMAN, Paul F., and David C. Hoaglin (1981). *Applications, Basics, and Computing of Exploratory Data Analysis.* Boston: Wadsworth.

VERGINO, Eileen S. (1989). "Soviet test yields." *EOS Transactions, American Geophysical Union, 70*(48): 1511.

WEISBERG, Sanford (1980). *Applied Linear Regression.* New York: Wiley.

WHITE, Halbert (1980)."A heteroscedasticity-consistent covariance matrix estimator and direct test for heteroscedasticity." *Econometrica, 48*: 817–838.

ZUPAN, Jeffrey M. (1973). *The Distribution of Air Quality in the New York Region.* Baltimore: Johns Hopkins University Press.

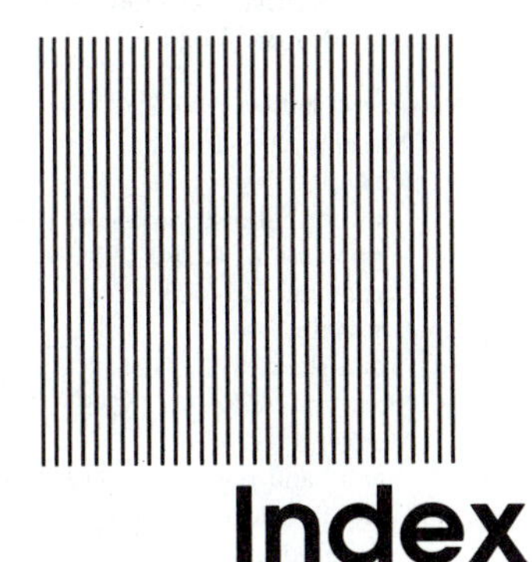

Index

Analysis of covariance (ANCOVA), 101
Analysis of variance (ANOVA)
  oneway, 92–95
  twoway, 95–101
Autocorrelation, 51, 111, 113, 118–124, 137, 141
Autoregressive integrated moving average (ARIMA), 141

Band regression, 117, 146–147, 181, 254
Beta weight, *see* Standardized regression coefficient
Bias, 110, 294–295, 304–313
Biweight estimation, 193–200
Bonferroni inequality, 132–133, 143–144
Bootstrap, 313–326, 329–331
Bounded influence, 207–212, 304–318
Bounded range, 141
Box-Cox transformation, 28
Boxplot, 8–10, 21–23, 27–28, 126

Causality, 29, 51
Centering, 152–153
Central Limit Theorem, 27
Collinearity (*also see* Multicollinearity), 82, 135
Change-score analysis, 105
Cluster analysis, 268–271
Coefficient of determination
  adjusted, 41–42
  from correlation matrix, 343
  in curvilinear regression, 181
  estimating communality, 271
  in logit regression, 233
  and multicollinearity, 133–136
  in nonlinear regression, 170
  in OLS, 38–42
  in regression through the origin, 50–51
  in robust and WLS regression, 198
Communality, 257, 271–273
Conditional effect plot, 153, 158–163, 231–233
Confidence interval
  for autocorrelation coefficient, 121
  in bivariate regression, 47–49
Confidence interval (*continued*)
  bootstrap, 319–325, 329–331
  boxplot, 9–10
  in logit regression, 229
  in multiple regression, 78–80
  in nonlinear regression, 171
  in robust regression, 198–200
Consistency, 111–112, 296
Constant error variance, *see* Homoscedasticity
Contaminated distribution, 307
Continuous variable, 290
Cook's D, 132–133, 143, 156, 158, 181, 211
Correlation
  of coefficients, 135–136, 171, 233, 247, 339
  and distribution shape, 63
  of errors with errors, *see* Autocorrelation
  of errors with variables, 112–113
  matrix, 113–114, 293
  population, 293

Correlation (*continued*)
regression from correlation matrix, 342–344
sample, 39–40
of variables, 34, 39–41, 113–114, 136
Correlogram, 121
Covariance
of coefficients, 339
of errors with errors, 111
of errors with variables, 112–113
matrix, 111, 293
population, 291–292
sample, 39, 291
of variables, 34, 39
C_P, 106
Cross-product matrix, 341, 347
Crosstabulation, 234, 247
Curvilinearity
in conditional effect plot, 158–163
in leverage plot, 137
in logit regression, 235
in residual versus predicted *Y* plot, 53, 137
in scatterplot matrix, 114–115, 136
and transformed variables, 57–58, 145–146, 148–163

Deviance residual, 235–242
Deviation score, 341
DFBETAS, 125–129, 142–143, 157–158, 280
DFFITS, 142–143
Diagonal matrix, 344
Differencing time series, 123
Discrete variable, 141, 290
Discrimination (logit analysis), 233–235
Dummy variable
and ANOVA, 92–101
as dependent variable, 218–223
and difference of means test, 86–87
intercept dummy variable, 85–88
and regression constant, 107
slope dummy variable, 88–92
Durbin-Watson test, 118, 120, 355–356

Effect coding, 99–101
Efficiency, 110, 215, 296, 304–312
Eigenvalue, 255–259, 268, 270, 272, 275, 345
Eigenvector, 255–256, 272, 345
Error term
in bivariate regression, 31–32, 51
in factor analysis, 250–252
in multiple regression, 110–113, 296–297
in robust regression, 189–190
Expected values, 31, 140, 289–291
Exploratory data analysis (EDA), 10, 28
Exponential models, 149–150, 163–165

F statistic
test ANOVA, 92–101

F statistic (*continued*)
test bivariate regression, 45–47
and multicollinearity, 82, 91–92, 136
test multiple regression, 80–81
reading *F* table, 64
in regression through the origin, 50–51
theoretical distribution, 298, 300, 351–353
Factor
common, 250
loading, 250–252, 257, 259–264, 288
maximum likelihood, 278–282
principal, 270–277, 281–282
rotation, *see* Rotation
score, 263–271, 275–277, 280, 282
unique, 250–251, 271
Fixed *X* assumption, 63, 110, 113, 140

Gaussian distribution, *see* Normal distribution
Gauss-Markov Theorem, 110–111
Gompertz curve, 166–173
Granularity, 16–17, 64

Hanning, 123
Hat matrix, 130, 201, 236–238, 339
Heteroscedasticity, 51, 53, 116–117, 137, 219, 246
Homoscedasticity, 111, 219
Huber estimation, 192–200

Identity matrix, 337
Influence, 51, 53, 125–133, 137, 152, 235–242
Information matrix, 247
Interaction
in ANOVA, 97–101, 107
slope dummy variable, 88–92
term in regression model, 84–85, 88–92, 227–228
due to transformed *Y* variable, 84, 161, 247
Interquartile range (IQR)
and boxplot, 9–10, 27
definition, 8
and normality, 8, 124–125, 316
Iteratively reweighted least squares (IRLS), 190, 195–198, 207
Inverse matrix, 337, 346–347
Inverse transformation, 57–58, 159

Jackknife, 323, 331
Jittering, 141, 274, 280

Ladder of powers, 18
Latent variable, 252, 273, 275
Leverage, 130–133, 144, 201, 206–212, 215, 236–238, 339
Leverage plot, 70–72, 127–129, 142
Likelihood function, 223–228
LISREL, 281

Logarithms, 17–18, 28, 150, 216
Logistic curve, 166–168
Logit, 220–223, 229–230, 246–247
Log-log curve, 149–150
Log odds, *see* Logit

M-estimation, 190–200
Masking, 215–216
Mean
of normal distribution, 4
population, 4, 289–290
running, 121–123
sample, 2–4
and skew, 8, 27, 124
Mean squared error (MSE), 73, 295
Median
and band regression, 117, 146–147, 181
and boxplot, 8–9
definition, 6–7
running, 122–123
and skew, 6, 8, 27, 124
and symmetry plot, 10–11
Median absolute deviation (MAD), 191
Monotonic curve, 151
Monte Carlo simulation, 303–312
Multicollinearity
consequences, 82, 113, 133–137, 339
and interaction term, 91–92, 107
and polynomial regression, 152–153
and principal components, 265–266
and stepwise regression, 84
Multiple comparison fallacy, 83–84

Nested models
F tests with OLS, 80–81
χ^2 tests with logit regression, 225–228
Nonconstant error variance, *see* Heteroscedasticity
Nonlinear regression, 163–174
Nonmonotonic curve, 151–153
Normal distribution, 4, 26, 297–298
Normal i.i.d. errors, 31–32
Normality
asymptotic, 296
checking for, 8, 15–17, 21–23, 27, 51–53, 124–125, 316
of errors, 51, 112–113, 137
of sampling distribution, 42, 306–307, 316, 319
Notched boxplot, 27–28

Odds, 220, 222–223, 230–231
Odds ratio, 230–231
Omitted variable, 51, 73, 113, 185, 203–206
Ordinary least squares (OLS)
assumptions, 110–113, 296–297
bivariate regression, 32–34
bootstrap performance, 314–325
M-estimation, 191–192
Monte Carlo performance, 304–312, 329
multiple regression, 67, 72

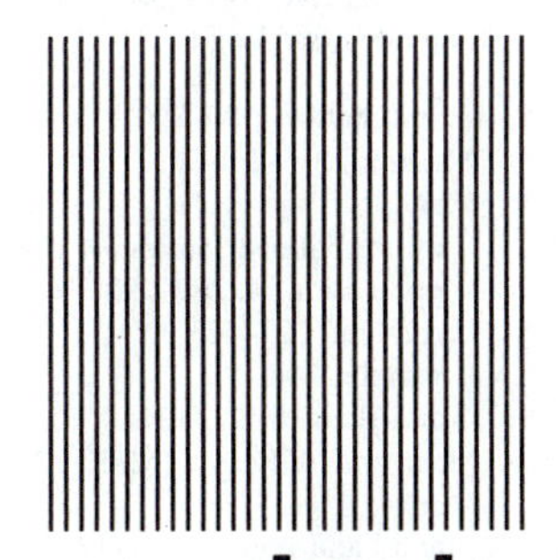

Index

Analysis of covariance (ANCOVA), 101
Analysis of variance (ANOVA)
oneway, 92–95
twoway, 95–101
Autocorrelation, 51, 111, 113, 118–124, 137, 141
Autoregressive integrated moving average (ARIMA), 141

Band regression, 117, 146–147, 181, 254
Beta weight, *see* Standardized regression coefficient
Bias, 110, 294–295, 304–313
Biweight estimation, 193–200
Bonferroni inequality, 132–133, 143–144
Bootstrap, 313–326, 329–331
Bounded influence, 207–212, 304–318
Bounded range, 141
Box-Cox transformation, 28
Boxplot, 8–10, 21–23, 27–28, 126

Causality, 29, 51
Centering, 152–153
Central Limit Theorem, 27
Collinearity (*also see* Multicollinearity), 82, 135
Change-score analysis, 105
Cluster analysis, 268–271
Coefficient of determination
adjusted, 41–42
from correlation matrix, 343
in curvilinear regression, 181
estimating communality, 271
in logit regression, 233
and multicollinearity, 133–136
in nonlinear regression, 170
in OLS, 38–42
in regression through the origin, 50–51
in robust and WLS regression, 198
Communality, 257, 271–273
Conditional effect plot, 153, 158–163, 231–233
Confidence interval
for autocorrelation coefficient, 121
in bivariate regression, 47–49
Confidence interval (*continued*)
bootstrap, 319–325, 329–331
boxplot, 9–10
in logit regression, 229
in multiple regression, 78–80
in nonlinear regression, 171
in robust regression, 198–200
Consistency, 111–112, 296
Constant error variance, *see* Homoscedasticity
Contaminated distribution, 307
Continuous variable, 290
Cook's D, 132–133, 143, 156, 158, 181, 211
Correlation
of coefficients, 135–136, 171, 233, 247, 339
and distribution shape, 63
of errors with errors, *see* Autocorrelation
of errors with variables, 112–113
matrix, 113–114, 293
population, 293

Correlation (*continued*)
  regression from correlation matrix, 342–344
  sample, 39–40
  of variables, 34, 39–41, 113–114, 136
Correlogram, 121
Covariance
  of coefficients, 339
  of errors with errors, 111
  of errors with variables, 112–113
  matrix, 111, 293
  population, 291–292
  sample, 39, 291
  of variables, 34, 39
C_P, 106
Cross-product matrix, 341, 347
Crosstabulation, 234, 247
Curvilinearity
  in conditional effect plot, 158–163
  in leverage plot, 137
  in logit regression, 235
  in residual versus predicted *Y* plot, 53, 137
  in scatterplot matrix, 114–115, 136
  and transformed variables, 57–58, 145–146, 148–163

Deviance residual, 235–242
Deviation score, 341
DFBETAS, 125–129, 142–143, 157–158, 280
DFFITS, 142–143
Diagonal matrix, 344
Differencing time series, 123
Discrete variable, 141, 290
Discrimination (logit analysis), 233–235
Dummy variable
  and ANOVA, 92–101
  as dependent variable, 218–223
  and difference of means test, 86–87
  intercept dummy variable, 85–88
  and regression constant, 107
  slope dummy variable, 88–92
Durbin-Watson test, 118, 120, 355–356

Effect coding, 99–101
Efficiency, 110, 215, 296, 304–312
Eigenvalue, 255–259, 268, 270, 272, 275, 345
Eigenvector, 255–256, 272, 345
Error term
  in bivariate regression, 31–32, 51
  in factor analysis, 250–252
  in multiple regression, 110–113, 296–297
  in robust regression, 189–190
Expected values, 31, 140, 289–291
Exploratory data analysis (EDA), 10, 28
Exponential models, 149–150, 163–165

F statistic
  test ANOVA, 92–101

F statistic (*continued*)
  test bivariate regression, 45–47
  and multicollinearity, 82, 91–92, 136
  test multiple regression, 80–81
  reading *F* table, 64
  in regression through the origin, 50–51
  theoretical distribution, 298, 300, 351–353
Factor
  common, 250
  loading, 250–252, 257, 259–264, 288
  maximum likelihood, 278–282
  principal, 270–277, 281–282
  rotation, *see* Rotation
  score, 263–271, 275–277, 280, 282
  unique, 250–251, 271
Fixed *X* assumption, 63, 110, 113, 140

Gaussian distribution, *see* Normal distribution
Gauss-Markov Theorem, 110–111
Gompertz curve, 166–173
Granularity, 16–17, 64

Hanning, 123
Hat matrix, 130, 201, 236–238, 339
Heteroscedasticity, 51, 53, 116–117, 137, 219, 246
Homoscedasticity, 111, 219
Huber estimation, 192–200

Identity matrix, 337
Influence, 51, 53, 125–133, 137, 152, 235–242
Information matrix, 247
Interaction
  in ANOVA, 97–101, 107
  slope dummy variable, 88–92
  term in regression model, 84–85, 88–92, 227–228
  due to transformed *Y* variable, 84, 161, 247
Interquartile range (IQR)
  and boxplot, 9–10, 27
  definition, 8
  and normality, 8, 124–125, 316
Iteratively reweighted least squares (IRLS), 190, 195–198, 207
Inverse matrix, 337, 346–347
Inverse transformation, 57–58, 159

Jackknife, 323, 331
Jittering, 141, 274, 280

Ladder of powers, 18
Latent variable, 252, 273, 275
Leverage, 130–133, 144, 201, 206–212, 215, 236–238, 339
Leverage plot, 70–72, 127–129, 142
Likelihood function, 223–228
LISREL, 281

Logarithms, 17–18, 28, 150, 216
Logistic curve, 166–168
Logit, 220–223, 229–230, 246–247
Log-log curve, 149–150
Log odds, *see* Logit

M-estimation, 190–200
Masking, 215–216
Mean
  of normal distribution, 4
  population, 4, 289–290
  running, 121–123
  sample, 2–4
  and skew, 8, 27, 124
Mean squared error (MSE), 73, 295
Median
  and band regression, 117, 146–147, 181
  and boxplot, 8–9
  definition, 6–7
  running, 122–123
  and skew, 6, 8, 27, 124
  and symmetry plot, 10–11
Median absolute deviation (MAD), 191
Monotonic curve, 151
Monte Carlo simulation, 303–312
Multicollinearity
  consequences, 82, 113, 133–137, 339
  and interaction term, 91–92, 107
  and polynomial regression, 152–153
  and principal components, 265–266
  and stepwise regression, 84
Multiple comparison fallacy, 83–84

Nested models
  F tests with OLS, 80–81
  χ^2 tests with logit regression, 225–228
Nonconstant error variance, *see* Heteroscedasticity
Nonlinear regression, 163–174
Nonmonotonic curve, 151–153
Normal distribution, 4, 26, 297–298
Normal i.i.d. errors, 31–32
Normality
  asymptotic, 296
  checking for, 8, 15–17, 21–23, 27, 51–53, 124–125, 316
  of errors, 51, 112–113, 137
  of sampling distribution, 42, 306–307, 316, 319
Notched boxplot, 27–28

Odds, 220, 222–223, 230–231
Odds ratio, 230–231
Omitted variable, 51, 73, 113, 185, 203–206
Ordinary least squares (OLS)
  assumptions, 110–113, 296–297
  bivariate regression, 32–34
  bootstrap performance, 314–325
  M-estimation, 191–192
  Monte Carlo performance, 304–312, 329
  multiple regression, 67, 72

Orthogonal matrix, 344
Outlier
and bootstrap, 319, 325–326
boxplot definition, 9–10, 27
deletion, 185
and influence statistics, 126
and jackknife, 331
and principal component plot, 267–268
and quantile-normal plot, 16–17

Parsimony, 72
Partial regression plot, *see* Leverage plot
Pearson χ^2 statistic, 235–242
Percentile
bootstrap confidence intervals from, 320–325, 329–330
a bounded-influence application, 207
definition, 7
Polynomial regression, 151–153
Power transformation, 17–23, 53–59, 145–146, 148–163, 173, 204
Predicted values
with dummy Y variable, 219–220
in linear regression, 32, 37–38
in logit regression, 221–223
Principal axis, *see* Factor, principal
Principal component
analysis, 249–273, 281–282
factor, 252
plot, 267–271
and principal factor analysis, 252, 271–272, 281
Principal factor, *see* Factor, principal
Probit regression, 246
Proportional leverage plot, 127–129, 142, 157–158
Pseudorandom number generator, 303, 329
Pseudostandard deviation (PSD), 315–316, 321–322

Quantile, 11–17, 27
Quantile-normal plot, 15–17
Quantile plot, 11–13
Quantile-quantile plot, 13–15
Quartile, 7, 10, 27

R^2, *see* Coefficient of determination
Random assignment, 69
Regression constant
suppressing, 49–51, 189, 247
in **X** matrix, 338
Regression through the origin, 49–51
Reliability, 266–267
Resampling, 313–315, 318–319, 329
Residual
in bivariate regression, 32–34, 36–38
scaled, 190–191
standard deviation, *see* Standard deviation of residual
standardized, *see* Standardized residual
studentized, *see* Studentized residual

Residual (*continued*)
uncorrelated with predicted Y, 63
uncorrelated with X variables, 113
variance, *see* Variance of residual
versus predicted Y plot, 51–53, 116–117, 156
zero mean, 113
Resistance, 8, 189, 191, 215
Ridge regression, 136
Robustness, 189–190, 215, 312
Rotation
oblique, 250–251, 262–264
orthogonal, 250, 259–262

Sampling distribution
definition, 289
properties, 294–296
of regression coefficients, 42–43, 296–297
theoretical, 297–299
Scalar, 334–335
Scatterplot
with boxplot, 35
matrix, 114–115, 254, 269, 274
oneway, 219
with regression line, 35–36
Scree graph, 258, 270, 275
Semilog curve, 148–149
Skew
definition, 6
and power transformation, 18–23, 53–56, 154–158
and quantile-normal plot, 16–17, 21–23
and symmetry plot, 10–11, 21–23
Slope
in bivariate regression, 30–31, 33–34, 294
in leverage plot, 71
regressing X on Y, 301
robust, 184–185, 201–203, 208–211
Smoothing, 121–123, 141
Standard deviation
of ith-case residual, 131–132
lack of resistance, 27
of normal distribution, 4
population, 4, 293
of residuals, 36, 106
sample, 3–4
Standard error
in bivariate regression, 42–43
bootstrap, 313–325
from correlation matrix, 344
of estimator, 295
jackknife, 323
Monte Carlo performance, 306–312
and multicollinearity, 134–136
in multiple regression, 77–80
in robust regression, 198–199
Standardized regression coefficients, 40–41, 76–77, 342–343
Standardized residual, 132–133

Standard normal distribution, 297–298
Standard score, 76, 252–253
Stepwise regression, 83–84
Studentized residual, 132–133
Sum of squares
from correlation matrix, 343
explained (ESS), 39
in nonlinear regression, 167, 170
residual (RSS), 33–34, 63
total (TSS), 3, 41
Symmetry plot, 10–11

t statistic
and multicollinearity, 82
reading t table, 64
test dummy variable, 87–90
test logit regression coefficient, 228–229
test regression coefficient, 43–45, 77–78
test robust regression coefficient, 198–200
theoretical distribution, 298–299, 350
Tolerance, 133–134, 233
Trace, 345
Transformed variable, *see* Power transformation
Transpose matrix, 334
Trimmed mean, 215
Type I and Type II error, 64, 106

Unbiased, *see* Bias
Uniqueness, 257, 271–272, 288

Variance
of estimator, 295
explained, 38–39
of ith-case residual, 131
population, 4, 292–293
residual, 106
sample, 3–4, 38–39, 292
Vector, 334

Weighted least squares (WLS), 186–189, 219, 238–239, 247

Y-intercept, *also see* Regression constant
in bivariate regression, 30–31, 33, 294
nonsensical, 63
regressing X on Y, 301
test difference between groups, *see* Dummy variable, intercept
test shift due to outlier, *see* Studentized residual

X pattern, 235–242
χ^2 statistic
test factor analysis, 278–279
test logit regression, 225–228
theoretical distribution, 297–299, 351

Z scores, *see* Standard scores